Undergraduate Texts in Mathematics

Undergraduate Texts in Mathematics

Series Editors

Pamela Gorkin
Mathematics, Bucknell University, Lewisburg, PA, USA

Jessica Sidman
Department of Mathematics and Statistics, Amherst College, Amherst, MA, USA

Advisory Editors

Colin Adams, *Williams College, Williamstown, MA, USA*
Jayadev S. Athreya, *University of Washington, Seattle, WA, USA*
Nathan Kaplan, *University of California, Irvine, CA, USA*
Jill Pipher, *Brown University, Providence, RI, USA*
Jeremy Tyson, *University of Illinois at Urbana-Champaign, Urbana, IL, USA*

Undergraduate Texts in Mathematics are generally aimed at third- and fourth-year undergraduate mathematics students at North American universities. These texts strive to provide students and teachers with new perspectives and novel approaches. The books include motivation that guides the reader to an appreciation of interrelations among different aspects of the subject. They feature examples that illustrate key concepts as well as exercises that strengthen understanding.

Andreas Klappenecker • Hyunyoung Lee

Discrete Structures

 Springer

Andreas Klappenecker
Dept. of Computer Science & Engineering
Texas A&M University
College Station, TX, USA

Hyunyoung Lee
Dept. of Computer Science & Engineering
Texas A&M University
College Station, TX, USA

ISSN 0172-6056　　　　　　　　ISSN 2197-5604　(electronic)
Undergraduate Texts in Mathematics
ISBN 978-3-031-73433-5　　　　ISBN 978-3-031-73434-2　(eBook)
https://doi.org/10.1007/978-3-031-73434-2

Mathematics Subject Classification: 68R01, 68R05, 05-01, 39-01

© The Editor(s) (if applicable) and The Author(s), under exclusive license to Springer Nature Switzerland AG 2025

This work is subject to copyright. All rights are solely and exclusively licensed by the Publisher, whether the whole or part of the material is concerned, specifically the rights of translation, reprinting, reuse of illustrations, recitation, broadcasting, reproduction on microfilms or in any other physical way, and transmission or information storage and retrieval, electronic adaptation, computer software, or by similar or dissimilar methodology now known or hereafter developed.
The use of general descriptive names, registered names, trademarks, service marks, etc. in this publication does not imply, even in the absence of a specific statement, that such names are exempt from the relevant protective laws and regulations and therefore free for general use.
The publisher, the authors and the editors are safe to assume that the advice and information in this book are believed to be true and accurate at the date of publication. Neither the publisher nor the authors or the editors give a warranty, expressed or implied, with respect to the material contained herein or for any errors or omissions that may have been made. The publisher remains neutral with regard to jurisdictional claims in published maps and institutional affiliations.

This Springer imprint is published by the registered company Springer Nature Switzerland AG
The registered company address is: Gewerbestrasse 11, 6330 Cham, Switzerland

If disposing of this product, please recycle the paper.

Preface

Discrete mathematics is concerned with the study of finite or countably infinite objects. It is a mélange of topics from logic, set theory, algebra, combinatorics, number theory, and other areas of mathematics rather than a mathematical discipline itself. Computer scientists pragmatically characterize it as the subject that provides the mathematical foundation for computing. In particular, it supplies the tools to analyze algorithms and data structures.

Since discrete mathematics consists of many different subjects, it uses a great variety of methods to solve problems. The aim of this book is to bring some coherency into these seemingly incongruent subjects. We begin by laying a solid foundation for the study of discrete structures by discussing logic, sets, and mathematical proofs. We systematically develop methods to find the closed form for finite sums. We deduce combinatorial methods from the foundations that we have laid.

The methods of discrete mathematics are often seemingly simple. For example, anyone can immediately grasp the pigeonhole principle, but may be dazzled by the sophisticated applications of the principle. Generally, it requires a lot of practice until one is able to master the methods. We have included 690 exercises of various levels of difficulty in this book. We encourage the reader to study the examples in the text and solve many of these exercises.

Chapter Synopses. The first chapter of this book discusses recreational problems that is meant to excite students. The solutions to these recreational problems illustrates the virtue of abstract methods. The reader should pay close attention to the proofs given in this chapter, as they illustrate styles of mathematical reasoning that will be examined in greater depth later.

The second chapter discusses propositional logic and a little bit of predicate logic, before exposing the student to the main principles of mathematical arguments.

The axiomatic method is illustrated by giving a brief introduction to Zermelo-Fraenkel set theory. As an application, we show how set theory can be used to derive profound limitations of computing.

The book contains an entire chapter devoted to proofs by induction, giving a thorough exposition to this fundamental proof technique.

In Chapter 5, we discuss equivalence relations and their application to the construction of number systems. We construct the set of integers and the set of rational numbers. Furthermore, we discuss the basics of modular arithmetic.

In Chapter 6, we discuss partial orders, strict orders, and cover relations. We give an introduction to lower and upper bounds, infima and suprema, and lattices.

The ubiquitous floor and ceiling functions are discussed in Chapter 7. These functions allow us for example to derive the number of digits of a positive integer that is written in base b, among other applications. This chapter eases the reader into topics in number theory.

Chapter 8 contains a more thorough discussion of topics in number theory. We discuss greatest common divisors and their applications in solving linear Diophantine equations. We discuss the RSA public key cryptosystem after introducing linear congruence equations and the Chinese remainder theorem.

The second part of the book concerns sums and asymptotic notations. We discuss the calculus of finite differences in Chapter 9. We will see that many sums can be turned into telescoping sums that are easy to evaluate.

We discuss asymptotic notations in Chapter 10. These notations allow us to characterize the limiting behavior of a function in terms of simpler functions. The main benefit is that bounds on the growth of function can be expressed in terms of functions that are easier to understand.

The last part of the book is concerned with combinatorial methods. In Chapter 11, we discuss fundamental counting principles. The chapter introduces combinatorial proofs that are often more insightful than inductive arguments. The combinatorial interpretation of falling factorials, binomial coefficients, Stirling numbers and other counting coefficients gives the student a deeper appreciation of their significance.

The next two chapters are concerned with generating functions and recurrence relations. Chapter 12 introduces the basic operations on generating functions and illustrates how they can simplify the solution to some counting problems. In Chapter 13, generating functions are used to solve recurrence relations.

Chapter 14 is concerned with undirected graphs. We discuss basic notions of graph theory and common examples of graphs. Simple graph-theoretic arguments are illustrated for connected graphs, trees, and planar graphs. A section on graph coloring provides an interesting problem that inspired many developments in graph theory.

In Chapter 15 we give an introduction of probability theory. We confine ourselves to countable sample spaces, so the entire theory can be developed without the need for measure-theoretic tools. We begin with combinatorial probability theory, which is essentially a variation on counting. Then we discuss the most important notions of elementary probability theory. A brief discussion of the probabilistic method concludes this chapter.

Preface

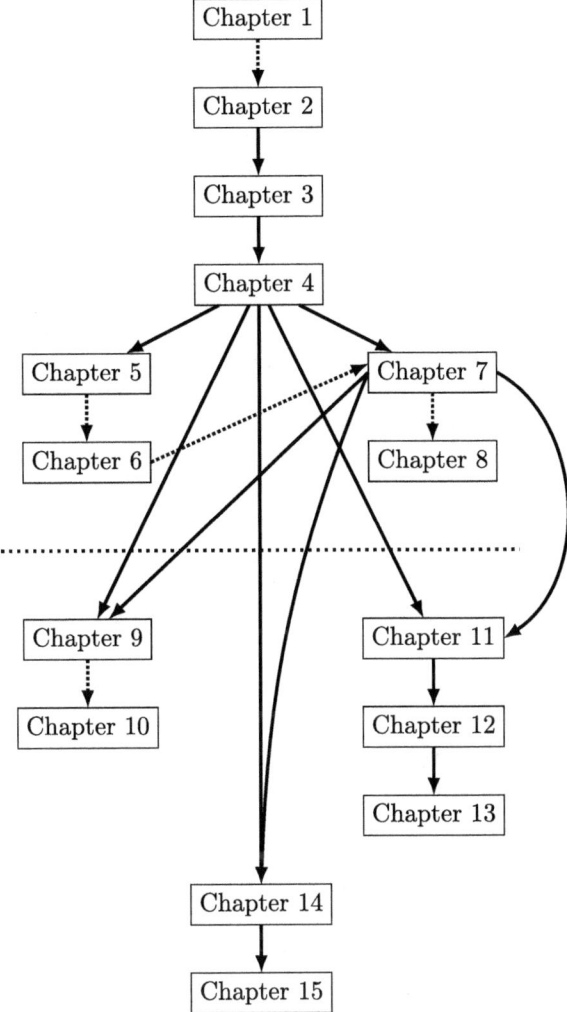

The main chapter dependencies are shown in this graph. A solid arrow Chapter A → Chapter B indicates that Chapter A should be studied before Chapter B. A dotted arrow means that it would be our recommendation to read the chapters in this order. We recommend to study Chapters 1–8 from the first part of the book before the later chapters.

For the Instructor. As an audience, we had undergraduate students in mind that have at least mastered precalculus. In a few instances, we made some remarks (and included a few exercises) for the benefit of students that have already taken calculus. These remarks and exercises are clearly marked, and can be ignored without loss of continuity. Furthermore, some more advanced material is contained in starred sections. The starred sections can be skipped at first reading without interrupting the flow.

The general outline of chapter dependencies is outlined in the chapter dependency graph. The order of the first four chapters is fairly fixed. Chapters 5, 6, and 7 are fairly independent, so they can be studied in any order. However, we recommend studying Chapter 5 before Chapter 6, as this is a natural progression in difficulty. Also, it should be noted that some exercises in Chapter 7 require material from Chapter 6. Chapter 8 is optional, as no other chapter depends on it. It should be evident from the chapter dependency graph that there is more flexibility in choosing the order of study for the later parts of the book.

We have taught this material in many different versions. For a semester course, we have taught for instance Chapters 1–7, Chapters 10–13, and Chapter 14. For more advanced students (such as an honors section), one could include the starred sections in Chapter 4 and end the course with Chapter 15 instead of Chapter 14. Of course, many other variations are possible.

When teaching in a quarter system, covering Chapters 1–7 followed by a selection of about two more advanced chapters is a good choice for a single term. One can accommodate a faster pace by omitting starred sections. The entire content of this text can be comfortably covered in two terms of the quarter system.

College Station, Texas, 2024   Andreas Klappenecker and Hyunyoung Lee

Acknowledgements. We thank our students who gave us feedback in numerous seminars and courses on discrete structures over the years. In particular, we would like to thank Andrew Nemec for many detailed comments. We also would like to thank the developers of `perusall.com`, especially Brian Lukoff, for allowing our students to directly comment on the manuscript. This feedback proved to be invaluable!

We would also like to thank the anonymous referees and the series editors of UTM for very valuable feedback. Above all, our heartfelt thanks go to Loretta Bartolini and Elizabeth Loew for their advice, guidance, and patience!

Contents

Contents		ix
Notation		xiii
I	**Discrete Structures**	**1**
1	**Introduction**	**3**
	1.1 Knight's Tour	3
	1.2 Notes	10
2	**Mathematical Arguments**	**11**
	2.1 Statements	11
	2.2 Logical Operations	13
	2.3 Logical Equivalence	20
	2.4 Logical Consequence	25
	2.5 Formal Arguments	28
	2.6 Predicates and Quantifiers	34
	2.7 Negations	39
	2.8 Proofs	42
	2.9 Notes	50
3	**Sets**	**51**
	3.1 Background and Motivation	51
	3.2 Fundamental Concepts	52
	3.3 Intersections and Unions	60
	3.4 Differences and Symmetric Differences	63
	3.5 Cartesian Products	67
	3.6 Relations	69
	3.7 Functions	72
	3.8 Numbers	78
	3.9 Cardinality	82
	3.10 Notes	88

4 Proofs by Induction — 89
- 4.1 Perfect Squares — 89
- 4.2 Bernoulli's Inequality — 93
- 4.3 Fibonacci Numbers — 95
- 4.4 Geometric Series — 97
- 4.5 Binomial Theorem — 99
- 4.6 Strong Induction — 101
- *4.7 Well-founded Induction — 106
- *4.8 Recursion — 117
- *4.9 Recursively Defined Sets — 122
- 4.10 Notes — 131

5 Equivalence Relations — 133
- 5.1 Generalities — 133
- 5.2 Integers — 139
- 5.3 Modular Arithmetic — 140
- 5.4 Rational Numbers — 142
- 5.5 Notes — 144

6 Partial Orders and Lattices — 145
- 6.1 Partial Orders — 145
- 6.2 Strict Order — 147
- 6.3 Cover Relations and Hasse Diagrams — 149
- 6.4 Dilworth's Theorem — 152
- 6.5 Lower and Upper Bounds — 157
- 6.6 Extensions of Partial Orders — 161
- 6.7 Monotonic Functions — 164
- 6.8 Lattices — 166
- 6.9 Notes — 169

7 Floor and Ceiling Functions — 171
- 7.1 Rounding Up and Down — 171
- 7.2 Divisibility and Primes — 177
- 7.3 Functions of Floors and Ceilings — 180
- 7.4 Notes — 184

8 Number Theory — 185
- 8.1 Divisibility — 185
- 8.2 The Greatest Common Divisor — 187
- 8.3 Linear Diophantine Equations — 191
- 8.4 Linear Congruence Equations — 196
- 8.5 The Chinese Remainder Theorem — 198
- 8.6 The RSA Public Key Cryptosystem — 202
- 8.7 Notes — 205

II Summation and Asymptotics — 207

9 Sums — 209
- 9.1 A Motivating Example — 209
- 9.2 Difference Calculus — 211
- 9.3 Falling Factorial Powers — 217
- 9.4 Stirling Numbers — 221
- 9.5 The Fundamental Theorem of Summation — 224
- *9.6 Analysis of Programs — 231
- 9.7 Notes — 235

10 Asymptotic Analysis — 237
- 10.1 Asymptotic Equality — 237
- 10.2 Limit Superior and Limit Inferior — 243
- 10.3 Asymptotically Tight Bounds — 250
- 10.4 Asymptotic Upper Bounds — 254
- 10.5 Asymptotic Lower Bounds — 259
- 10.6 Analysis of Algorithms — 260
- 10.7 Notes — 264

III Combinatorics — 265

11 Counting — 267
- 11.1 Fundamental Counting Principles — 267
- 11.2 Permutations and Combinations — 273
- 11.3 Combinatorial Proofs — 276
- 11.4 Selections with Repetitions — 284
- 11.5 Set Partitions — 288
- 11.6 The Inclusion-Exclusion Principle — 290
- 11.7 Pigeonhole Principle — 296
- 11.8 Notes — 299

12 Generating Functions — 301
- 12.1 The Basic Concept — 301
- 12.2 Operations on Generating Functions — 303
- 12.3 Elementary Generating Functions — 311
- 12.4 Giving Change — 315

13 Recurrence Relations — 319
- 13.1 Recurrence Relations — 319
- 13.2 A Motivating Example — 323
- 13.3 Fibonacci Sequence — 325
- 13.4 Partial Fractions — 328
- 13.5 Reciprocal Polynomials — 331
- 13.6 Linear Homogeneous Recurrence Relations — 333

13.7 Characteristic Polynomials	**336**
13.8 Inhomogeneous Linear Recurrence Relations	**339**
13.9 Catalan Numbers	**343**
13.10 Notes	**347**

14 Graphs — **349**
- 14.1 Undirected Graphs — **349**
- 14.2 Common Graphs — **354**
- 14.3 Connected Graphs — **358**
- 14.4 Trees — **362**
- 14.5 Planar Graphs — **364**
- 14.6 Graph Coloring — **367**
- 14.7 Hamiltonian Cycles and Paths — **374**

15 Probability Theory — **381**
- 15.1 Probability Spaces — **381**
- 15.2 Combinatorial Probability — **385**
- 15.3 Conditional Probabilities — **389**
- 15.4 Independence — **394**
- 15.5 Random Variables — **398**
- 15.6 Expectation — **401**
- 15.7 The Probabilistic Method — **406**
- 15.8 Notes — **411**

Bibliography — **413**

Index — **419**

Notation

Logic

$A \wedge B$	conjunction, A and B
$A \vee B$	disjunction, A or B
$\neg A$	negation, not A
$A \to B$	implication, A implies B
$A \leftrightarrow B$	equivalence, A if and only if B
$v[\![A]\!]$	a valuation v of the Boolean formula A
$A \equiv B$	logical equivalence, so $v[\![A]\!] = v[\![B]\!]$ for all valuations v
$\{P_1, \ldots, P_n\} \models A$	A is a logical consequence of the premises $P_1, \ldots, P_n$
$\{P_1, \ldots, P_n\} \vdash A$	A can be deduced from the premises $P_1, \ldots, P_n$
$\forall n\, P(n)$	the predicate $P(n)$ holds for all n in the universe
$\exists n\, P(n)$	the predicate $P(n)$ holds for some n in the universe

Sets

$\mathbf{N}_0$	the set of nonnegative integers $\{0, 1, 2, 3, \ldots\}$
$\mathbf{N}_1$	the set of positive integers $\{1, 2, 3, \ldots\}$
$\mathbf{Q}$	the set of rational numbers
$\mathbf{R}$	the set of real numbers
$\mathbf{Z}$	the set of integers
$\varnothing$	the empty set
$A \subseteq B$	A is a subset of the set B
$A \subsetneq B$	A is a proper subset of the set B
$A \cap B$	intersection of the sets A and B
$A \cup B$	union of the sets A and B
$A - B$	set of elements in A that are not in the set B, same as $A \smallsetminus B$
$A \smallsetminus B$	set of elements in A that are not in the set B, same as $A - B$
$A^\complement$	complement of A with respect to a universe U, so $A^\complement = U \smallsetminus A$
$P(A)$	power set of A, the set of all subsets of A,
2^A	alternate notation for power sets, $2^A = P(A)$
$|A|$	cardinality of the set A

Functions

$f\colon A \to B$	function with domain A and codomain B
$\operatorname{dom}(f)$	domain of the function f
$\operatorname{ran}(f)$	range of the function, $\operatorname{ran}(f) = \{f(x) \mid x \in \operatorname{dom}(f)\}$
$f \upharpoonright X$	restriction of the function to X, also denoted as $f\big|_X$
f^{-1}	inverse of the function f or preimage of f
$g \circ f$	composition of functions, $g \circ f(x) = g(f(x))$
i_A	identity map on a set A

Sums and Products

$\sum_{k=1}^{n} f(n)$	summation, $\sum_{k=1}^{n} f(n) = f(1) + f(2) + \cdots + f(n)$
Δ	difference operator
Δ^{-1}	summation operator
$\prod_{k=1}^{n} f(n)$	product, $\prod_{k=1}^{n} f(n) = f(1)f(2) \cdots f(n)$

Number Theory

$a \mid b$	integer a divides the integer b
$a \nmid b$	integer a does not divide the integer b
$a \equiv b \pmod{n}$	the integer $a - b$ is a multiple of n

Combinatorics

$k!$	factorial, $k! = 1 \cdot 2 \cdots (k-1) \cdot k$
$n^{\underline{k}}$	falling factorial power, $n^{\underline{k}} = n(n-1) \cdots (n-k+1)$
$n^{\overline{k}}$	rising factorial power, $n^{\overline{k}} = n(n+1) \cdots (n+k-1)$
$\binom{n}{k}$	binomial coefficient
$\binom{n}{k_1, k_2, \ldots, k_m}$	multinomial coefficient
$\left\{ {n \atop k} \right\}$	Stirling number of the second kind, also denoted as $S(n,k)$

Graph Theory

$V(G)$	vertex set of a graph G
$E(G)$	edge set of a graph G
$N(v)$	neighborhood of the vertex v
$N[v]$	closed neighborhood of the vertex v
$\deg v$	degree of the vertex v
$\delta(G)$	minimal degree of a graph G
$\Delta(G)$	maximal degree of a graph G
$d(u,v)$	distance between vertices u and v
$d(G)$	diameter of the graph G
$\alpha(G)$	independence number of G
$\chi(G)$	chromatic number of G
$\omega(G)$	clique number of G
$k(G)$	number of components of G
E_n	empty graph with n nodes
P_n	path graph with n nodes
C_n	cycle graph with n nodes
K_n	complete graph with n nodes
$K_{m,n}$	complete bipartite graph with $m+n$ nodes
Q_n	hypercube graph with 2^n nodes
$\overline{G}$	complementary graph of G
$G \square H$	Cartesian product of the graphs G and H

Probability Theory

Ω	a sample space, assumed to be finite or countable
$\Pr$	probability measure, sometimes denoted by μ
$\Pr[A]$	probability of event A
$\Pr[A \mid B]$	conditional probability of A given B
$E[X]$	expected value of X

Part I

Discrete Structures

Chapter 1

Introduction

> *Problems worthy*
> *of attack*
> *prove their worth*
> *by hitting back.*
>
> — Piet Hein, *Grooks*

Many areas of mathematics have a long history. For instance, calculus has been studied for hundreds of years. By contrast, courses on discrete structures emerged more recently during the past 50 years. They were designed to provide a mathematical support of computer science. Yet, bits and pieces of this subject have a longer history. We will begin with a playful recreational problem that fascinated Euler. The problem asks whether it is possible to move a knight on a chessboard such that it will visit each square of the chessboard precisely once. We will show that the question whether or not a solution exists depends on the size of the chessboard.

Euler was prompted by this and other recreational problems to invent graph theory. Even though graphs are not absolutely necessary to solve the knight's tour problem, they make the reasoning about the problem simpler. This motivational chapter will show you how to formalize the problem. We will also illustrate how to argue about the existence or nonexistence of solutions to the problem.

1.1 Knight's Tour

A knight can move on a chessboard either two squares in horizontal direction and one square in vertical direction or alternatively one square in horizontal direction and two squares in vertical direction. Thus, there are up to eight possible **knight moves** from a given square, as shown in Fig. 1.1. In other words, the knight moves according to the rules of the game of chess.

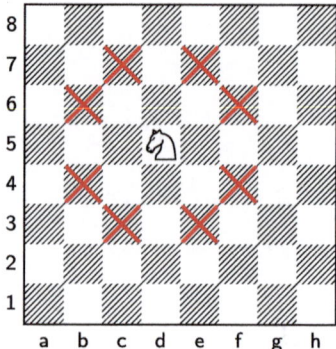

Figure 1.1: All possible eight knight's moves starting from the square d5 are marked with a cross

A sequence of knight moves that visits every square of the chessboard precisely once is called a **knight's tour**. If the end square of the knight's tour is within a knight's move of the beginning square, then the knight's tour is called **closed**. A famous ancient chess puzzle asks to find a closed knight's tour on the chessboard.

Let us first study the problem on smaller chessboards. A chessboard with 3 rows and 4 columns is the smallest size that allows one to find knight's tours. It is not difficult to find a knight's tour for the 3×4 rectangular chessboard, see Fig. 1.2.

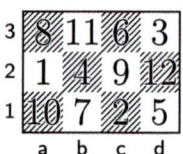

Figure 1.2: A knight's tour on a 3×4 chessboard starting at a2 and ending at d2. This tour is not closed, though

The knight's tour given in Fig. 1.2 is not unique. In Exercise 1.1, you are encouraged to find more knight's tours on a 3×4 chessboard. You might have difficulties finding a closed knight's tour for this size chessboard for the reason explained in Exercise 1.2.

One can work out the knight problems on the boards alone, but a drawback is that the board lacks the structure of the moves. A good approach is to formulate the possible moves with the help of a graph, a useful datatype that we will discuss in more detail later.

Before explaining the concept of a graph, let us recall the notion of a set. A **set** is an unordered collection of distinct objects, called the **elements** of the set. A finite set can be defined by listing its elements in curly braces. For instance, the set $A = \{a, b, c, d\}$ contains four elements, namely a, b, c and d.

1.1 Knight's Tour

The order of the elements does not matter, so $\{b, c, a, d\}$ is another way of specifying the same set. We write $a \in A$ to denote that a is an element of the set A. A set B is called a **subset** of a set A if and only if every element of B is an element of A. An n-element subset B of A is a subset of A that contains precisely n elements. For instance, $B = \{a, d\}$ is a 2-element subset of A.

A **graph** is a pair (V, E) of sets such that the set E consists of 2-element subsets of V. The elements of V are called **vertices** or **nodes** of the graph, and the elements of E are called its **edges**. We represent a vertex by a point in the plane and edges by line segments or curves connecting the points of the two vertices representing the edge.

The **knight graph** of an $n \times m$ chessboard consists of a set of nm vertices, where each vertex represents one square of the chessboard. Two vertices are connected by an edge if and only if the corresponding squares on the chessboard are one knight's move apart. Thus, the knight graph models the possible movements of a knight on the chessboard.

Example 1.1. The knight graph (V, E) of a 4×4 chess board has a set V of 16 vertices modeling the squares of the chess board and a set E of 24 edges modeling the knight moves. The set V of vertices is given by

$$V = \{a1, a2, a3, a4, b1, b2, b3, b4, c1, c2, c3, c4, d1, d2, d3, d4\},$$

and the set E of edges by

$$\begin{aligned}E = \{&\{a1,b3\}, \{a1,c2\}, \{a4,b2\}, \{a4,c3\}, \{a2,c1\}, \{a2,c3\}, \{a2,b4\}, \{a3,c2\},\\&\{a3,c4\}, \{a3,b1\}, \{b1,d2\}, \{b1,c3\}, \{b2,d1\}, \{b2,d3\}, \{b2,c4\}, \{b3,c1\},\\&\{b3,d2\}, \{b3,d4\}, \{b4,c2\}, \{b4,d3\}, \{c1,d3\}, \{c2,d4\}, \{c3,d1\}, \{c4,d2\}\}.\end{aligned}$$

The edge $\{a1,c2\}$ formalizes, for instance, that the squares a1 and c2 can reach each other through a knight's move. As aforementioned, the vertices of a graph are often illustrated by a dot and edges by a line segment or a curve. Figure 1.3 shows the knight graph (V, E) of the 4×4 chess board in this diagrammatic form.

A **walk** in a graph (V, E) is a sequence of edges $e_1, \ldots, e_k$ in E such that $e_i = \{v_i, v_{i+1}\}$. Notice that the edge e_i and the subsequent edge e_{i+1} share the vertex v_{i+1}. So a walk models a sequence of vertices

$$(v_1, v_2, \ldots, v_k, v_{k+1})$$

that are connected by traveling along the edges from $e_1, e_2, \ldots, e_k$. In a walk, a vertex might be visited multiple times.

Example 1.2. One walk on the knight graph of the 4×4 chessboard is, for instance, given by

$$(\{a1,c2\}, \{c2, b4\}, \{b4, d3\}, \{d3, b4\}, \{b4, a2\}).$$

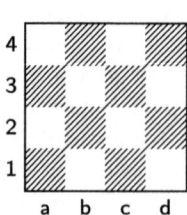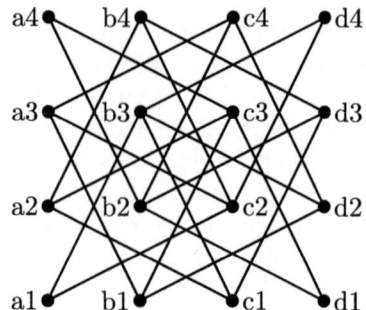

Figure 1.3: A 4 × 4 chessboard and the corresponding knight graph. The vertices b2 and d3 are connected by an edge, since these squares on the chessboard are one knight's move apart

It describes a walk consisting of five knight moves. The first move starts at the square a1, then moves to c2. From c2, the knight moves to b4, then d3, and back to b4. In the final move, the knight reaches the square a2. A less verbose description of this walk is given by (a1, c2, b4, d3, b4, a2).

A **path** in a graph is a walk with pairwise distinct vertices. Thus, a path is a walk

$$(v_1, v_2, \ldots, v_k, v_{k+1})$$

that does not revisit vertices, so v_i is not the same as v_j when the indices i and j are distinct.

A **Hamiltonian path** in a graph is a path that visits *every* vertex of the graph exactly once. A knight's tour corresponds to a Hamiltonian path in the knight graph.

A **cycle** in a graph is a walk such that the vertices are pairwise distinct except that the first and the last vertex are the same. A cycle of a graph is called a **Hamiltonian cycle** if and only if it visits every vertex of the graph exactly once. A closed knight's tour corresponds to a Hamiltonian cycle in the knight graph.

Proposition 1.3. *There does not exist a closed knight's tour on a 4 × 4 chessboard.*

Proof. How can we go about showing that a closed knight's tour does not exist? Well, we can show that there is going to be a problem with any purported knight's tour, which will force us to conclude that no knight's tour can exist. This method of proof is called a proof by contradiction and will be studied in more detail in the next chapter.

Seeking a contradiction, let us assume that a closed knight's tour exists, so there must exist a Hamiltonian cycle in the 4 × 4 knight graph. Then each vertex is adjacent to precisely two edges of the Hamiltonian cycle. The vertices a1 and d4 are forced to use the dashed edges in the knight graph illustrated as follows.

1.1 Knight's Tour

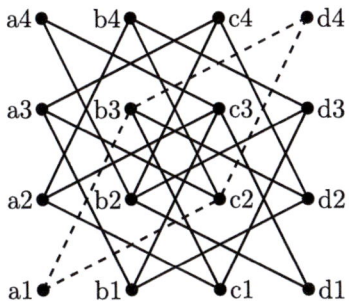

Then the dashed cycle must be a subpath of the Hamiltonian cycle, which is impossible as {a1, b3, c2, d4} cannot be connected to any other nodes on the Hamiltonian path. □

The previous proposition showed that a closed knight's tour is impossible on a 4×4 chessboard. We will now take advantage of the obstacle to the Hamiltonian cycle on the 4×4 board to systematically construct a closed knight's tour on the original 8×8 board.

Theorem 1.4. *There exists a closed knight's tour on an 8×8 chessboard.*

Proof. We will subdivide the 8×8 board into four 4×4 boards. On each 4×4 board, a knight can take the following paths:

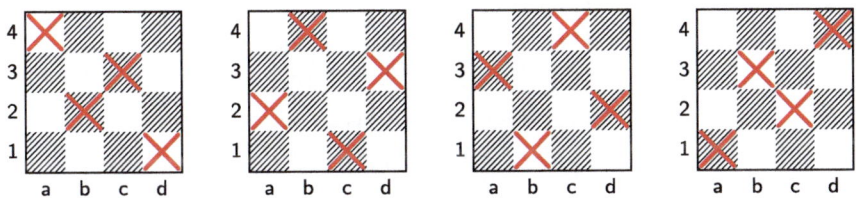

Notice that the four paths partition the 4×4 board. The idea is to walk in one quadrant according to one of these patterns and then connect it to another such pattern in the same or in another quadrant. Specifically, we get the following four paths

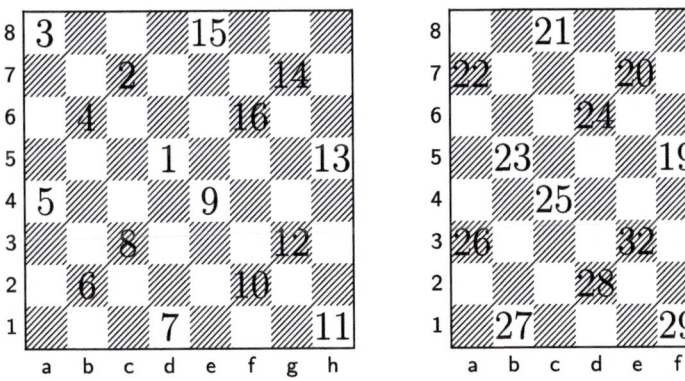

and

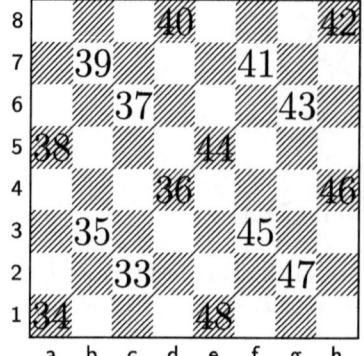 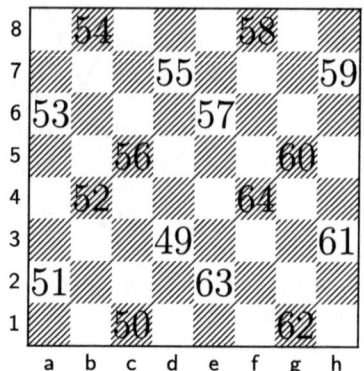

These four paths partition the entire board. Putting it all together, we get the following closed knight's tour:

	a	b	c	d	e	f	g	h
8	3	54	21	40	15	58	17	42
7	22	39	2	55	20	41	14	59
6	53	4	37	24	57	16	43	18
5	38	23	56	1	44	19	60	13
4	5	52	25	36	9	64	31	46
3	26	35	8	49	32	45	12	61
2	51	6	33	28	63	10	47	30
1	34	27	50	7	48	29	62	11

This solution is not unique. In fact, there are 26,534,728,821,063 other closed knight tour solutions as well. □

Despite the fact that there are so many knight's tours on an 8 × 8 board, most people find it challenging to discover one. In general, it is a good idea to solve smaller versions of the problem as we did at the beginning of this section, to gain insight into the problem.

If you lack the patience to solve it by hand, then you might want to solve some small cases with the help of a computer. A general strategy to solve such combinatorial problems is given by backtracking. The idea is to start from a square on the chessboard and incrementally build up a knight's tour from this square. At each step, we make a knight's move to another square that has not been visited yet. We keep going on in this fashion until it is impossible to move to another square that has not been visited yet. If we found a tour, then we can print it; otherwise, we backtrack to a previous square that offers still alternative moves and try the next move.

1.1 Knight's Tour

For example, suppose that we want to find a knight's tour on a 3×4 chessboard starting from the square a3. Then after nine knight moves, one arrives at the configuration shown on the left in the next figure.

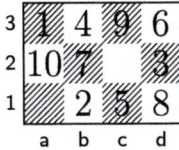

This partial tour cannot be completed, as there are no further knight moves possible from position 10. One has to backtrack five steps to find a square that offers an alternative move. Instead of moving from position 5 at the square c1 to the square d3, as we did previously, we move instead to the square at a2. The resulting partial tour is shown on the far right in the next figure.

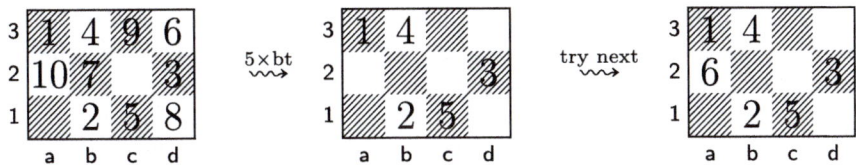

Starting from the partial tour shown on the far right, one notices after a few steps that this partial tour cannot be completed either. Since there are no other knight's moves available from the square at position 5, we have to backtrack further to the square at position 4 and try an alternative knight's move from there.

It is an instructive exercise to write a recursive program that implements a backtrack search for a knight's tour, see Exercise 1.6.

EXERCISES

1.1. Find at least two more knight's tours on a 3×4 chessboard. These tours should be essentially different (that is, don't just differ in direction).

1.2. Show that there does not exist a closed knight's tour on a 3×4 chessboard. Hint: Mark the columns a and d with crosses and the columns b and c with circles. Where can the knight move if it sits on a field with a cross (respectively, circle)? Now consider a putative closed knight's tour.

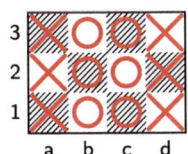

1.3. (a) Give a simple argument showing that there cannot exist a knight's tour on a 3×3 chessboard. (b) Give the knight's graph of the 3×3 chessboard and explain what additional insights are provided by this graph representation.

1.4. Suppose that you are given an $m \times n$ chessboard such that the product nm is an even integer. Show that the knight's tour problem cannot be solved when the first and the last square of the tour have the same color.

1.5. Suppose that you are given an $m \times n$ chessboard such that nm is an odd integer. Show that there cannot exist a closed knight tour.

1.6. Write a recursive program that searches for a (open) knight's tour by backtracking for an $n \times m$ chessboard. The input should be a starting square on the chessboard.

1.2 Notes

Finding a knight's tour is a recreational chess problem. It was studied by Euler, who gave an open knight's tour [26]. It is a special case of the Hamiltonian path problem on the knight's graph. The knight's tour problem can be solved in linear time. By contrast, deciding whether a general graph has a Hamiltonian path is NP-complete. NP-complete problems have solutions that are easy to verify, but there are no known efficient (polynomial time) algorithms that can solve NP-complete problems. Thus, it seems unlikely that one can find a Hamiltonian path in polynomial time.

Chapter 2

Mathematical Arguments

> *My dear friend, I'd advise, in sum,*
> *First, the Collegium Logicum.*
> *There your mind will be trained,*
> *As if in Spanish boots, constrained,*
> *So that painfully, as it ought,*
> *It creeps along the way of thought, ...*
>
> — Johann Wolfgang von Goethe, *Faust I*

In this chapter, we will discuss the structure of mathematical arguments. We begin with an overview of propositional logic. We first introduce propositions and logical operators. Then we discuss tautologies, satisfiability, and logical equivalence. We discuss a propositional calculus that gives us a first glimpse into the method of formal proofs. We sketch the rudiments of predicate logic including universal and existential quantifiers, predicates, and their interpretations. We conclude this chapter with a discussion of basic proof techniques. Furthermore, we will learn how to properly negate statements and how to leverage this knowledge in proofs by contradiction.

2.1 Statements

In mathematics, a **statement** is a sentence that is either true or false. Statements are also called **assertions** or **propositions**. For example, the sentence

There exists a closed knight's tour on an 8×8 chessboard

is a true statement. We showed in Chap. 1, Proposition 1.3, that the sentence

There exists a closed knight's tour on a 4×4 chessboard

is a false statement.

Sometimes it is difficult to decide whether a mathematical statement is true or false. The innocuous assertion

The number 2^e is irrational

is a mathematical statement, since it is certainly either true or false. However, so far nobody has been able to settle whether 2^e is a rational number of not. There are many similar mathematical statements that suffer from the same predicament.

In general, how do we go about proving a mathematical statement? We assume some assertions that are believed to be true as **axioms**. Then we use logical reasoning to derive the statement from these axioms. Euclid nicely illustrated this deductive method in his *Elements*, a most influential textbook written around 300BC that developed the foundations of Euclidean geometry and the beginnings of number theory.

It should be self-evident that one needs to understand the rules of logic to construct a valid mathematical argument. Therefore, we give a brief discussion of symbolic logic. We conclude this chapter by discussing some proof strategies that illustrate how to apply the logical reasoning in mathematical arguments.

EXERCISES

2.1. Which of the following sentences are mathematical statements? Justify your answers.
(a) The number π is the smallest irrational real number.
(b) The number π is an irrational number.
(c) The quadratic equation $x^2 - 4x + 2 = 0$ does not have a solution in the integers.
(d) The inequality $123 - 100 > 23$ holds.
(e) The positive integer x is a prime number.

2.2. Which of the following sentences are statements? Justify your answers.
(a) Is 17 a prime?
(b) $12x + 4 = 10$.
(c) Does there exist a knight's tour on a 31×31 chessboard?
(d) The positive integer x is the sum of two squares.
(e) Every even integer greater than 2 can be expressed as the sum of two primes.

2.3. For each of the following statements, determine whether the statement is true or false. Explain why.
(a) The real number $0.111\cdots$ is equal to $1/9$.
(b) The real number $0.1212121212\cdots$ is not a rational number.
(c) The greatest common divisor of 1111 and 11111111 is equal to 1.
(d) $(-1)(-1) = 1$.
(e) The set of positive integers $\{1, 2, 3, \ldots\}$ is infinite.

2.4. Research the following statement: There exist infinitely many pairs of prime numbers that differ by 2. It is evidently a statement, but is it true?

2.2 Logical Operations

2.5. Consider the statement "This statement is false." Is it true or false? Discuss.

2.2 Logical Operations

Logical operations allow one to combine several statements into a new compound statement. We discuss the disjunction, conjunction, negation, implication, and equivalence operations. You need to be aware of the precise meaning of these operations so that you can properly follow a mathematical argument.

Disjunction. Let A and B be statements. When is "A or B" true? In English, the word "or" is sometimes used in the inclusive sense "A or B, or both" and sometimes in the exclusive sense "either A or B, but not both." For example, if a restaurant offers complimentary coffee or tea after a meal, they don't expect you to choose both. If they offer milk or sugar with your coffee, then it is meant that you can select milk, sugar, or both.

In mathematics, one cannot have this kind of ambiguity. Mathematicians agreed that the **disjunction** "A or B" is always understood in the inclusive way "A or B, or both." For instance, if we consider a real number x, then

$$x \leqslant 0 \quad \text{or} \quad x \geqslant 0$$

is a true statement, even though $x = 0$ fulfills both $x \leqslant 0$ and $x \geqslant 0$.

In symbolic logic, we write $A \vee B$ for the mathematical statement A or B. The statements A and B can each be true or false, so there are four possibilities in all for the arguments of the disjunction. Writing T for true and F for false, we can express the behavior of the disjunction by the truth table

A	B	$A \vee B$
F	F	F
F	T	T
T	F	T
T	T	T

Note that "A or B" is false only when both the statement A and the statement B are false.

You might be wondering why we are concerned with statements that are false. Well, one proof technique first negates the statement and then tries to find a contradiction. So false statements occur quite often within mathematical arguments, even though we are ultimately concerned with establishing the truth of mathematical assertions that are stated in theorems, propositions, and lemmas.

Conjunction. The **conjunction** "A and B" is true if and only if both A is true and B is true. The conjunction is also written as $A \wedge B$. The truth table is given by

A	B	$A \wedge B$
F	F	F
F	T	F
T	F	F
T	T	T

For example, one might want to establish that $2 < e < 3$, which amounts to prove the conjunction of the inequalities

$$2 < e \quad \text{and} \quad e < 3.$$

Negation. Let A be a statement. Then "not A" or "$\neg A$" is the **negation** of the statement. For instance, if A is the statement "We do serve breakfast until 11:00am," then the negation $\neg A$ is "We do not serve breakfast until 11:00am." The statement A is true if and only if the negated statement $\neg A$ is false. The truth table is given by

A	$\neg A$
F	T
T	F

Implication. The **implication** "A implies B" is particularly important in proofs. It is often written in the form "If A, then B." We will also denote it as $A \to B$. We call A the **hypothesis** and B the **conclusion**.

If the hypothesis A is true and the conclusion B is true, then "A implies B" ought to be true. If the hypothesis A is true and the conclusion B is false, then "A implies B" ought to be false, since we do not want to conclude something that is false from a true hypothesis.

However, it is not so clear what the truth value of the implication should be if the hypothesis is false. Mathematicians agreed on the convention that an implication is true if its hypothesis is false. If the hypothesis A is false, then we say that the implication $A \to B$ is **vacuously true**. Thus, the truth table of the implication is given by

A	B	$A \to B$
F	F	T
F	T	T
T	F	F
T	T	T

It might seem strange that one can conclude anything from a false statement. However, even in everyday English we sometimes make a point by drawing an even more absurd conclusion from an absurd statement, such as *If this mafioso is innocent, then I will eat my hat*. Exercise 2.14 explains why mathematicians did not really have the option to define the implication in a different way.

Another important point is that an implication is not based on cause and effect. For example, if A is the statement "1+1=2" and B is the statement

2.2 Logical Operations

"blueberries are blue," then the implication $A \to B$ is true. Evidently, the correctness of $1 + 1 = 2$ does not cause blueberries to be blue. So merely the truth of the statements A and B determine whether the implication $A \to B$ is true.

Equivalence. If an implication $A \to B$ and its converse $B \to A$ are both true, then we say that A and B are **equivalent** and we write $A \leftrightarrow B$. The truth table of an equivalence is given by

A	B	$A \leftrightarrow B$
F	F	T
F	T	F
T	F	F
T	T	T

We see that $A \leftrightarrow B$ is true if and only if A and B have the same truth value. In the literature, an equivalence is sometimes also called a **bi-implication**.

Boolean Formulas. A **Boolean variable** is a variable that can take on the values T for true and F for false. We can construct **Boolean formulas** as follows.

B1. Every Boolean variable is a Boolean formula.

B2. If A is a Boolean formula, then so is $\neg A$.

B3. If A and B are Boolean formulas, then $(A \lor B)$, $(A \land B)$, $(A \to B)$, and $(A \leftrightarrow B)$ are Boolean formulas.

A Boolean formula is formed by applying **B1**–**B3** a finite number of times.

Example 2.1. We claim that

$$((\neg A \land B) \to (\neg A \lor B))$$

is a Boolean formula. Indeed, the Boolean variables A and B are Boolean formulas by **B1**. It follows that $\neg A$ is a Boolean formula by **B2**. By **B3**, $(\neg A \land B)$ and $(\neg A \lor B)$ are Boolean formulas. Using **B3** once more, we can conclude that $((\neg A \land B) \to (\neg A \lor B))$ is a Boolean formula.

By building up Boolean formulas starting from Boolean variables, their negations, and more and more complicated compound formulas, we can unravel the syntactic structure of the Boolean formula. However, this also allows us to form the truth table that assigns truth values to the Boolean variables. After a truth value is assigned to a Boolean variable, we can deduce the value of its negation. If the truth values are known for subformulas A and B, then we can deduce the truth value for expressions such as $(A \lor B)$, $(A \land B)$, $(A \to B)$, and $(A \leftrightarrow B)$.

Let us illustrate how to derive a truth table for the Boolean formula $((\neg A \wedge B) \rightarrow (\neg A \vee B))$ of the aforementioned example. The formula contains two Boolean variables, so there are a total for four assignments of truth values to these two Boolean variables A and B. The value of A determines $\neg A$. Knowing the truth values of $\neg A$ and B, we can deduce the truth values of $(\neg A \wedge B)$ and $(\neg A \vee B)$. After determining the truth values of $(\neg A \wedge B)$ and $(\neg A \vee B)$, we can deduce the truth value of the entire formula

$$((\neg A \wedge B) \rightarrow (\neg A \vee B)).$$

The resulting truth table is given by

A	B	$\neg A$	$(\neg A \wedge B)$	$(\neg A \vee B)$	$((\neg A \wedge B) \rightarrow (\neg A \vee B))$
F	F	T	F	T	T
F	T	T	T	T	T
T	F	F	F	F	T
T	T	F	F	T	T

We observe that regardless of the assignment of truth values to A and B, the Boolean formula $((\neg A \wedge B) \rightarrow (\neg A \vee B))$ always evaluates to true.

The definition of Boolean formulas mainly governs the syntactical aspect of the formulas. In other words, it determines whether a string of symbols forms a valid Boolean formula. We are able to give meaning to a Boolean formula by presenting its truth table. However, this approach does not scale well, as it is limited to Boolean formulas with a small number of Boolean variables. For this reason, we will introduce the notion of a valuation in the next section. A valuation is a useful abstraction of a row of a truth table, but it is more convenient to use when we reason about the meaning of Boolean formulas.

Knights and Knaves. The inimitable Raymond Smullyan popularized a great variety of amusing logic puzzles that are perfect to hone your skills in logical reasoning, see [67]. Imagine an island that has two kinds of inhabitants called knights and knaves. The knights have the property that they always speak the truth and the knaves have the peculiar property that they invariably lie. The problem is that one cannot tell the knights and knaves apart.

Now let us suppose you meet on the island two people $\mathcal{A}$ and $\mathcal{B}$. Person $\mathcal{A}$ says, "I am a knave but $\mathcal{B}$ is not." What can you conclude about $\mathcal{A}$ and $\mathcal{B}$?

For brevity, let us write A for the statement "$\mathcal{A}$ is a knight." and B for the statement "$\mathcal{B}$ is a knight." So $\mathcal{A}$ made the statement $\neg A \wedge B$. This statement is true if $\mathcal{A}$ is a knight, but it is false if $\mathcal{A}$ is a knave. However, we can combine both facts into the single statement

$$A \leftrightarrow (\neg A \wedge B), \tag{2.1}$$

which is always true by definition of the equivalence operator. The solution to the puzzle can now be found by considering the truth table of the statement

2.2 Logical Operations

(2.1). We first find the truth values of the subexpression $\neg A$, then of $\neg A \wedge B$, and finally of the statement $A \leftrightarrow (\neg A \wedge B)$,

A	B	$\neg A$	$\neg A \wedge B$	$A \leftrightarrow (\neg A \wedge B)$
F	F	T	F	T
F	T	T	T	F
T	F	F	F	F
T	T	F	F	F

The statement $A \leftrightarrow (\neg A \wedge B)$ is true if and only if A and B are both false. In other words, we can conclude that $\mathcal{A}$ and $\mathcal{B}$ are both knaves.

You should recognize that the solution was quite simple after we succeeded with the formalization of the problem. We used truth tables in our argument, since this illustrates the operations once more. You are encouraged to use truth tables until you become fluent with the meaning of the logical operations. We will learn more elegant proof techniques in the remainder of this chapter.

EXERCISES

2.6. Let A denote the statement "the supermarket is open," B the statement "I go shopping," and C the statement "the pharmacy is open." Translate the following statements into English phrases.
(a) $A \wedge C$,
(b) $A \vee C$,
(c) $A \to B$,
(d) $B \leftrightarrow (A \wedge C)$.

2.7. Let A denote the statement "Albert is happy," C the statement "Albert cooks pasta," and E the statement "Emmy is happy." Translate the following statements into English phrases.
(a) $(C \wedge \neg E)$,
(b) $(C \to (A \wedge E))$,
(c) $(E \leftrightarrow \neg A)$,
(d) $((A \wedge E) \vee \neg C)$.

2.8. Let C denote the statement "it is cloudy," R the statement "it is rainy," and S the statement "it is sunny." Formalize each of the following sentences:
(a) If it is cloudy, then it is not sunny.
(b) It is cloudy and rainy.
(c) It is rainy or it is cloudy and not sunny.
(d) It is sunny if and only if it is not cloudy.

2.9. Argue that each of the following expressions is a Boolean formula. Start with the Boolean variables occurring in the expression. Then show how the properties **B1**, **B2**, and **B3** are used to construct the entire given Boolean formula.
(a) $((A \to B) \to A)$

(b) $((\neg A \lor B) \land A)$
(c) $((A \land \neg B) \lor (A \land B))$
(d) $((\neg A \land \neg B) \leftrightarrow (A \land B))$

2.10. Explain why each of the following expressions is not a Boolean formula
(a) $\cdots (((A \to A) \to A) \to A) \to \cdots$
(b) $(((A \to B) \lor C)$

2.11. (a) How many rows are in the truth table of a Boolean formula f with n variables?
(b) How can you systematically generate all possible truth assignments for the n Boolean variables of F in lexicographic order?

2.12. Find the truth tables of the Boolean formulas
(a) $((A \to B) \to A)$,
(b) $(A \to (B \to A))$,
(c) $((A \to B) \to B)$,
(d) $(A \to (B \to B))$.

2.13. Find the truth table of the Boolean formula
$$((A \to B) \leftrightarrow \neg(A \land \neg B)).$$

2.14. In this exercise, you will demonstrate that there is really just one sensible choice for the implication operation. If the hypothesis A is true and the conclusion B is true, then the implication $A \to B$ should be true. If the hypothesis A is true and the conclusion B is false, then $A \to B$ should be false. However, we have four choices $\xrightarrow{a}, \ldots, \xrightarrow{d}$ for the implication when the hypothesis is false.

A	B	$A \xrightarrow{a} B$	$A \xrightarrow{b} B$	$A \xrightarrow{c} B$	$A \xrightarrow{d} B$
F	F	F	T	F	T
F	T	F	F	T	T
T	F	F	F	F	F
T	T	T	T	T	T

Give a compelling reason why $\xrightarrow{d}$ is the only sensible choice for the implication. [Hint: Express the equivalence $\leftrightarrow$ in terms of the newly defined implication.]

2.15. We have introduced five logical operators, namely the unary negation operation $\neg$, and the four binary operations conjunction $\land$, disjunction $\lor$, implication $\to$, and equivalence $\leftrightarrow$. There exist a total of $2^4 = 16$ binary logical operators, see Table 2.1. Show that all 16 binary logical operations can be expressed in terms of the five logical operations that we have introduced so far by completing the "equivalent formulation" column in terms of expressions using the operations $\{\neg, \land, \lor, \to, \leftrightarrow\}$.

2.16. Rewrite the following expressions using only operators from the given set S.

2.2 Logical Operations

Notation	Equivalent formulation	Truth table	Name
$A \perp B$	(1)	$FFFF$	Falsehood
$A \wedge B$	(2)	$FFFT$	Conjunction
$A \nrightarrow B$	(3)	$FFTF$	Nonimplication
$A \mathbin{\text{L}} B$	(4)	$FFTT$	Left projection
$A \nleftarrow B$	(5)	$FTFF$	Converse nonimplication
$A \mathbin{\text{R}} B$	(6)	$FTFT$	Right projection
$A \oplus B$	(7)	$FTTF$	Exclusive or
$A \vee B$	(8)	$FTTT$	Inclusive or
$A \barwedge B$	(9)	$TFFF$	Nor
$A \leftrightarrow B$	(10)	$TFFT$	Equivalence
$A \bar{\text{R}} B$	(11)	$TFTF$	Right complementation
$A \leftarrow B$	(12)	$TFTT$	Converse implication
$A \bar{\text{L}} B$	(13)	$TTFF$	Left complementation
$A \rightarrow B$	(14)	$TTFT$	Implication
$A \bar{\wedge} B$	(15)	$TTTF$	Nand
$A \top B$	(16)	$TTTT$	Tautology

Table 2.1: A table listing all 16 logical binary operators. The values of the truth tables are listed by respectively evaluating the operators for the arguments (A, B) in the order $(F, F), (F, T), (T, F)$, and (T, T)

(a) $A \nrightarrow B$ using $S = \{\wedge, \vee, \neg\}$,
(b) $A \barwedge B$ using $S = \{\neg, \rightarrow\}$,
(c) $A \bar{\wedge} B$ using $S = \{\neg, \rightarrow\}$,
(d) $A \leftarrow B$ using $S = \{\wedge, \vee, \neg\}$.
In each case, prove your claim using a truth table.

2.17. Suppose Peter visits the island of knights and knaves. Peter plans to ask every person he meets the question "Are you a knave?" Explain to Peter why his question is useless.

2.18. Suppose that $\mathcal{A}$ and $\mathcal{B}$ are inhabitants of the island of knights and knaves. Suppose that $\mathcal{A}$ says "I am a knave or $\mathcal{B}$ is a knight" and $\mathcal{B}$ does not say anything. What can you say about $\mathcal{A}$ and $\mathcal{B}$?

2.19. Suppose that we visit the island of knights and knaves and meet $\mathcal{A}$ and $\mathcal{B}$. If $\mathcal{A}$ says "We are both knaves" and $\mathcal{B}$ says nothing, what can you conclude about $\mathcal{A}$ and $\mathcal{B}$?

2.20. Suppose that we visit the island of knights and knaves. We meet the friendly inhabitants $\mathcal{A}$ and $\mathcal{B}$. Then $\mathcal{A}$ tells us that "$\mathcal{B}$ is a knight" and $\mathcal{B}$ asserts that "the two of us are of opposite type." What are the types of $\mathcal{A}$ and $\mathcal{B}$?

2.21. Suppose that you visit the island of knights and knaves. You meet four people $\mathcal{A}, \mathcal{B}, \mathcal{C}$, and $\mathcal{D}$. First $\mathcal{A}$ declares, "If $\mathcal{B}$ is a knave, then $\mathcal{D}$ is a knave."

Then $\mathcal{B}$ adds, "If $\mathcal{D}$ is a knight, then $\mathcal{A}$ is a knave". Finally, $\mathcal{C}$ utters, "$\mathcal{A}$ is a knave or $\mathcal{B}$ is a knight". The last person $\mathcal{D}$ does not say anything and looks rather grim. Who is a knight and who is a knave?

2.22. Suppose that you visit the island of knights and knaves. You meet a group of five people and ask them: How many of you are knaves? The first answers that one of them is a knave, the second answers that two are knaves, the third answers that three of them are knaves, the fourth answers that there are four knaves, and the fifth person answers—unsurprisingly—that there are five knaves. At first, you might think that this question was quite useless, given the variety of answers. At second thought, you realize that you can actually figure out how many knaves there are. So how many are there? [Hint: You can easily obtain the answer by logical reasoning.]

2.3 Logical Equivalence

We call two Boolean formulas equivalent if and only if they evaluate to the same truth values for all possible truth assignments. We begin by introducing the concept of a valuation that generalizes a row of a truth table to an arbitrary number of Boolean variables.

A **valuation** v is a function from the set of Boolean formulas to the set of truth values such that it assigns to propositional variables a truth value and is compatible with the usual interpretation of the logical operators, namely

V1. $v[\![\neg A]\!] = \neg v[\![A]\!]$,

V2. $v[\![A \vee B]\!] = v[\![A]\!] \vee v[\![B]\!]$,

V3. $v[\![A \wedge B]\!] = v[\![A]\!] \wedge v[\![B]\!]$,

V4. $v[\![A \to B]\!] = v[\![A]\!] \to v[\![B]\!]$,

V5. $v[\![A \leftrightarrow B]\!] = v[\![A]\!] \leftrightarrow v[\![B]\!]$.

The valuations are conceptually simple. Essentially, the truth values are assigned to the variables and consistently extended to Boolean formulas.

Example 2.2. Suppose that we are given the Boolean formula $A \leftrightarrow (B \wedge \neg C)$. If a valuation v assigns the truth values

$$v[\![A]\!] = F, \quad v[\![B]\!] = T, \quad v[\![C]\!] = F,$$

then we can deduce $v[\![\neg C]\!] = \neg v[\![C]\!] = T$. Furthermore,

$$v[\![B \wedge \neg C]\!] = v[\![B]\!] \wedge v[\![\neg C]\!] = T \wedge T = T.$$

Therefore, we can conclude that the valuation v must assign to the Boolean formula $A \leftrightarrow (B \wedge \neg C)$ the truth value

$$v[\![A \leftrightarrow (B \wedge \neg C)]\!] = v[\![A]\!] \leftrightarrow (v[\![B \wedge \neg C]\!]) = F \leftrightarrow T = F.$$

2.3 Logical Equivalence

As long as we restrict ourselves to one particular Boolean formula, then after assigning truth values to the variables that occur in the formula, the truth value of the formula is determined by this assignment.

A Boolean formula A is called a **tautology** if and only if $v[\![A]\!] = T$ holds for all valuations v. In other words, A is a tautology if and only if T is the value of A in every row of its truth table.

A Boolean formula A is called **satisfiable** if and only if $v[\![A]\!] = T$ holds for some valuation v. Put differently, the Boolean formula A is satisfiable if and only if in the truth table of A, there exists some row such that A evaluates to T.

We call two Boolean formulas A and B **logically equivalent** if and only if $v[\![A]\!] = v[\![B]\!]$ holds for all valuations v. We write $A \equiv B$ if and only if A and B are logically equivalent. For example, the double negation $\neg\neg A$ is logically equivalent to A, so $\neg\neg A \equiv A$.

Proposition 2.3. *Let A and B be statements. Then $A \to B \equiv \neg A \vee B$.*

Proof. For a valuation v, we get the truth value $v[\![A \to B]\!] = v[\![A]\!] \to v[\![B]\!] = F$ if and only if $v[\![A]\!] = T$ and $v[\![B]\!] = F$. On the other hand, a valuation v satisfies $v[\![A]\!] = T$ and $v[\![B]\!] = F$ if and only if $v[\![\neg A \vee B]\!] = F$. Therefore, $v[\![A \to B]\!] = v[\![\neg A \vee B]\!]$ for all valuations v. □

Since the result is used so often, we will give another proof that illustrates a different style of argument. The proof rests on the fact that two Boolean formulas A and B are logically equivalent if and only if $A \leftrightarrow B$ is a tautology. Indeed, we have equality $v[\![A]\!] = v[\![B]\!]$ for all valuations v if and only if

$$v[\![A \leftrightarrow B]\!] = T$$

holds for all valuations v, which is equivalent to saying that $A \leftrightarrow B$ is a tautology.

Proof. It suffices to show that $(A \to B) \leftrightarrow (\neg A \vee B)$ is a tautology. Indeed, consider the truth table

A	B	$A \to B$	$\neg A \vee B$	$(A \to B) \leftrightarrow (\neg A \vee B)$
F	F	T	T	T
F	T	T	T	T
T	F	F	F	T
T	T	T	T	T

Therefore, $(A \to B) \leftrightarrow (\neg A \vee B)$ is a tautology, which means that $A \to B \equiv \neg A \vee B$, as claimed. □

The **contrapositive** of an implication $A \to B$ is the implication $\neg B \to \neg A$. Remarkably, the contrapositive of an implication expresses the same logical operation as the implication itself, as the next proposition shows.

Proposition 2.4. *An implication $A \to B$ and its contrapositive $\neg B \to \neg A$ are logically equivalent, $A \to B \equiv \neg B \to \neg A$.*

Proof. It suffices to show that the truth tables of both expressions are the same. This is indeed the case as the following table shows:

A	B	$A \to B$	$\neg A$	$\neg B$	$\neg B \to \neg A$
F	F	T	T	T	T
F	T	T	T	F	T
T	F	F	F	T	F
T	T	T	F	F	T

We notice that the implication $A \to B$ and its contrapositive $\neg B \to \neg A$ are false precisely when A is true and B is false. □

The **converse** of an implication $A \to B$ is the implication $B \to A$. The truth of one implication does not say anything about the truth of the other. For example, the implication

$$\text{If } 2+2 = 5, \text{ then } 2 \text{ is even}$$

is a true statement, since the hypothesis is false (and the conclusion is true). The converse

$$\text{If } 2 \text{ is even, then } 2+2 = 5$$

is a false statement, since the hypothesis is true, but the conclusion is false. Simply put, the implication $F \to T$ is true, but its converse $T \to F$ is false. Thus, $A \to B \not\equiv B \to A$.

The Boolean operations have a very rich structure when we consider the Boolean formulas up to logical equivalence. We begin by recording some important properties of the conjunction operation.

Proposition 2.5. *Let A, B, and C be Boolean formulas. Then*
(a) $A \wedge (B \wedge C) \equiv (A \wedge B) \wedge C$ *(associative law)*
(b) $A \wedge B \equiv B \wedge A$ *(commutative law)*
(c) $A \wedge A \equiv A$ *(idempotence law)*
(d) $A \wedge T \equiv A$ *(identity law)*
(e) $A \wedge F \equiv F$ *(domination law)*

Proof. One can use truth tables to prove these laws, see Exercise 2.28. □

The disjunction operation has very similar laws.

Proposition 2.6. *Let A, B, and C be Boolean formulas. Then*
(a) $A \vee (B \vee C) \equiv (A \vee B) \vee C$ *(associative law)*
(b) $A \vee B \equiv B \vee A$ *(commutative law)*
(c) $A \vee A \equiv A$ *(idempotence law)*
(d) $A \vee F \equiv A$ *(identity law)*

2.3 Logical Equivalence

(e) $A \vee T \equiv T$ (domination law)

Proof. One can use truth tables to verify these laws, see Exercise 2.29. □

The similarities between the conjunction and disjunction operations are not accidental. The next proposition shows that the negation of a conjunction can be expressed in terms of a disjunction with negated arguments.

Proposition 2.7 (De Morgan's Laws). *Let A and B be Boolean formulas. Then*
(a) $\neg(A \wedge B) \equiv \neg A \vee \neg B$,
(b) $\neg(A \vee B) \equiv \neg A \wedge \neg B$.

Proof. We will first show that the negation of the conjunction A and B is logically equivalent to the disjunction of the negated statements. Indeed, consider the truth table

A	B	$A \wedge B$	$\neg(A \wedge B)$	$\neg A$	$\neg B$	$\neg A \vee \neg B$
F	F	F	T	T	T	T
F	T	F	T	T	F	T
T	F	F	T	F	T	T
T	T	T	F	F	F	F

Since the truth values of $\neg(A \wedge B)$ and $\neg A \vee \neg B$ are the same for all arguments, this proves our claim (a).

By substituting the negated statements for A and B, we can infer from $\neg(A \wedge B) \equiv \neg A \vee \neg B$ the logical equivalence

$$A \vee B \equiv \neg(\neg A \wedge \neg B).$$

Applying the not operator to both sides, we obtain $\neg(A \vee B) \equiv \neg A \wedge \neg B$, which proves our claim (b). □

Suppose that C is a Boolean formula containing A as a subformula. Given a logical equivalence $A \equiv B$, we can substitute the expression B for A in C and obtain a new formula C'. Since A and B take on the same truth values for all valuations, we can conclude that $C \equiv C'$. We will call this the **substitution technique**.

In the proof of the next proposition, we will illustrate how to use this substitution technique repeatedly to prove the claims (b) and (c) after establishing that conjunction distributes over disjunction.

Proposition 2.8. *Let A, B, and C be Boolean formulas. We have*
(a) $A \wedge (B \vee C) \equiv (A \wedge B) \vee (A \wedge C)$ *(conjunction distributes over disjunction)*
(b) $A \vee (B \wedge C) \equiv (A \vee B) \wedge (A \vee C)$ *(disjunction distributes over conjunction)*
(c) $A \wedge (A \vee B) \equiv A$ *(absorption law for conjunction)*
(d) $A \vee (A \wedge B) \equiv A$ *(absorption law for disjunction)*

Proof. (a) For a valuation v, we get the truth value $v[\![A \wedge (B \vee C)]\!] = T$ if and only if $v[\![A]\!] = T$ and $v[\![B \vee C]\!] = T$ if and only if $v[\![A]\!] = T$ and either $v[\![B]\!] = T$ or $v[\![C]\!] = T$. In other words, we have $v[\![A \wedge (B \vee C)]\!] = T$ if and only if $v[\![A]\!] = T$ and $v[\![B]\!] = T$ holds or $v[\![A]\!] = T$ and $v[\![C]\!] = T$ holds. The latter statement is equivalent to $v[\![A \wedge B]\!] = T$ or $v[\![A \wedge C]\!] = T$, which is equivalent to $v[\![(A \wedge B) \vee (A \wedge C)]\!] = T$. Therefore, for any valuation v, we have the equality

$$v[\![A \wedge (B \vee C)]\!] = v[\![(A \wedge B) \vee (A \wedge C)]\!],$$

so the claimed logical equivalence $A \wedge (B \vee C) \equiv (A \wedge B) \vee (A \wedge C)$ holds.

(b) We can deduce the second distributive law using the distributive law from part (a) and repeated application of de Morgan's law,

$$\begin{aligned}
A \vee (B \wedge C) &\equiv \neg\neg(A \vee (B \wedge C)) && \text{by double negation} \\
&\equiv \neg(\neg A \wedge \neg(B \wedge C)) && \text{by de Morgan's law} \\
&\equiv \neg(\neg A \wedge (\neg B \vee \neg C)) && \text{by de Morgan's law} \\
&\equiv \neg((\neg A \wedge \neg B) \vee (\neg A \wedge \neg C)) && \text{by the distributive law (a)} \\
&\equiv \neg(\neg A \wedge \neg B) \wedge \neg(\neg A \wedge \neg C) && \text{by de Morgan's law} \\
&\equiv (A \vee B) \wedge (A \vee C) && \text{by de Morgan and double negation}
\end{aligned}$$

(c) We can show the absorption law for conjunction using a series of known logical equivalences. Indeed,

$$\begin{aligned}
A \wedge (A \vee B) &\equiv (A \vee F) \wedge (A \vee B) && \text{by the identity law} \\
&\equiv A \vee (F \wedge B) && \text{by the distributive law} \\
&\equiv A \vee F && \text{by the domination law} \\
&\equiv A && \text{by the identity law}
\end{aligned}$$

(d) We encourage the reader to prove the last claim in several different ways to apply the proof techniques that we have learned, see Exercise 2.32. □

EXERCISES

2.23. Let $v[\![A]\!] = F$, $v[\![B]\!] = T$, and $v[\![C]\!] = T$. Which of the following Boolean formulas evaluate to true under the valuation v?
(a) $(A \to (B \wedge C))$,
(b) $((\neg A \wedge B) \leftrightarrow C)$,
(c) $A \to \neg C$,
(d) $((A \wedge B) \vee \neg C)$.

2.24. Use a truth table to prove that the implication $A \to B$ and its converse $B \to A$ are **not** logically equivalent.

2.25. Show that $(A \to B) \wedge (B \to A)$ and $A \leftrightarrow B$ are logically equivalent.

2.4 Logical Consequence

2.26. Show that $(A \land (A \rightarrow B)) \rightarrow B$ is a tautology. This is the basis for the modus ponens inference rule: If A and $A \rightarrow B$ are true, then we can conclude that B must be true.

2.27. Show that there exists a Boolean expression that uses only the operators of the form $\lor$ and $\leftrightarrow$ that is logically equivalent to $A \land B$.

2.28. Prove the claims of Proposition 2.5 using truth tables, that is, show that the conjunction operation is an associative and commutative operation satisfying the idempotence, identity, and domination laws.

2.29. Prove the claims of Proposition 2.6 using truth tables, that is, show that the disjunction operation is an associative and commutative operation satisfying the idempotence, identity, and domination laws.

2.30. Show that the associative laws hold for the conjunction and the disjunction operations
(a) $A \land (B \land C) \equiv (A \land B) \land C$,
(b) $A \lor (B \lor C) \equiv (A \lor B) \lor C$.
Do not use truth tables. Instead, show that for any valuation, the left-hand side yields the same truth value as the right-hand side of the putative logical equivalence.

2.31. Show that $((A \lor B) \land (\neg A \lor C)) \rightarrow (B \lor C)$ is a tautology. This is the basis of the resolution inference rule: If the premises $A \lor B$ and $\neg A \lor C$ are true, then $B \lor C$ must be true.

2.32. Prove the absorption law from Proposition 2.8 (d) using
(a) a truth table
(b) logical equivalences that exploit de Morgan's law, double negation, and the absorption law from Proposition 2.8 (c).
(c) known logical equivalences (such as distributive, identity, and domination laws), but without the use of de Morgan's law.

2.4 Logical Consequence

In this section, we will start to formalize the notion of valid logical arguments. These arguments must entail the conclusion from one or more premises. A valid argument must ensure that its conclusion is true whenever the premises are true. We will formalize this using the notion of logical consequence.

Let S be a set of Boolean formulas. We say that a Boolean formula C is a **logical consequence** of S if and only if for all valuations v, the condition $v[\![P]\!] = T$ for all P in S implies $v[\![C]\!] = T$. We write

$$S \models C$$

to express that C is a logical consequence of S. In other words, we write $S \models C$ if and only if there does not exist a valuation v such that $v[\![C]\!] = F$ when $v[\![P]\!] = T$ holds for all formulas P in S.

One of the most pressing questions one might have about a mathematical argument is whether the steps of the argument are valid. The logical consequence paves the way to answer this question for arguments that are formulated in the logic that we have developed so far. Given the **premises** or **hypotheses** $P_1, \ldots, P_n$, an argument seeks to assert a **conclusion** C. We call this argument **valid** if and only if $\{P_1, \ldots, P_n\} \models C$. In other words, a valid argument will ensure that the conclusion is true when its premises are true.

As an example, let us establish the validity of the logical argument known as **modus ponens**. This is an argument form that has two premises and one conclusion. The two premises of modus ponens are of the form A and $A \to B$. If the premises are true, then modus ponens concludes that B must be true. As a particular example, consider the premises $A =$ "it rains" and $A \to B =$ "if it rains, then the ground is wet." Assuming that these premises are true, modus ponens allows us to conclude that $B =$ "the ground is wet" must be true.

Example 2.9 (Validity of Modus Ponens). For all Boolean formulas A and B, we have
$$\{A, A \to B\} \models B,$$
so the conclusion B is the logical consequence of the premises A and $A \to B$. Indeed, seeking a contradiction, let us assume that there exists a valuation v such that the premises evaluate to true, but the conclusion evaluates to false. In symbols, we can express this in the form $v[\![A]\!] = T$, $v[\![A \to B]\!] = T$, and $v[\![B]\!] = F$. However, $v[\![A]\!] = T$ and $v[\![B]\!] = F$ implies that $v[\![A \to B]\!] = F$, contradicting our assumption. Therefore, all valuations assigning true to the formulas in the set $\{A, A \to B\}$, must assign true to the formula B, whence $\{A, A \to B\} \models B$ as claimed.

In the exercises, you are encouraged to establish in a similar manner the validity of other forms of logical arguments.

Notation. If the set S contains a single element B, then we write $B \models A$ instead of $\{B\} \models A$. Therefore, two Boolean formulas A and B are logically equivalent, $A \equiv B$, if and only if $A \models B$ and $B \models A$. If S is the empty set, then we write $\models A$ instead of $\emptyset \models A$. In other words, $\models A$ is another way of expressing that A is a tautology.

The next theorem gives an alternative approach to prove logical consequence. By combining all the premises and the conclusion into a single formula, it suffices to establish that the resulting formula is a tautology.

Theorem 2.10. *Suppose that $\{P_1, P_2, \ldots, P_n\}$ is a finite set of Boolean formulas and C is a Boolean formula. Then*
$$\{P_1, P_2, \ldots, P_n\} \models C$$
if and only if $(P_1 \wedge P_2 \wedge \cdots \wedge P_n) \to C$ is a tautology.

Proof. By definition, $\{P_1, P_2, \ldots, P_n\} \models C$ if and only if there does not exist a valuation v such that $v[\![P_1]\!] = T, \ldots, v[\![P_n]\!] = T$, and $v[\![C]\!] = F$. This is

2.4 Logical Consequence

equivalent to the nonexistence of a valuation v such that $v[\![P_1 \wedge \cdots \wedge P_n]\!] = T$ and $v[\![C]\!] = F$. In other words, there does not exist a valuation v such that $v[\![P_1 \wedge \cdots \wedge P_n \to C]\!] = F$, which is simply a different way of expressing that the formula $(P_1 \wedge P_2 \wedge \cdots \wedge P_n) \to C$ is a tautology. □

Example 2.11. Given the premises $A \vee B$ and $\neg A \vee C$, we can conclude $B \vee C$. This is the so-called resolution rule. The resolution rule is a valid argument, since $((A \vee B) \wedge (\neg A \vee C)) \to (B \vee C)$ is a tautology by Exercise 2.31, so $\{A \vee B, \neg A \vee C\} \models B \vee C$ holds by the previous theorem. Automated theorem provers are often based on the resolution rule.

In principle, we can always use a truth table to establish that

$$\{P_1, P_2, \ldots, P_n\} \models C$$

is a valid argument by proving that $(P_1 \wedge P_2 \wedge \cdots \wedge P_n) \to C$ is a tautology, regardless of the complexity of the argument. However, this is impractical, as the number of rows in the truth table grows exponentially with the number of Boolean variables that occur in the premises and conclusion.

In general, we will verify the logical consequence for a few simple rules such as the resolution rule or the modus ponens. We will then compose an argument as a sequence of steps, where each step uses either an axiom or a rule whose correctness we have established. The advantage of this modular approach is that it is easier to verify the correctness of such a proof. In the next section, we will give a first taste of such formal proofs.

EXERCISES

2.33. Let A, B, and C be Boolean formulas. Show that the following three arguments are valid.
(a) $(A \vee (B \vee C)) \models ((A \vee B) \vee C)$ (the associative rule)
(b) $(A \vee A) \models A$ (the contraction rule)
(c) $A \models B \vee A$ (the expansion rule)

2.34. Let A, B, and C be Boolean formulas. We proved that the modus ponens $\{A, A \to B\} \models B$ is a valid argument. Use a similar style of argument to show that the following arguments are valid.
(a) $\{A \to B, \neg B\} \models \neg A$ (modus tollens)
(b) $\{A \to B, B \to C\} \models A \to C$ (hypothetical syllogism)

2.35. Show that the following arguments are valid:
(a) $\{A \to B, \neg A \to B\} \models B$,
(b) $\{A \to B, A \to \neg B\} \models \neg A$.

2.36. Let A, B, and C be Boolean formulas. Show that the following arguments are valid:
(a) $(B \to C) \models ((A \wedge B) \to C)$ (strengthening the hypothesis),
(b) $(A \to B) \models (A \to (B \vee C))$ (weakening the conclusion).

2.37. Determine whether the following is a valid argument:
$$A \models (A \wedge B).$$

2.38. Determine whether the following is a valid argument:
$$\{(A \vee B), (A \vee \neg C)\} \models (A \wedge C).$$

2.39. Determine whether the following is a valid argument:
$$\{(A \rightarrow (B \vee C)), (B \rightarrow C)\} \models (A \rightarrow C).$$

2.40. Determine whether the following argument is valid:
$$\{(A \rightarrow (B \vee \neg C)), (B \rightarrow (A \wedge C))\} \models (A \rightarrow C).$$

2.41. Let $P_1, P_2, \ldots, P_n$ and A be Boolean formulas. Show that $\{P_1, P_2, \ldots, P_n\} \models A$ if and only if $\{P_1, P_2, \ldots, P_n\} \cup \{\neg A\}$ is not satisfiable.

2.42. What is a potential practical problem when formulating an argument about propositional logic in the form $\{P_1, P_2, \ldots, P_n\} \models C$?

2.5 Formal Arguments

In this section, we will study a **propositional calculus** that allows us to formally deduce a conclusion C from premises $P_1, \ldots, P_n$ using an argument that consists of several steps. In each step, we use an axiom, the premises, or some already established result and apply some inference rule until we can conclude C. This mimics the style of mathematical arguments, but is more detailed and more formal.

A propositional calculus is a special type of formal system. A **formal system** consists of a formal language, a set of axioms, and a set of inference rules. We will now introduce these three components for the propositional calculus.

Formal Language. We will confine ourselves in our formal language to Boolean formulas that contain operators from the set $\{\neg, \rightarrow\}$. This does not restrict expressiveness, since all other logical operators can be defined with the help of negation and implication, see Exercises 2.43 and 2.45. However, our exposition is simplified by restricting to these two logical operators.

Axioms. Let A, B, and C denote Boolean formulas. Then we have the following three axiom schemes:

A1. $A \rightarrow (B \rightarrow A)$,

A2. $(A \rightarrow (B \rightarrow C)) \rightarrow ((A \rightarrow B) \rightarrow (A \rightarrow C))$,

2.5 Formal Arguments

A3. $(\neg B \to \neg A) \to (A \to B)$.

We call these axiom schemes rather than axioms, since each of these three rules will lead to an infinite number of axioms by substituting Boolean formulas for A, B, and C. One should note that **A1**, **A2**, and **A3** are tautologies, see Exercise 2.46. This means that we do not make a mistake by assuming them as facts.

Inference Rule. We will be allowed to use the modus ponens rule of inference in our arguments. If A and B are Boolean formulas, then the modus ponens rule allows us to infer from A and $A \to B$, the Boolean formula B. We will denote this inference rule in the form

R1. $\dfrac{A, A \to B}{B}$.

In the previous section, we established that modus ponens is a valid argument, so we will not make erroneous conclusions using this inference rule.

We will denote the formal system consisting of the Boolean formulas, the axiom schemes **A1**, **A2**, **A3**, and the inference rule **R1** by the letter H, honoring the famous mathematician David Hilbert, who championed such formal systems.

Formal Proofs. We will now use the axioms and inference rule to derive formal proofs in the propositional calculus H. Let $P = \{P_1, P_2, \ldots, P_n\}$ be a set of premises and C a desired conclusion. We will write $P \vdash_H C$ if and only if we can deduce the conclusion C from the premises using the axioms and the inference rule of the formal system H. In other words, we write $P \vdash_H C$ if and only if we can formulate a sequence of $S_1, S_2, \ldots, S_m$ of Boolean formulas such that for all indices k, one of the following three choices applies

F1. S_k is an axiom from one of the axiom schemes **A1**, **A2**, or **A3**,

F2. S_k is a premise in P, or

F3. $S_k = B$ can be obtained using the modus ponens inference rule from two prior steps $S_i = A$ and $S_j = A \to B$,

and additionally S_m coincides with the desired conclusion C.

Notation. If P consists of a single premise $P = \{A\}$, then we write $A \vdash_H C$ instead of the more cumbersome $\{A\} \vdash_H C$. If the set P of premises is empty, then we write shortly $\vdash_H C$ instead of $\emptyset \vdash_H C$.

We will use a very small example to illustrate this style of formal argument.

Example 2.12. We will formally prove that $A \vdash_H (A \to A)$. In other words, we will deduce $(A \to A)$ from the assumption A. We can take advantage of any axioms, but we are limited to the modus ponens inference rule in this argument. The argument goes as follows:

(1) A premise
(2) $A \to (A \to A)$ by axiom **A1**
(3) $(A \to A)$ **R1**(1), (2)

On the right-hand side, the comments give the justification for this step in the proof.

The style of the argument is very different from the mathematical arguments from the previous section. The argument does not involve truth values or valuations, but merely manipulations at a syntactic level.

You might have noticed that we should be able to derive $(A \to A)$ without the assumption A. After all, $(A \to A)$ is a tautology. We can indeed drop the assumption A, as our second example shows.

Example 2.13. Let us formally prove that

$$\vdash_H A \to A$$

holds. Since there are no premises, it means that we should be able to derive it from the axioms using the modus ponens rule. Indeed, we observe that

(1) $A \to ((A \to A) \to A)$ by axiom **A1**
(2) $(A \to ((A \to A) \to A)) \to ((A \to (A \to A)) \to (A \to A))$ by axiom **A2**
(3) $(A \to (A \to A)) \to (A \to A)$ by **R1**(1), (2)
(4) $A \to (A \to A)$ by axiom **A1**
(5) $A \to A$ by **R1**(4), (3)

This proof might not be exactly intuitive. However, a formal proof has the advantage that one can easily verify each single step—well, at least in principle. The comments are crucial, as otherwise the verification becomes cumbersome.

In the third example, we will prove that implication is a transitive operation, which means that if $A \to B$ and $B \to C$, then $A \to C$. This is also known as the rule of hypothetical syllogism. This fact is frequently used in other proofs.

Example 2.14. We will formally prove that

$$\{A \to B, B \to C\} \vdash_H A \to C.$$

Indeed, we have

(1) $A \to B$ premise
(2) $B \to C$ premise
(3) $(B \to C) \to (A \to (B \to C))$ by axiom **A1**
(4) $A \to (B \to C)$ by **R1**(2), (3)
(5) $(A \to (B \to C)) \to ((A \to B) \to (A \to C))$ by axiom **A2**
(6) $(A \to B) \to (A \to C)$ by **R1**(4), (5)
(7) $A \to C$ by **R1**(1), (6)

2.5 Formal Arguments

When developing this proof, we did not proceed by assuming (1), (2), and then successively deducing steps (3) through (7). Rather, we guessed that the instance (5) of the axiom **A2** might be a promising way to deduce $A \to C$. Then we found how to obtain $A \to (B \to C)$ through the steps (2), (3), and (4). Fleshing out the rest of the proof was then straightforward.

One can in principle continue in this fashion and prove many basic facts. However, the style of argument is a bit too cumbersome, as the expressions can become very long. Thanks to the influential logician Jacques Herbrand, there is a more convenient way available, known as the deduction theorem.

Theorem 2.15 (Deduction Theorem). *Let A and B be Boolean formulas, and P a set of Boolean formulas. Then*

$$P \cup \{A\} \vdash_H B \quad \text{if and only if} \quad P \vdash_H A \to B.$$

The interested reader can find a proof of the deduction theorem for instance in Goldrei [29, pages 97–98] or other introductory texts on symbolic logic.

The next example illustrates how to use this theorem.

Example 2.16. We will now give a formal proof that

$$\{A \to B, A \to (B \to C)\} \vdash_H A \to C$$

By the deduction theorem, this statement is equivalent to

$$\{A \to B, A \to (B \to C), A\} \vdash_H C$$

However, proving the latter claim is simple, as

(1)	A	premise
(2)	$A \to B$	premise
(3)	B	by **R1**(1), (2)
(4)	$A \to (B \to C)$	premise
(5)	$B \to C$	by **R1**(1), (4)
(6)	C	by **R1**(3), (5)

Since we were able to move A into the set of premises, we were able to use the modus ponens more effectively.

In the exercises, you will show that the axioms **A1**–**A3** are tautologies. We have shown in the previous section that the modus ponens rule is a valid argument. One might worry that the composition of the formal proof might somehow lead to an invalid argument. The next theorem puts these worries to rest.

Theorem 2.17 (Soundness Theorem). *Let P be a set of Boolean formulas and C a Boolean formula. If $P \vdash_H C$, then $P \models C$. In other words, if we can find a formal proof in the formal system H that deduces C from the set of premises P, then this constitutes a valid argument.*

Proof. Suppose that the conclusion C can be deduced from the set of premises P in the formal system H; thus, we suppose that

$$P \vdash_H C$$

holds. This means that there exists a sequence of Boolean formulas

$$S_1, S_2, \ldots, S_m$$

satisfying **F1**, **F2**, or **F3** for each index k in the range $1 \leq k \leq m$ and $S_m = C$.

Let v be any valuation such that $v[\![A]\!] = T$ holds for all premises $A \in P$. We claim that $v[\![S_k]\!] = T$ holds for all k in the range $1 \leq k \leq m$.

We use a proof by contradiction to prove this claim. Seeking a contradiction, let us assume that k is the smallest index in the range $1 \leq k \leq m$ such that $v[\![S_k]\!] = F$. The step S_k in the proof can arise from **F1**, **F2**, or **F3**. So let us examine each of these three cases.

(1) If S_k results from **F1**, then S_k is an axiom, hence it is a tautology by Exercise 2.46, which means that $v[\![S_k]\!]$ must evaluate to true.
(2) If S_k results from **F2**, then S_k is a premise, so satisfies $v[\![S_k]\!] = T$ by assumption.
(3) If $S_k = B$ results from **F3**, then it was obtained from prior steps $S_i = A$ and $S_j = A \to B$ using the modus ponens rule **R1**. Since the indices i and j are smaller than k, we have $v[\![S_i]\!] = T$ and $v[\![S_j]\!] = T$. However, the modus ponens is a valid argument by Example 2.9. Therefore, we must have $v[\![S_k]\!] = T$.

In all three cases, we get a contradiction to $v[\![S_k]\!] = F$. Therefore, we can conclude that $v[\![S_k]\!] = T$ holds for all k in the range $1 \leq k \leq m$. In particular, we can conclude that $v[\![S_m]\!] = v[\![C]\!] = T$.

Since v was an arbitrary valuation, this means that $P \models C$, as claimed. $\square$

The propositional calculus satisfies a rather remarkable property. One can show that the three axiom schemes and modus ponens alone suffice to prove every valid argument.

Theorem 2.18 (Completeness Theorem). *Let P be a set of Boolean formulas and C a Boolean formula. If $P \models C$, then $P \vdash_H C$. In other words, every valid argument of propositional logic can be proved in the formal system H.*

The proof of the completeness theorem is more involved than the proof of the soundness theorem. We will not discuss a proof here, as this would lead us too far astray. The interested reader can find a proof in most textbooks on logic, such as Goldrei [29].

Other Argument Forms. There are some other argument forms besides modus ponens that are commonly used. Since we can use negation and implication to define all other logical operators, see Exercises 2.43 and 2.45, we will express the subsequent argument forms using a broader selection of logical operators.

2.5 Formal Arguments

Modus Tollens	Disjunctive Syllogism
$A \to B$	$A \vee B$
$\neg B$	$\neg A$
$\overline{\neg A}$	$\overline{B}$

Reductio ad Absurdum	Resolution Rule
$A \to B$	$A \vee B$
$A \to \neg B$	$\neg A \vee C$
$\overline{\neg A}$	$\overline{B \vee C}$

Weakening of a Conjunction   **Weakening to a Disjunction**

$$\dfrac{A \wedge B}{A} \quad \text{or} \quad \dfrac{A \wedge B}{B} \qquad \dfrac{A}{A \vee B} \quad \text{or} \quad \dfrac{B}{A \vee B}$$

One can easily verify the validity of inference rules such as modus tollens and the like. By the completeness theorem, we know that there must exist some formal proof in H. The deduction theorem allows us to give a short formal proof of the modus tollens.

Proposition 2.19 (Modus Tollens). *If A and B are Boolean formulas, then*

$$\{A \to B, \neg B\} \vdash_H \neg A.$$

Proof. We have

(1) $\vdash_H (A \to B) \to (\neg B \to \neg A)$ by axiom **A3**
(2) $A \to B \vdash_H \neg B \to \neg A$ by the deduction theorem
(3) $\{A \to B, \neg B\} \vdash_H \neg A$ by the deduction theorem,

as claimed. □

One example should suffice to illustrate this argument form.

Example 2.20. Consider the argument: *If it rains in Texas, then it pours. It does not pour. Therefore, it does not rain in Texas.* If we denote the statement "It rains in Texas" by A and "It pours" by B, then we see that the argument is of the form

$A \to B$	*If it rains in Texas, then it pours.*
$\neg B$	*It does not pour.*
$\overline{\neg A}$	*It does not rain in Texas.*

This example is a typical use of the modus tollens inference rule.

EXERCISES

2.43. Show that
(a) $A \vee B \equiv \neg A \rightarrow B$
(b) $A \wedge B \equiv \neg(A \rightarrow \neg B)$
(c) $A \leftrightarrow B \equiv \neg((A \rightarrow B) \rightarrow \neg(B \rightarrow A))$
Thus, disjunction, conjunction, and equivalence can be expressed in terms of implication and negation.

2.44. Show that a Boolean function $f(X_1, X_2, \ldots, X_n)$ in n Boolean variables can be expressed in **canonical disjunctive normal form**. This means that $f(X_1, X_2, \ldots, X_n)$ is logically equivalent to a disjunction of minterms. A minterm is a conjunction of n literals $Y_1 \wedge Y_2 \wedge \cdots \wedge Y_n$, where $Y_k \in \{X_k, \neg X_k\}$ for all integers k in the range $1 \leqslant k \leqslant n$. [Hint: Inspect the truth table of f.]

2.45. Show that $\{\neg, \rightarrow\}$ is a **functionally complete set** of operators. This means that for each Boolean formula f there exists a logically equivalent formula f' that contains just negation and implication operators. [Hint: Use the fact that a Boolean formula can be written in disjunctive normal form.]

2.46. Prove that
(a) $A \rightarrow (B \rightarrow A)$,
(b) $(A \rightarrow (B \rightarrow C)) \rightarrow ((A \rightarrow B) \rightarrow (A \rightarrow C))$,
(c) $(\neg B \rightarrow \neg A) \rightarrow (A \rightarrow B)$,
are tautologies.

2.47. Show that $\{A \rightarrow (B \rightarrow C), B\} \vdash_H A \rightarrow C$.

2.48. Show that $(A \rightarrow (B \rightarrow C)) \vdash_H (B \rightarrow (A \rightarrow C))$ holds using the deduction theorem once.

2.49. Reprove $\vdash_H (\neg A \rightarrow (A \rightarrow B))$ without using hypothetical syllogism in your argument.

2.50. Show that $\vdash_H ((\neg A \rightarrow A) \rightarrow (\neg A \rightarrow B))$. You can use the previous exercise.

2.51. Reprove $\{(A \rightarrow B), (A \rightarrow (B \rightarrow C))\} \vdash_H (A \rightarrow C)$ without using the deduction theorem.

2.52. Show that
$$\neg\neg A \vdash_H A.$$

2.6 Predicates and Quantifiers

The propositional logic that we have discussed so far is not expressive enough to formulate most mathematical proofs. We will now study first-order languages that generalize propositional logic and form an appropriate framework for most mathematical arguments. We will informally explain the two main concepts on which first-order languages are based.

2.6 Predicates and Quantifiers

Predicates. We often like to express statements that depend on one or more parameters. For example, let $P(n)$ have the interpretation

n is a prime number.

Then $P(2)$ expresses the statement "2 is a prime number" and $P(4)$ expresses the statement "4 is a prime number." Thus, one obtains a statement that can be true or false after substituting an integer for the argument of the predicate.

In general, a **predicate** or **propositional function** is a function that has propositions as values. A big advantage of predicates is that they allow us to formulate propositions that are dependent on each other. For example, suppose that $A(n)$ means $n > 0$. Then we can express by the implication

$$A(n) \to A(n+1)$$

the fact that if n is positive, then $n+1$ must be positive. This implication is true for all integers n.

The parameters of a predicate can take on values from the **universe of discourse**, or shortly called the **universe**. The universe comprises elements that one wants to refer to in the predicates. For example, the universe could be the set of integers, the set of real numbers, the set of all people, or the like.

A predicate can have several arguments. A predicate $P(x_1, \ldots, x_n)$ with n arguments is called a **predicate of degree** n or an **n-ary predicate**. An interpretation of a predicate of degree n decides for which n-tuples of elements from the universe the predicate is supposed to be true. In the language of set theory of the next chapter, a predicate is then interpreted as a characteristic function of an n-ary relation.

Example 2.21. Suppose that the universe of discourse is the set of points in the Euclidean plane. If the predicate $C(x, y)$ has the interpretation $x^2 + y^2 \leq 1$, then $C(x, y)$ evaluates to true for any point (x, y) that belongs to the unit circle about $(0, 0)$, and false for any point that lies outside this circle (Fig. 2.1).

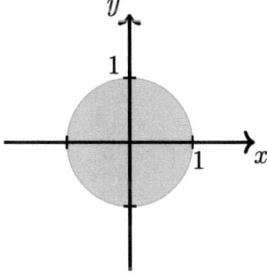

Figure 2.1: The points in the Euclidean plane for which the predicate $C(x, y)$ when interpreted as $x^2 + y^2 \leq 1$ evaluates to true are shaded in gray

Example 2.22. Suppose that the universe of discourse is the set of points in the Euclidean plane. If the predicate $U(x, y)$ is interpreted to mean $y \geq 0$, then $U(x, y)$ evaluates to true for all points that lie in the closed upper half-plane and to false for all points in the open lower half-plane (Fig. 2.2).

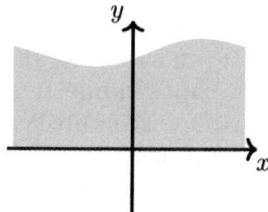

Figure 2.2: The points in the Euclidean plane for which the predicate $U(x, y)$ when interpreted as $y \geq 0$ evaluates to true are shaded in gray

Quantifiers. We frequently want to express that a predicate is true for all elements of the universe. Suppose that a predicate $P(n)$ has the interpretation $(n!)^2 \geq n^n$, where $n!$ is the factorial function that is the product of the first n positive integers, $n! = 1 \cdot 2 \cdot \ldots \cdot n$. A typical mathematical statement is

for all positive integers n, the inequality $(n!)^2 \geq n^n$ holds.

If the universe is the set of positive integers, then we can express this in predicate logic as $\forall n\, P(n)$. We call $\forall$ the **universal quantifier** and n the variable **bound** by the universal quantifier. The statement $\forall n\, P(n)$ is true if and only if $P(n)$ is true for all n in the universe. Note that the truth depends on the interpretation of the predicate and the choice of the universe.

We can also express that a predicate $P(n)$ is true for some[1] element of the universe. We express this in the form $\exists n\, P(n)$. We call the symbol $\exists$ the **existential quantifier** and n the variable bound by the existential quantifier.

For instance, if the predicate $P(n)$ has the interpretation "n is a prime," and the universe is the set of positive integers, then the statement

$$\exists n(P(n) \land P(n+1))$$

expresses that there exists a prime n such that $n + 1$ is a prime as well. This happens to be a true statement, since $P(2) \land P(3)$ is true.

The arguments of a predicate that are not bound by a quantifier and require values are called **free variables**.

Example 2.23. Let $P(n)$ denote the predicate with the interpretation "n is a prime." The value of n ranges over the set of positive integers as a universe.

---

[1] In mathematics, "some element" is always understood in the sense of "at least one element".

2.6 Predicates and Quantifiers

In this interpretation, $P(n)$ means that n is an integer such that 1 and n are the only divisors of n, and $n > 1$. We can express $P(n)$ in the form

$$\forall a \forall b \left((n = a \cdot b) \to ((a = 1) \lor (b = 1))\right) \land (n > 1),$$

where both a and b have the set of positive integers as a universe. Here a and b are bound variables and n is a free variable.

A formula may have a variable that occurs both as a bound variable in one subexpression and as a free variable in another. For instance, the variable x in the expression

$$(\forall x P(x)) \land Q(x),$$

is bound to the universal quantifier in the predicate $P(x)$ in the first expression, but free in $Q(x)$. It is a good practice to rename the bound variable, since the expression

$$(\forall y P(y)) \land Q(x),$$

is equivalent, but less confusing.

If an expression is quantified by two universal quantifiers such as

$$\forall a \forall b P(a, b),$$

then the order among the universal quantifiers does not matter, so we could have written $\forall b \forall a P(a,b)$. We can also freely exchange existential quantifiers, so it does not matter whether we write $\exists a \exists b Q(a,b)$ or $\exists b \exists a Q(a,b)$. However, the order among existential and universal quantifiers matters, as the following example illustrates.

Example 2.24. Let $H(s,c)$ denote the predicate with the interpretation that "the student s owns the computer c." The statement

$$\forall c \exists s H(s,c)$$

expresses the fact that every computer is owned by a student. On the other hand, interchanging the quantifiers yields the statement

$$\exists s \forall c H(s,c),$$

which means that there exists a student who owns all computers.

EXERCISES

2.53. Let $C(a,b,c)$ and $S(a,b,c)$ be predicates with the interpretation $a^3+b^3 = c^3$ and $a^2 + b^2 = c^2$, respectively. How many values of (a,b,c) make the predicates true for the given universe?
(a) $C(a,b,c)$ over the universe U of nonnegative integers.
(b) $C(a,b,c)$ over the universe U of positive integers.
(c) $S(a,b,c)$ over the universe $U = \{1,2,3,4,5\}$.
(d) $S(a,b,c)$ over the universe U of positive integers.

2.54. Suppose that the universe of discourse is the set of real numbers. Translate the following formal statements into mathematical statements in plain English.
(a) $\forall x(x^2 \geq 0)$
(b) $\forall x \exists y(x < y)$
(c) $\forall x \forall z \exists y(x < z) \rightarrow ((x < y) \wedge (y < z))$
(d) $\forall x \forall y \forall z((x < y) \rightarrow (x + z < y + z))$
(e) $\forall x \forall y((x < y) \vee (x = y) \vee (x > y))$.

2.55. Express each of the following statements in first-order predicate logic. Specify the universe of discourse and predicates when needed.
(a) Every integer is a rational number.
(b) The square of any real number does not equal -1.
(c) There exists an integer that is equal to its own square.
(d) There do not exist positive integers x, y, and z such that $x^3 + y^3 = z^3$
(e) For every positive real number x, there exists a positive real number y such that $x = y^2$.

2.56. Determine whether the following statements are true or false and justify your answers. Suppose that the universe of discourse is the set of positive integers for each quantifier.
(a) $\forall a \exists b\,(a = b)$.
(b) $\forall a \exists b\,(a = b + 2)$.
(c) $\forall a \exists b\,(a \leq b)$.
(d) $\exists a \forall b\,(a < b)$.
(e) $\exists a \forall b\,(a \leq b)$.

2.57. Suppose that the universe $U = \{1,2,3\}$ and $P(n)$ a predicate over the universe U. For each of the following statements, give an equivalent expression that does not use quantifiers (but may use conjunctions, disjunctions, and negations).
(a) $\forall n P(n)$
(b) $\exists n P(n)$
(c) $\neg \forall n P(n)$
(d) $\neg \exists n P(n)$
For expressions (c) and (d), express the statement without negated compound expressions.

2.7 Negations

Finding the negation of statements is frequently needed. In this section, we summarize the facts that we already know about the negation of formulas and extend the results to quantified expressions and other operators.

The double negation law allows one to simplify a statement A that is preceded by two negations.

N1. $\neg\neg A \equiv A$.

The de Morgan's laws govern the negation of conjunction and disjunction of two statements A and B.

N2. $\neg(A \wedge B) \equiv \neg A \vee \neg B$.
N3. $\neg(A \vee B) \equiv \neg A \wedge \neg B$.

Example 2.25. Suppose that A is the statement "the triangle $\triangle DEF$ is an isosceles triangle" and B the statement "the triangle $\triangle DEF$ is a right triangle." Then $A \wedge B$ means

> The triangle $\triangle DEF$ is an isosceles triangle and a right triangle.

When we negate this statement, we have to be careful where we insert the word "not" to obtain the correct negation $\neg(A \wedge B)$ of the statement. One correct version is "It is not the case that the triangle $\triangle DEF$ is an isosceles triangle and a right triangle." De Morgan's law leads to the more transparent version "The triangle $\triangle DEF$ is not an isosceles triangle or is not a right triangle."

Sometimes we will need to negate an implication. Since $A \to B \equiv \neg A \vee B$, the negation yields

N4. $\neg(A \to B) \equiv \neg(\neg A \vee B) \equiv A \wedge \neg B$

by de Morgan's law. Thus, the negation of an implication is the conjunction of the hypothesis with the negated conclusion.

 The negation of an implication can never be an implication! This is a common error that beginners are prone to make.

Example 2.26. Consider the statement "If the dessert is tiramisu, then she eats the dessert." The negation of this implication is given by

> The dessert is tiramisu and she does not eat the dessert.

This is the conjunction of the hypothesis "the dessert is tiramisu" with the negation of the conclusion "she eats the dessert."

The negation of an equivalence $A \leftrightarrow B$ is given by

N5. $\neg(A \leftrightarrow B) \equiv A \leftrightarrow \neg B$.

Example 2.27. The negation of the statement "She is happy if and only if the dessert is tiramisu" has the negation

> She is happy if and only if the dessert is not tiramisu.

Convince yourself that this fits the pattern outlined in **N5**.

The negation of statements that involve existential or universal quantifiers is fairly straightforward as well. If $P(x)$ is a predicate, then

N6. $\neg \forall x\, P(x) \equiv \exists x\, \neg P(x)$,
N7. $\neg \exists x\, P(x) \equiv \forall x\, \neg P(x)$.

We justify the previous two laws in the next proposition.

Proposition 2.28. *The negation laws **N6** and **N7** for quantified expressions are valid.*

Proof. Let D denote the universe. The statement $\neg \forall x P(x)$ expresses that $P(x)$ does not hold for all $x \in D$. This is equivalent to saying that there exists an x in the domain D such that $P(x)$ does not hold, which is equivalent to $\exists x \neg P(x)$. Therefore, the law **N6** holds.

By the first claim, we have $\neg \forall x Q(x) \equiv \exists x \neg Q(x)$. For $Q(x) = \neg P(x)$, this yields $\neg \forall x \neg P(x) \equiv \exists x \neg(\neg P(x)) \equiv \exists x\, P(x)$. Negating both sides yields the law **N7**. □

Example 2.29. Let the universe of discourse be the set of all primes, and let $P(x)$ be the predicate with the interpretation "x is an odd integer." Then $\forall x P(x)$ means "all primes are odd integers." The negation of this statement is $\neg \forall x P(x)$, which means "not all primes are odd integers." By **N6**, this can also be expressed in the form $\exists x \neg P(x)$, which can be interpreted as "there exists a prime that is not an odd integer" or simply as "there exists a prime that is an even integer."

Example 2.30. Let the universe of discourse be the set of all right triangles in the Euclidean plane. Let $P(x)$ be the predicate with the interpretation "x is an equilateral triangle". Then $\neg \exists x P(x)$ means that "there does not exist a right equilateral triangle." By **N7**, we can also formulate this in the equivalent form $\forall x \neg P(x)$, which means "all right triangles are not equilateral."

The formalism shines when we want to determine the negation of expressions with multiple quantifiers. In such cases, we rely on repeated application of the laws **N6** and **N7**. The next example illustrates this for the negation of a statement with three quantifiers.

Example 2.31. Suppose that we want to negate the expression

$$\forall x \exists y \forall z\, C(x, y, z).$$

2.7 Negations

Then repeatedly applying the rules for negation yields

$$\begin{aligned} \neg \forall x \exists y \forall z\, C(x,y,z) &\equiv \exists x \neg \exists y \forall z\, C(x,y,z) \\ &\equiv \exists x \forall y \neg \forall z\, C(x,y,z) \\ &\equiv \exists x \forall y \exists z\, \neg C(x,y,z). \end{aligned}$$

The beauty of the formalism is that the negation of long nested compound statements becomes rather straightforward.

EXERCISES

2.58. Negate the following statements and simplify such that negations are either eliminated or occur only directly before predicates.
(a) $\forall x \exists y (P(x) \to Q(y))$,
(b) $\forall x \exists y (P(x) \wedge Q(y))$,
(c) $\forall x \forall y \exists z ((P(x) \vee Q(y)) \to R(x,y,z))$,
(d) $\exists x \forall y (P(x,y) \leftrightarrow Q(x,y))$,
(e) $\exists x \exists y (\neg P(x) \wedge \neg Q(y))$.

2.59. For each of the following statements, find the negation of the statement. Simplify such that the negations are eliminated (but you are allowed to use $\neg$).
(a) For all integers x, x^2 is nonnegative.
(b) For all integers a and b, if $a < b$, then $f(a) < f(b)$.
(c) For all integers a, there exists an integer b such that $a + b = 1001$.
(d) There exists an integer a such that for all integers b, $a + b = 1001$.
(e) For all positive integers a, there exists a positive integer b such that $b < a$.

2.60. A student uses the following statement in a proof: "If x is a real number, then $x \leqslant x^2$." The statement is evidently false. Negate this statement. For this purpose, (a) formalize the statement, (b) negate the formalized statement and rewrite it in a form that does not explicitly contain any negations, and (c) translate the negated statement back into English.

2.61. Negate the following statement from a calculus book: "There exists a constant $C \in \mathbf{R}$ such that $|f(x)| \leqslant C$ for all $x \in \mathbf{R}$." For this purpose, (a) first formalize the statement with quantifiers, (b) negate the formalized statement and rewrite it in a form such that no negations occur before a quantifier, and (c) translate the negated statement back into English.

2.62. One of the main difficulties in a calculus course is to grasp the concept of a limit. If the definition of the limit is given in terms of an epsilon-delta style, then this involves three quantifiers and an implication. In this case, logic can help to unravel the definition.

Let D be a subset of the set $\mathbf{R}$ of real numbers. A function $f \colon D \to \mathbf{R}$ has a **limit** L at $c \in D$, denoted $L = \lim_{x \to c} f(x)$, if and only if

> for all $\epsilon > 0$ there exists a $\delta > 0$ such that for all x in D satisfying $0 < |x - c| < \delta$, we have $|f(x) - L| < \epsilon$.

(a) Reformulate the definition of a limit in predicate logic.
(b) Negate the resulting statement.
(c) Rewrite the condition for the nonexistence of a limit in plain English.

2.8 Proofs

In this section, you will learn how to do certain proofs. The goal of a proof is to convince someone that a mathematical statement is true. Ideally, a proof will also explain why the statement is true. We will focus on three different proof techniques: direct proofs, proofs by contraposition, and proofs by contradiction.

Terminology. Before we proceed, it might be helpful to clarify some mathematical terminology. Mathematical statements are often called theorems, propositions, lemmas, corollaries, or scholia. The name helps the reader judge how important a statement is. By convention, a mathematical statement is usually called a proposition. If we want to emphasize the importance of a proposition, then we call it a theorem. A lemma is a statement that is used in the course of a proof of another proposition or theorem. If a proposition is a simple consequence of another proposition or theorem that hardly requires a proof, then it is called a corollary. A commentary on an important theorem is called a scholium.

Direct Proofs. Many mathematical statements are of the form $A \to B$. In a direct proof of $A \to B$, we assume that the hypothesis A is true and deduce the conclusion B. This is a valid proof technique, since $A \to B$ is vacuously true when A is false.

> **Direct Proof** To prove that a claim "If A, then B" holds, we can proceed as follows.
> *Proof* Suppose that A holds.
> $\vdots$
> Deduce B using axioms, rules of inference, and prior results.

An integer n is called **even** if and only if there exists an integer k such that $n = 2k$. An integer n is called **odd** if and only if there exists an integer k such that $n = 2k + 1$.

Proposition 2.32. *If n is an odd integer, then n^2 is an odd integer as well.*

Proof. If n is an odd integer, then by definition there must exist an integer k such that $n = 2k + 1$. It follows that $n^2 = (2k+1)^2$. Expanding the right-hand side yields
$$n^2 = 4k^2 + 4k + 1 = 2(2k^2 + 2k) + 1.$$

2.8 Proofs

Since $2k^2 + 2k$ is an integer, it follows that n^2 is an odd integer, which proves the claim. □

Remark 2.33. How exactly did we prove this result? We were given the hypothesis *n is an odd integer*. Our goal was to prove that *n^2 is an odd integer*. It is natural to use the definition of an odd integer to get started. In general, it is often a good idea to make things explicit, so we expressed n in the form $n = 2k + 1$. Since our goal was to prove a property about n^2, we simply investigated whether $n^2 = (2k+1)^2$ will have that property. In this case, the remaining steps were driven by the goal to show that n^2 is of the form $2\ell + 1$. Our explicit calculation proved that $\ell = 2k^2 + 2k$ will do the trick.

An integer n is called a **perfect square** if and only if there exists an integer k such that $n = k^2$. In our next proposition, we are supposed to show that a given odd integer can be expressed as a difference of two perfect squares. It might not be apparent what these two perfect squares are. A good idea is to look at some examples to get a feel for the problem.

$$
\begin{aligned}
3 &= 2^2 - 1^2 = 4 - 1, \\
5 &= 3^2 - 2^2 = 9 - 4, \\
7 &= 4^2 - 3^2 = 16 - 9.
\end{aligned}
$$

You probably see a pattern in these examples. So how do we get started with the proof? By hypothesis, n is an odd integer, so we use the definition of an odd integer, as we did in the previous proposition. For instance, $3 = 2 \times 1 + 1$, $5 = 2 \times 2 + 1$, and $7 = 2 \times 3 + 1$. Compare this with the differences of squares that we have guessed earlier and you should be all set to write down the proof.

Proposition 2.34. *If n is an odd integer, then it is the difference between two perfect squares.*

Proof. If n is an odd integer, then there exists by definition an integer k such that $n = 2k + 1$. Then adding $0 = k^2 - k^2$ to $n = 2k + 1$ yields

$$
\begin{aligned}
n = 2k + 1 &= k^2 + 2k + 1 - k^2 \\
&= (k+1)^2 - k^2,
\end{aligned}
$$

which is a difference between two perfect squares, as claimed. □

Given two integers a and b, we say that a **divides** b if and only if there exists an integer q such that $b = aq$. We will sometimes use the shorthand $a \mid b$ to denote that a divides b.

Proposition 2.35. *Let $a, b,$ and c be integers. Suppose that a divides both b and c. Then a must divide $mb + nc$ for all integers m and n. In particular, a divides both $b + c$ and $b - c$.*

Proof. By assumption, there exist integers p and q such that $b = ap$ and $c = aq$. Therefore, $mb + nc = map + naq = a(mp + nq)$, so $a \mid mb + nc$. □

The number 1 is a **unit**. A positive integer $p > 1$ whose only positive divisors are 1 and itself is called a **prime number**. An integer $c > 1$ that is not a prime number is called a **composite number**. The difference between two successive prime numbers is called a **prime gap**. For example, if we look at the first few prime numbers,

$$2, 3, 5, 7, 11, 13, 17, 19, 23, 29, 31, 37,$$

then the biggest gap between any pair of successive primes is 6. The next proposition shows that in general it is possible to find arbitrarily long gaps between primes. The argument is a simple direct proof.

Proposition 2.36. *There exist arbitrarily long gaps between prime numbers.*

Proof. For all integers k in the range $2 \leqslant k \leqslant n+1$, the integer $(n+1)! + k$ is divisible by k. Therefore, there exists a sequence of n consecutive composite numbers, for instance, none of the numbers

$$(n+1)! + 2, \ldots, (n+1)! + n + 1$$

is prime. It follows that there are arbitrarily long gaps between prime numbers. □

Direct proofs are used in all areas of mathematics, not just number theory. We will give one example from calculus that is very useful in its own right. Let $|x|$ denote the absolute value of a real number x. So

$$|x| = \begin{cases} x & \text{if } x \geqslant 0, \\ -x & \text{if } x < 0. \end{cases}$$

The triangle inequality is a simple observation that is often useful.

Proposition 2.37 (Triangle Inequality). *If a and b are real numbers, then*

$$|a + b| \leqslant |a| + |b|.$$

Proof. We observe that the square of the left-hand side is equal to

$$\begin{aligned} |a+b|^2 &= (a+b)^2 = a^2 + 2ab + b^2 \\ &= |a|^2 + 2ab + |b|^2 \\ &\leqslant |a|^2 + 2|a|\,|b| + |b|^2 = (|a| + |b|)^2 \end{aligned}$$

By taking the square root of both sides, we obtain the claim. □

2.8 Proofs

Proofs by Contraposition. If we want to prove that an implication $A \to B$ is true, then it suffices to show that $\neg B \to \neg A$ is true, since the two implications are logically equivalent. This is called a proof by contraposition.

> **Proof by Contraposition** To prove that a claim "If A, then B" holds, we can proceed as follows.
> *Proof (By Contraposition)*
> Suppose that $\neg B$ holds.
> $\vdots$
> Deduce $\neg A$ using axioms, rules of inference, and prior results.

An example will make this proof strategy clear.

Proposition 2.38. *If n^2 is an even integer, then n is an even integer.*

Proof. (By Contraposition.) Let n be an integer. Suppose that n is not even. Therefore, n is odd. By Proposition 2.32, it follows that n^2 is an odd integer. The claim follows, since we proved its contrapositive. □

Proposition 2.39. *Let a, b, and c be integers. If a does not divide bc, then a does not divide b and a does not divide c.*

Proof. (By Contraposition.) We are going to prove the contrapositive of the claim: if a divides b or a divides c, then a divides bc. We will discuss the two cases (a) a divides b and (b) a divides c separately.
(a) Suppose that a divides b. This means that there exists an integer q such that $b = aq$. Then $bc = aqc$, so $a \mid bc$.
(b) Suppose that a divides c. This means that there exists an integer p such that $c = ap$. Then $bc = bap = abp$, so $a \mid bc$.
We can conclude that a divides bc. Therefore, we have proved the claim by contraposition. □

Proposition 2.40. *Let n be a positive integer. If $4^n - 1$ is prime, then n is odd.*

Proof. We prove the contrapositive: if n is even, then $4^n - 1$ is composite.
If n is even, then there must exist a positive integer k such that $n = 2k$. Then $4^n - 1 = 4^{2k} - 1 = (4^k - 1)(4^k + 1)$ is composite, which proves the claim by contraposition. □

Proofs by Contradiction. In a proof by contradiction, our goal is to prove a mathematical statement A by assuming its negation $\neg A$ as a hypothesis and deriving a contradiction. Put differently, we prove A by showing that $\neg A \to F$ is true. Since the implication $\neg A \to F$ can only evaluate to true when $\neg A$ is false, it must be the case that the statement A is true.

> **Proof by Contradiction** To prove that a claim A holds, we can proceed as follows.
> **Proof** Seeking a contradiction, suppose that $\neg A$ holds.
> $\vdots$
> Deduce falsity F.

If you want to prove a statement A using a proof by contradiction, then you need to make sure that your hypothesis is the negation $\neg A$ of the statement that you would like to prove. A common mistake is to start with a statement that is not precisely the negation of A.

Let us have a look at a simple example. A rational number is a fraction of two integers such as $1/3$ or $2/17$. The claim is that there cannot exist a smallest positive rational number. This means that the set of positive rational numbers is qualitatively different from the set of positive integers, where 1 is evidently the smallest positive integer.

In general, it is a good idea to warn the reader that the proof will be by contradiction, for instance, by starting your proof with the phrase "Seeking a contradiction, we assume."

Proposition 2.41. *There does not exist a smallest positive rational number.*

Proof. Seeking a contradiction, we assume that there exists a smallest positive rational number s. Then $s/2$ is a positive rational number that is smaller than s, which contradicts our choice of s. □

Remark 2.42. Let us revisit the structure of the previous argument. The claim A of the proposition was that *there does not exist a smallest positive rational number*. Then the negation $\neg A$ of the statement is *there exists a smallest positive rational number*. We called this putative smallest number s. We can divide s by a positive integer greater than 1 to obtain a smaller number, which leads to a contradiction.

Let us prove a particularly useful fact.

Proposition 2.43. *Any integer $m > 1$ can be factored into a product of primes.*

Proof. Seeking a contradiction, let us assume that m is the smallest integer $m > 1$ that cannot be factored into a product of primes. In particular, m cannot be a prime and must be composite. Thus, $m = ab$ for some positive

2.8 Proofs

integers $a, b > 1$. Since a and b are numbers that are smaller than m, we can write them in the form

$$a = a_1 a_2 \cdots a_k, \quad \text{and} \quad b = b_1 b_2 \cdots b_\ell$$

such that $a_1, a_2, \ldots, a_k$ and $b_1, b_2, \ldots, b_\ell$ are primes. This would mean that $m = a_1 a_2 \cdots a_k b_1 b_2 \cdots b_\ell$ is a product of primes, contradicting our assumption. $\square$

Remark 2.44. Let $P(m)$ denote the predicate that is true if and only if m is an integer that can be factored into a product of primes. The claim of the previous proposition is that $\forall m P(m)$, where m extends over all integers that are greater than 1. Then the negation is the statement $\neg \forall m\, P(m) \equiv \exists m\, \neg P(m)$. In words, we begin the proof by assuming that there exists an integer $m > 1$ that cannot be factored into a product of primes. However, we used an important trick. We assumed that m is the smallest integer $m > 1$ that cannot be factored into primes. This made it easier to find a contradiction.

Proofs by contradiction go back to antiquity. The next theorem is contained in Euclid's elements, but we give Kummer's elegant variant of Euclid's proof.

Theorem 2.45 (Euclid). *There are infinitely many prime numbers.*

Proof. Seeking a contradiction, suppose that there exist only a finite number of primes $p_1, p_2, \ldots, p_k$. Let $P = p_1 p_2 \cdots p_k$ be their product. The integer $P - 1$ can be factored into primes by Proposition 2.43. Each prime p_m dividing $P - 1$ also divides P, hence by Proposition 2.35, it must divide the difference

$$P - (P - 1) = 1,$$

which is absurd. $\square$

Let us conclude this chapter with a particularly striking application of a proof by contradiction. The game of hex was invented by the Danish mathematician, scientist, and poet Piet Hein and later rediscovered by John Nash. It is a board game that is played by two players on an $n \times n$ rhombus-shaped grid with hexagonal fields. Two opposing sides of the game board are labeled white and the other two are labeled black. The four corner hexagons belong to either side. A 4×4 game board is depicted on the left of Fig. 2.3.

One player plays white playing pieces and the other plays black. The players take turns placing a single playing piece of their color into an empty hexagon cell anywhere on the game board. The goal of each player is to connect the two sides of their color with a connected path of playing pieces of their color. The first player to connect the two regions of their color wins. Figure 2.3 on the right shows a game that ends with White winning after a total of 7 moves.

One can show that the game of hex cannot end in a tie. Nash proved in 1952 that the first player has a winning strategy assuming optimal play. We give his argument here.

Theorem 2.46. *The first player in a hex game has a winning strategy.*

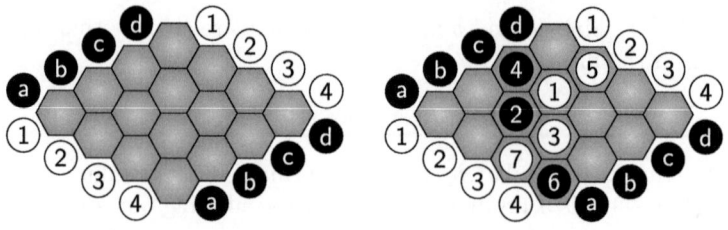

Figure 2.3: The figure on the left shows a 4 × 4 hex board with the lower left and upper right sides in white and the upper left and lower right in black. We can use algebraic notation to specify the hexagons of the game board. For instance, a1 is the left-most hexagon and d4 is the rightmost hexagon. The figure on the right illustrates one game. White begins with c2, then Black plays b2, White plays b3, then Black c1. White plays d2 and Black responds with a4. White's final move a3 completes the path between the two white regions, so White wins

Proof. Since the hex game cannot end in a tie, one of the players must have a winning strategy. Seeking a contradiction, let us assume that the second player has a winning strategy. Then the first player A can place a playing piece T anywhere on the board and from then on pretend to be the second player; this means that A will play according to the winning strategy. If the winning strategy requires the first player to place a playing piece on the field where T is placed, then A can place during this move A's piece on any other field, as this cannot hurt A's chances. This means that the second player cannot win anymore, contradicting our assumption that the second player has a winning strategy. □

This strategy-stealing argument works because the first move cannot hurt the first player. What does it really mean that the first player can steal the strategy? The nexus of the argument is that the second player simply cannot have a winning strategy!

The theorem proves the existence of a winning strategy for the first player, but it does not provide a recipe for such a strategy. Working out an explicit winning strategy turns out to be difficult. Even six decades later, all explicitly known winning strategies for hex are confined to small boards of size $n \leq 10$, despite a significant amount of research. In fact, we cannot hope that good algorithms for hex will be developed, since it is know that hex is PSPACE-complete (which means that it ranks among the most difficult problems that can be solved on a computer with a polynomial amount of space, and it is unlikely that such problems can be solved in polynomial time). You can learn more about PSPACE-complete problems in a course on algorithms.

2.8 Proofs

EXERCISES

2.63. Suppose that m and n are consecutive integers. Use a direct proof to show that their sum $m + n$ is an odd integer.

2.64. Suppose that n is an odd integer. Use a direct proof to show that $n^2 - 1$ is divisible by 8.

2.65. Use a direct proof to show that $n^3 - n$ is divisible by 6 for all integers n.

2.66. Let n be an integer. Give a direct proof that $n(n+1)(n+2)(n+3)$ is divisible by 24.

2.67. Let n be a positive even integer. Show that

$$\frac{1!2!3!\cdots(2n)!}{n!}$$

is a perfect square.

2.68. Using a direct proof, show that if a and b are real numbers, then the reverse triangle inequality

$$|a - b| \geq ||a| - |b||$$

holds.

2.69. Using a direct proof, show that if a and b are real numbers, then

$$|a| + |b| \leq |a + b| + |a - b|.$$

2.70. Let m and n be integers. Use a proof by contraposition to show that if their sum $m + n > 100$, then $m > 40$ or $n > 60$.

2.71. Let n be a positive integer. Use a proof by contraposition to show that if $n \geq ab$ for some positive integers a and b, then $a \leq n^{1/2}$ or $b \leq n^{1/2}$.

2.72. Let m be an integer. Use a proof by contraposition to show that if $5m + 7$ is even, then m is odd.

2.73. Let n be a positive integer. Use a proof by contraposition to show that if $2^n - 1$ is prime, then n must be prime. [A prime M_n of the form $M_n = 2^n - 1$ is called a **Mersenne prime**.]

2.74. Prove by contradiction that the equation $42m + 70n = 1000$ does not have an integer solution.

2.75. Prove by contradiction that $\sin x + \cos x \geq 1$ holds for all real numbers x in the range $0 \leq x \leq \pi/2$.

2.9 Notes

You will need to read many proofs and practice regularly to master the art of proofs. The books by Eccles [23] and Velleman [77] are recommended for further reading about proof techniques. We gave a brief introduction to symbolic logic, and we hope that you want to study this important topic in more depth. Fortunately, there are many excellent texts on first-order predicate logic, see for example Ebbinghaus et al. [22], Enderton [25], Gallier [28], Goldrei [29], Huth and Ryan [37], Mendelson [60], Smullyan [68] and [69]. We recommend these texts for further reading. A delightful collection of knight and knave puzzles is given in [67]. There is a wide range of applications of symbolic logic and a very long history that starts with Aristotle's logical works. The formalization of arguments that started with Aristotle were brought to perfection through Frege, Hilbert, Łukasiewicz, Russell, and Whitehead and many others. We strongly recommend that you explore `metamath.org` to see many more examples of formal proofs. It would be very remiss of us not to mention at least some limitations of formal methods. For example, it is undecidable whether a formula of first-order logic is true under all possible interpretations.

Chapter 3

Sets

> *No one shall expel us from the Paradise that Cantor has created.*
> — David Hilbert, *On the Infinite, Math. Ann. 95*

Set theory was invented by Georg Cantor in a remarkable series of six seminal papers. In this chapter, we will learn notation, terminology, and basic operations of set theory. We will follow the axiomatic method by Zermelo with the amendments suggested by Fraenkel and Skolem, so that we can avoid the consistency problems of "naive set theory." The axioms of this set theory are commonly known as ZFC, where the acronym honors Zermelo with Z and Fraenkel with F, and the letter C indicates the inclusion of the axiom of choice. This axiom schema forms the basis of most areas of modern mathematics.

3.1 Background and Motivation

A set is a collection of distinct objects. The objects contained in a set are called its elements. For instance,

$$\{1, 2, 3\}$$

is the set containing three elements, namely 1, 2, and 3. Another set is $\{A, B, C, D, E\}$ that contains the five elements A, B, C, D, and E.

Sets were implicitly used by mathematicians since antiquity. However, they were not seriously studied in their own right before Cantor conceived his theory of sets. Set theory quickly flourished and soon became an established mathematical area. Nowadays, sets are commonly used as a foundation for most mathematical theories.

Sets can have a finite number of elements or an infinite number of elements. One of the key insights by Cantor was that infinite sets can come in different sizes (or, in technical terms, cardinalities). One of the many consequences of this fact is a simple proof for the existence of functions on the natural numbers that cannot be computed. So Cantor's insight into the infinite has tangible benefits to the theory of computation!

The pioneering work by Cantor was not without some flaws. He did not delineate well enough what can and cannot constitute a set, so his theory lead to contradictions. Fortunately, it is possible to fix these flaws, and the subsequent sections detail the improved version of set theory that was developed by Zermelo, Fraenkel, and others.

For those who are familiar with Cantor's set theory, we point out some apparent differences. In ZFC, every element is a set itself—so there are no so-called urelements that are not sets themselves. This allows for a more coherent development of the theory, but the difference is not essential. The unrestricted set comprehension of Cantor's set theory is replaced by a more restricted version, which eliminates troublesome antinomies.

3.2 Fundamental Concepts

The central concept of mathematics is a **set**. A set is an abstraction of a container. The objects contained in it are called its **elements**. We write

$$a \in A$$

to denote that a is an element of the set A. If a is not an element of A, then we write $a \notin A$. Put differently, $a \notin A$ is the negation of $a \in A$.

The first axiom asserts that there exists a set that does not contain any elements.

> **S1.** There exists a set that does not have any elements.

We call it the **empty set** and denote it by $\emptyset$, or by $\{\}$. The membership relation $a \in \emptyset$ is always false. The empty set occurs more often than you might think. Its role in set theory is similar to the role of 0 in the integers.

The second axiom asserts that the membership relation completely determines the equality of sets.

> **S2.** Two sets are **equal** if and only if they have the same elements.

In the Zermelo–Fraenkel set theory, every element is a set. Our third axiom schema asserts that we can form a finite set given its elements.

> **S3.** Let $a_1, \ldots, a_n$ be sets. Then there exists a set S such that $x \in S$ if and only if $x = a_1$ or $x = a_2$ or $\cdots$ or $x = a_n$.

3.2 Fundamental Concepts

The set S is commonly denoted by listing its elements in curly braces,

$$S = \{a_1, a_2, \ldots, a_n\}.$$

For example, $\{a, b, c\}$ is the set containing the elements a, b, c, and no other elements.

It should be stressed that listing elements more than once does not change the set. For instance, $\{a, a, b\}$ denotes the same set as $\{a, b\}$. We drive this point home in the next example.

Example 3.1. The four sets

$$S_1 = \{\emptyset\}, S_2 = \{\emptyset, \emptyset\}, S_3 = \{\emptyset, \emptyset, \emptyset\}, S_4 = \{\emptyset, \emptyset, \emptyset, \emptyset\}$$

all contain the empty set $\emptyset$ as an element. Since none of them contain any other element, they actually all specify the same set! So, despite appearances, the sets S_1, S_2, S_3, and S_4 each contain a single element. Listing an element multiple times is simply redundant.

Although we have only learned a few axioms of ZFC set theory, they already allow us to specify many different sets. Starting from the empty set and the set builder notation, we can form many different sets.

We call a set $\{a\}$ with one element a **singleton set**. This terminology is helpful in the next example.

Example 3.2. The empty set $\emptyset$ and the singleton set $\{\emptyset\}$ containing the empty set as an element are different sets. Indeed, the empty set $\emptyset$ does not contain any element, but the singleton set $\{\emptyset\}$ contains one element, namely the empty set. It follows from **S2** that $\emptyset$ is not equal to $\{\emptyset\}$.

By the axiom **S2**, the order of the elements does not matter, so $\{a, b, c\}$ denotes the same set as $\{b, c, a\}$.

Example 3.3. We can form a set S containing as elements the empty set $\emptyset$, the singleton set $\{\emptyset\}$, and the set $\{\emptyset, \{\emptyset\}\}$ containing two elements. Apparently, all three elements of this set are different, as the first contains no element, the second contains one element, and the third contains two elements. So even if we do not repeat elements, there are already six different ways to list the elements of S, namely

$$S_1 = \{\emptyset, \{\emptyset\}, \{\emptyset, \{\emptyset\}\}\}, \quad S_2 = \{\emptyset, \{\emptyset, \{\emptyset\}\}, \{\emptyset\}\},$$
$$S_3 = \{\{\emptyset\}, \emptyset, \{\emptyset, \{\emptyset\}\}\}, \quad S_4 = \{\{\emptyset\}, \{\emptyset, \{\emptyset\}\}, \emptyset\},$$
$$S_5 = \{\{\emptyset, \{\emptyset\}\}, \emptyset, \{\emptyset\}\}, \quad S_6 = \{\{\emptyset, \{\emptyset\}\}, \{\emptyset\}, \emptyset\}.$$

Each set $S_1, S_2, \ldots, S_6$ is equal to S, as the order of elements does not matter by axiom **S2**.

If we form sets starting with the empty set and set builder notations, we can quickly end up with a bewildering number of curly braces. This can be a bit tedious to decipher. Therefore, it is customary to introduce abbreviations. For instance, the next example reformulates the previous example with abbreviations, which dramatically increase the readability.

Example 3.4. If we denote the empty set $\emptyset$ as 0, the singleton set $\{\emptyset\}$ as 1, and the set $\{\emptyset, \{\emptyset\}\}$ with two elements as 2, then the sets from the previous example can be stated in the more readable form $S_1 = \{0, 1, 2\}$, $S_2 = \{0, 2, 1\}$, $S_3 = \{1, 0, 2\}$, $S_4 = \{1, 2, 0\}$, $S_5 = \{2, 0, 1\}$, and $S_6 = \{2, 1, 0\}$.

The idea of defining a nonnegative integer n by a set with n elements goes back to von Neumann. We will explore his construction in depth later on. One can apply the same principle to other objects. The next example briefly sketches one possible set-theoretic representation of ASCII characters.

Example 3.5. If we want to reason about sets of ASCII characters, then we can use von Neumann's idea and represent the upper case A with ASCII code 65 by a set with 65 elements, the lower case a with ASCII code 97 by a set with 97 elements, and so on. This allows us to form sets of characters such as $\{A, B, C\}$.

A set A is called a **subset** of a set B if and only if all elements of A are also elements of B. We write $A \subseteq B$ to denote that A is a subset of B. If A is a subset of B, but not equal to B, then we call A a **proper subset** of B. We write $A \subsetneq B$ to denote that A is a proper subset of B.

Example 3.6. If $A = \{1, 2\}$, $B = \{2, 3\}$ and $C = \{1, 2, 3\}$, then $A \subseteq C$ and $B \subseteq C$, so A and B are subsets of C. In fact, they are both proper subsets of C. However, A is not a subset of B, as 1 is an element of A that is not contained in B. Furthermore, B is not a subset of A, as 3 is an element of B that is not contained in A.

A convenient way to prove that two sets A and B are equal is to show that $A \subseteq B$ and $B \subseteq A$ holds. The next proposition establishes this simple fact.

Proposition 3.7. *If A and B are two sets such that $A \subseteq B$ and $B \subseteq A$, then $A = B$.*

Proof. Since $A \subseteq B$, $x \in A$ implies $x \in B$. Since $B \subseteq A$, $x \in B$ implies $x \in A$. Therefore, $x \in A$ if and only if $x \in B$, which means that $A = B$. $\square$

Every set A has itself as a subset, since the definition of a subset does not require that a subset is a proper subset. The next proposition shows that every set has the empty set as a subset. This is obvious if you have a good grounding in logic, but perhaps a bit surprising for a novice.

Proposition 3.8. *The empty set is a subset of each set.*

3.2 Fundamental Concepts

Proof. Seeking a contradiction, we assume that there exists a set A such that the empty set $\emptyset$ is not a subset of A. This means that there must exist an element $x \in \emptyset$ such that $x \notin A$. However, this is absurd since the empty set does not contain any element x. □

The most common way to specify a subset of a set A is by defining a property $S(x)$ that an element x of A may or may not have.

> **S4.** Let $S(x)$ be a property of x. For a set A, there exists a set B such that $x \in B$ if and only if x is an element of A that has property $S(x)$.

The set B is commonly denoted by

$$\{x \in A \mid S(x)\}.$$

The main point of the construction is that a subset of A is obtained by filtering out all elements x that do not satisfy $S(x)$. What remains is evidently a subset of A, and it contains precisely the elements x of A that do satisfy the property $S(x)$.

There are some technicalities to keep in mind. We need to be able to specify the property $S(x)$ using a formula in the first-order language of set theory that has only x as a free variable. Furthermore, the formula $S(x)$ cannot reference the set B, so that circular definitions are avoided. In other words, $S(x)$ is constructed using the equality $=$ and the membership relation $\in$, the logical operators $\neg$, $\wedge$, $\vee$, $\rightarrow$, and $\leftrightarrow$, as well as the quantifiers $\forall$ and $\exists$ using the rules of first-order languages.

The next example illustrates this principle.

Example 3.9. Let A denote the set

$$A = \{\emptyset, \{\emptyset\}, \{\emptyset, \{\emptyset\}\}, \{\emptyset, \{\emptyset\}, \{\emptyset, \{\emptyset\}\}\}\}$$

with four distinct elements. The set A contains the empty set $\emptyset$, the singleton set $\{\emptyset\}$, the set $\{\emptyset, \{\emptyset\}\}$ with two elements, and the set $\{\emptyset, \{\emptyset\}, \{\emptyset, \{\emptyset\}\}\}$ with three elements. If $S(x)$ is the property $(\{\emptyset\} \in x)$ then

$$\{x \in A \mid S(x)\} = \{x \in A \mid \{\emptyset\} \in x\} = \{\{\emptyset, \{\emptyset\}\}, \{\emptyset, \{\emptyset\}, \{\emptyset, \{\emptyset\}\}\}\}$$

is the subset of A containing as elements $\{\emptyset, \{\emptyset\}\}$ and $\{\emptyset, \{\emptyset\}, \{\emptyset, \{\emptyset\}\}\}$. Both elements evidently contain $\{\emptyset\}$. Missing from this subset of A are the elements $\emptyset$ and $\{\emptyset\}$ of A that do not contain $\{\emptyset\}$ as an element.[1]

---

[1] Indeed, the empty set $\emptyset$ does not contain any elements, and the set $\{\emptyset\}$ contain just one element, namely $\emptyset$, but this element is different from $\{\emptyset\}$.

One does not need to express the property $S(x)$ in such a formal way. It is sufficient to specify the property $S(x)$ unambiguously in a way that can be expressed—in principle—by a formal statement. The next example illustrates this point.

Example 3.10. Let A denote the set

$$A = \{\emptyset, \{\emptyset\}, \{\emptyset, \{\emptyset\}\}, \{\emptyset, \{\emptyset\}, \{\emptyset, \{\emptyset\}\}\}\}$$

with four distinct elements. If $S(x)$ is the property that (x has an even number of elements), then

$$\{x \in A \mid S(x)\} = \{\emptyset, \{\emptyset, \{\emptyset\}\}\}$$

is the subset of A containing the empty set with 0 elements, and the set $\{\emptyset, \{\emptyset\}\}$ with two elements.

Axiomatic set theory is very expressive and subsumes many areas of mathematics such as analysis and number theory. As we have already mentioned, it is possible to interpret nonnegative integers as sets. Properties of numbers such as $S(x) = $ (x is a prime) can be expressed in principle using the formal language of set theory, even though the expression would be long and difficult to understand. Therefore, the next example is still within the purview of axiomatic set theory, even though everything is expressed by a convenient shorthand.

Example 3.11. Let $S = \{2, 3, \ldots, 1947\}$ be the set of integers in the range from 2 to 1947. The set

$$\{x \in S \mid x \text{ is a prime factor of } 1947\}$$

is equal to the set $\{3, 11, 59\}$.

Remark 3.12. The idea behind set comprehension has also inspired set and list comprehension in programming languages, including Haskell, Python, Ruby, and many others.

So let us get back to set theory. In the next axiom, we postulate that one can collect all of the subsets of a set into a set.

> **S5.** For a set A, there exists a set that contains all subsets of A as elements.

We call this set the **power set** of A and denote it by $P(A)$.

The power set $P(A)$ contains the empty set $\emptyset$ as well as the set A and any set in between these two extremes.

Example 3.13. We offer three small examples.

(a) The power set of the empty set is given by $P(\emptyset) = \{\emptyset\}$.

3.2 Fundamental Concepts

(b) The power set of a singleton set is given by $P(\{a\}) = \{\varnothing, \{a\}\}$.

(c) The power set of a set with two elements is given by

$$P(\{a,b\}) = \{\varnothing, \{a\}, \{b\}, \{a,b\}\}.$$

Indeed, it must contain the empty set and the entire set $\{a,b\}$, as well as the smaller subsets $\{a\}$ and $\{b\}$ of $\{a,b\}$. There are no other subsets.

Example 3.14. When constructing the power set of a larger set such as $S = \{a,b,c\}$, it can be helpful to notice that each subset of S either contains the element c or does not contain c. The subsets of S that do not contain c are given by the four elements of $P(\{a,b\})$, namely

$$\varnothing, \{a\}, \{b\}, \{a,b\}.$$

On the other hand, the subsets of S that do contain c are obtained by adding c to each of these four sets, resulting in

$$\{c\}, \{a,c\}, \{b,c\}, \{a,b,c\}.$$

Therefore, the power set $P(\{a,b,c\})$ is given by

$$P(\{a,b,c\}) = \{\varnothing, \{a\}, \{b\}, \{a,b\}, \{c\}, \{a,c\}, \{b,c\}, \{a,b,c\}\}.$$

So $P(\{a,b,c\})$ contains twice as many elements as $P(\{a,b\})$.

So far, we have asserted the existence of sets such as the empty set, finite collections of sets, power sets, and subsets using specification. However, there are some sets that cannot exist, as the next theorem shows. The irksome "set of all sets" confounded set theorists until the proper axioms of set theory were found.

Theorem 3.15. *A set having all sets as elements cannot exist.*

Proof. Seeking a contradiction, we suppose that U is a set that has all sets as elements. We can then form the subset

$$C = \{x \in U \mid x \notin x\}.$$

of U. Since C is both an element of U and a subset of U, we have two choices: either the element C belongs to the subset C or it does not belong to it. Let us carefully inspect both cases.

(a) If $C \in C$, then we have $C \notin C$ by definition of C, which is absurd.
(b) If $C \notin C$, then we have $C \in C$ by definition of C, which is also absurd.

In both cases, we can conclude that both $C \in C$ and $C \notin C$ hold, and this is a contradiction. Therefore, the set U cannot exist. $\square$

Some readers might not feel comfortable with the definition
$$C = \{x \in U \mid x \notin x\}.$$
The construction used here is valid, since any formula from predicate logic using set membership and equality can be used as a specification, so there is nothing out of the ordinary to use the membership relation $x \in x$ or its negation.

We remark here that sets containing themselves as elements are ruled out by the following Axiom of Foundation in the Zermelo–Fraenkel set theory.

> **S6.** Every nonempty set A must contain an element $B \in A$ such that the sets A and B do not have any element in common.

This axiom was introduced to set theory by von Neumann. Even though it might not be immediately obvious from its formulation, this axiom rules out self-referential formulations of sets (such as sets containing themselves, see Exercise 3.14). Furthermore, the axiom **S6** implies there cannot exist infinite descending membership chains,[2] such as
$$A_1 \ni A_2 \ni A_3 \ni \cdots,$$
see Exercise 3.15. The main purpose of axiom **S6** is to rule out weird sets that do not occur in common mathematical practice.

EXERCISES

3.1. For each of the elements $e_1 = \emptyset$, $e_2 = \{\emptyset\}$ and $e_3 = \{\{\emptyset\}\}$, determine whether or not they are elements of the sets
(a) $A_1 = \{\emptyset\}$,
(b) $A_2 = \{\emptyset, \{\emptyset\}\}$,
(c) $A_3 = \{\{\emptyset\}, \{\{\emptyset\}\}\}$.
You have to consider nine potential membership cases.

3.2. For each of the sets $S_1 = \emptyset$, $S_2 = \{\emptyset\}$ and $S_3 = \{\{\emptyset\}\}$, determine whether or not they are subsets of the sets
(a) $A_1 = \{\emptyset\}$,
(b) $A_2 = \{\emptyset, \{\emptyset\}\}$,
(c) $A_3 = \{\{\emptyset\}, \{\{\emptyset\}\}\}$.
You have to consider nine potential subset cases.

3.3. Prove that the subset relation is transitive, i.e., show that $A \subseteq B$ and $B \subseteq C$ implies $A \subseteq C$.

3.4. Express each of the following statements in your own words:

---

[2] In the terminology of the next chapter, the membership relation is a well-founded relation.

3.2 Fundamental Concepts

(a) $x \in S$
(b) $\{x\} \subseteq S$
(c) $\{x\} \in P(S)$

Then show that all three statements are equivalent.

3.5. Show that if S is a set, then $S \in P(S)$.

3.6. Show that the empty set is unique, meaning that if A and B are both empty sets, then $A = B$.

3.7. Show that a set, which is a subset of every set, must be the empty set.

3.8. Let $\mathbf{Z}$ denote the set of all integers. Describe the set

$$E = \{(x, y) \in \mathbf{Z} \times \mathbf{Z} \mid x^2/4 + y^2 \leq 1\}$$

by set enumeration, that is, by explicitly listing its elements in curly braces.

3.9. Let $\mathbf{Z}$ denote the set of integers. Determine the power set $P(S)$ of the set $S = \{x \in \mathbf{Z} \mid x^2 < 4\}$.

3.10. Suppose that a set A has n elements. Show that $P(A)$ has 2^n elements by giving a counting argument. [Hint: Notice that each element of A can be included or excluded from a subset.]

3.11. Let A and B be sets. Show that $P(A) = P(B)$ implies $A = B$.

3.12. Let A be a set containing elements a, b, and c. Show that

$$\{\{a\}, \{a, b\}, \{a, b, c\}\}$$

is an element of $P(P(A))$.

3.13. Show that a set cannot contain its power set.

3.14. The Axiom of Foundation **S6** states that every nonempty set A must contain an element $B \in A$ such that the sets A and B do not have any element in common. Show that the axiom of foundation implies that a set A cannot satisfy $A \in A$. [Hint: Consider the set $\{A\}$.]

3.15. Show that the Axiom of Foundation **S6** implies that there cannot exist a sequence $A_1, A_2, A_3, \ldots$ of sets such that

$$A_{n+1} \in A_n$$

holds for all positive integers n.

3.3 Intersections and Unions

Given two sets A and B, we can form the subset

$$\{x \in A \mid x \in B\}.$$

We will denote this set by $A \cap B$ and call it the **intersection** of the sets A and B. The intersection of A and B contains all elements that are contained in both sets.

We say that two sets A and B are **disjoint** if and only if $A \cap B = \emptyset$.

Example 3.16. If $A = \{1,2,3,4,5\}$, $B = \{2,4,6,8\}$ and $C = \{1,3,5,7\}$, then

$$A \cap B = \{2,4\}, \quad A \cap C = \{1,3,5\}, \quad B \cap C = \emptyset.$$

The sets B and C are disjoint, since B contains only even integers and the set C contains only odd integers.

We postulate[3] that there exists a set U that contains precisely the elements of the sets A and B. In other words, $x \in U$ if and only if $x \in A$ or $x \in B$. We call U the **union** of A and B and denote it by $A \cup B$.

Example 3.17. If $A = \{1,2,3\}$ and $B = \{2,4,6,8\}$ then

$$A \cup B = \{1,2,3,4,6,8\}.$$

The union of A and B contains the elements of both sets A and B, but no other elements (Fig. 3.1).

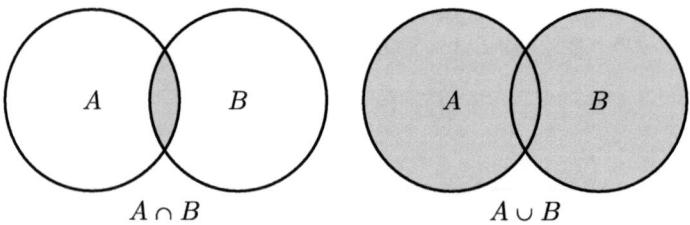

Figure 3.1: The two figures show the Venn diagrams of the sets A and B. The figure on the left illustrates the intersection $A \cap B$ as the shaded region in the intersection of the two circles. The shaded part in the figure on the right illustrates the union $A \cup B$

Proposition 3.18. *Set intersection and set union are associative operations, that is, for all sets A, B, and C, we have*

$$A \cap (B \cap C) = (A \cap B) \cap C,$$
$$A \cup (B \cup C) = (A \cup B) \cup C.$$

---

[3]This is a special case of axiom S7 below.

3.3 Intersections and Unions

Proof. We have $x \in A \cap (B \cap C)$ if and only if $x \in A$ and $x \in (B \cap C)$. The latter condition holds if and only if $x \in A$ and $x \in B$ and $x \in C$. The condition $x \in A$ and $x \in B$ and $x \in C$ holds if and only if $x \in (A \cap B)$ and $x \in C$, and the latter condition is equivalent to $x \in (A \cap B) \cap C$. Therefore, $A \cap (B \cap C) = (A \cap B) \cap C$.

The proof for the associativity of unions is similar. □

One can use a similar style of proof to show that set intersection and set union are commutative operations, so

$$A \cap B = B \cap A \quad \text{and} \quad A \cup B = B \cup A$$

hold for all sets A and B. The operations are idempotent,

$$A \cap A = A \quad \text{and} \quad A \cup A = A.$$

The empty set is a zero element for the intersection operation

$$A \cap \emptyset = \emptyset \cap A = \emptyset,$$

and a neutral element for the union operation

$$A \cup \emptyset = \emptyset \cup A = A.$$

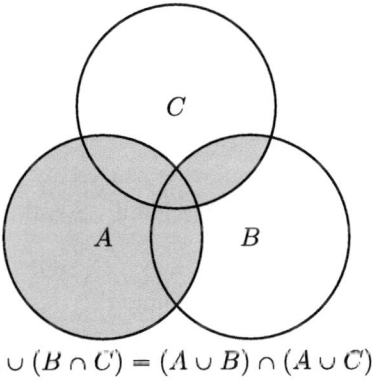

$$A \cup (B \cap C) = (A \cup B) \cap (A \cup C)$$

Figure 3.2: Venn diagrams can be useful to check set identities. The shaded part of the figure illustrates the set $A \cup (B \cap C)$. Since the same area can be obtained by shading $(A \cup B) \cap (A \cup C)$, we have a visual "proof" that the two sets are the same. Of course, this does not replace a formal proof, but it can serve as a motivation

It is interesting to see how the two set operations interact.

Proposition 3.19. *Set intersection and set union satisfy the distributive laws*

$$\begin{aligned} A \cap (B \cup C) &= (A \cap B) \cup (A \cap C), \\ A \cup (B \cap C) &= (A \cup B) \cap (A \cup C), \end{aligned}$$

for all sets A, B, and C.

Proof. We have $x \in A \cap (B \cup C)$ if and only if $x \in A$ and $x \in (B \cup C)$ if and only if $x \in A$ and ($x \in B$ or $x \in C$). This is equivalent to ($x \in A$ and $x \in B$) or ($x \in A$ and $x \in C$) and this holds if and only if $x \in (A \cap B) \cup (A \cap C)$.

We leave the proof of the second assertion to the reader. A visual "proof" motivating the correctness of the assertion is given in Fig. 3.2. □

General Intersections and Unions. Set intersections and unions can be extended to an arbitrary number of sets. The sets are collected in a family S. The **intersection** $\bigcap S$ of a nonempty family S is defined to be the set

$$\bigcap S = \{x \in X \mid \text{for all } A \in S, x \in A\},$$

where X is an element of S. The intersection of S contains the elements that are common to all members of the family S.

Example 3.20. The intersection of the sets $\{1, 2, 3\}, \{1, 3, 4\}, \{2, 3, 4\}$ is given by

$$\bigcap \{\{1, 2, 3\}, \{1, 3, 4\}, \{2, 3, 4\}\} = \{3\},$$

as 3 is the only element that is common to all three sets.

The next axiom postulates the existence of the union of sets.

> **S7.** For any family S of sets, there exists a set U such that $x \in U$ if and only if $x \in A$ for some $A \in S$.

We call U the **union** of the sets in the family S and denote it by $\bigcup S$.

Example 3.21. Let $A = \{1, 2\}$, $B = \{2, 3\}$, and $C = \{1, 3\}$. The family $S = \{A, B, C\}$ has the intersection and union

$$\bigcap S = \emptyset \quad \text{and} \quad \bigcup S = \{1, 2, 3\},$$

respectively.

The intersection and union of a family of two sets coincide with the previous definitions of intersection and union:

$$A \cap B = \bigcap \{A, B\} \quad \text{and} \quad A \cup B = \bigcup \{A, B\}.$$

Remark 3.22. Since the intersection and union of a set family are at different level than the notation in the binary case, many people prefer to write

$$\bigcap_{A \in S} A \quad \text{for} \quad \bigcap S \quad \text{and} \quad \bigcup_{A \in S} A \quad \text{for} \quad \bigcup S.$$

EXERCISES

3.16. Suppose that $A = \{1, 2, 3\}$, $B = \{2, 3, 5\}$, and $C = \{3, 4\}$. Determine the following sets: (a) $A \cap B$, (b) $A \cup B$, (c) $B \cap C$, (d) $B \cup C$, (e) $A \cap C$, (f) $A \cup C$.

3.17. Let A denote the set of even integers, B the set integers that are multiples of 3, C the set of positive integers that are powers of 6. Describe the sets (a) $A \cap B \cap C$ and (b) $A \cup B$.

3.18. Show that $A \subseteq B$ if and only if $A \cap B = A$.

3.19. Show that $A \subseteq B$ if and only if $A \cup B = B$.

3.20. Show that $P(A) \cup P(B) \subseteq P(A \cup B)$.

3.21. Show that $P(A \cup B) \subseteq P(A) \cup P(B)$ implies that either $A \subseteq B$ or $B \subseteq A$.

3.22. Let S be a set. Show that $\bigcup P(S) = S$.

3.23. Let A and B be finite sets. We denote by $|A|$ the number of elements in A. Show that
$$|A \cup B| + |A \cap B| = |A| + |B|.$$

3.4 Differences and Symmetric Differences

In this section, we introduce the set difference operation that singles out elements that belong to one set but not another. The set difference between a fixed large set and a subset is called a complement, which has a convenient notation. The final operation that we introduce is the symmetric difference that has nicer properties than the set difference.

Difference. Given two sets A and B, we define the **difference** between the sets A and B as the set $A - B$ defined by
$$A - B = \{x \in A \mid x \notin B\}.$$

If the sets A and B carry an additive operation, then the alternative notation $A \setminus B$ is used to denote the set difference to avoid notational ambiguity.

Let us have a look at some properties of the set difference. The set B does not need to be a subset of A in the difference $A - B$, but only the elements in the intersection $A \cap B$ will matter, since
$$A - B = A - (A \cap B).$$

The set difference can characterize subsets, since
$$A - B = \emptyset \quad \text{if and only if} \quad A \subseteq B;$$

and disjoint sets, because

$$A - B = A \quad \text{if and only if} \quad A \cap B = \emptyset.$$

The two most significant properties of the set difference are given in the next proposition.

Proposition 3.23. *For all sets A, B, and C, the de Morgan's laws*

$$\begin{aligned} C - (A \cap B) &= (C - A) \cup (C - B), \\ C - (A \cup B) &= (C - A) \cap (C - B) \end{aligned}$$

hold.

Proof. We have $x \in C - (A \cap B)$ if and only if $x \in C$ and $x \notin A \cap B$. The latter condition is equivalent to $x \in C$ and not $(x \in A \cap B)$ or equivalently $x \in C$ and not $(x \in A$ and $x \in B)$. We can rewrite this condition in the logically equivalent form $x \in C$ and $(x \notin A$ or $x \notin B)$, which can also be expressed in the form $(x \in C$ and $x \notin A)$ or $(x \in C$ and $x \notin B)$. In other words, this condition is equivalent to $x \in (C - A)$ or $x \in C - B$, which can be rewritten as $x \in (C - A) \cup (C - B)$. Therefore, we can conclude that $C - (A \cap B) = (C - A) \cup (C - B)$.

We leave the proof of the second de Morgan's law to the reader. □

Complement. We will often encounter the situation that all sets in an argument are subsets of a given set U, which we can consider the "universe." In that case, it is convenient to define for each set A a **complement** $A^\complement$ in U by

$$A^\complement = U - A.$$

The main point of the notation is to suppress the set U. Of course, the universe U needs to be clear from the context. In this case, the de Morgan's laws read

$$(A \cap B)^\complement = A^\complement \cup B^\complement \quad \text{and} \quad (A \cup B)^\complement = A^\complement \cap B^\complement.$$

Furthermore, we have the identities

$$A \cup A^\complement = U \quad \text{and} \quad A \cap A^\complement = \emptyset,$$

as well as the double complement identity $(A^\complement)^\complement = A$. The complement notation often allows one to simplify proofs. For example, let the universe U contain both A and B. Then

$$A - B = A \cap B^\complement.$$

This is convenient, since it allows one to translate set differences into set intersections that are more straightforward to manipulate.

3.4 Differences and Symmetric Differences

Symmetric Difference. The **symmetric difference** between A and B is the set $A \triangle B$ defined by

$$A \triangle B = (A - B) \cup (B - A).$$

The symmetric difference between A and B consists of all elements that belong to either A or to B but not to both. The Venn diagrams in Fig. 3.3 might be helpful to understand the difference between set difference and symmetric difference.

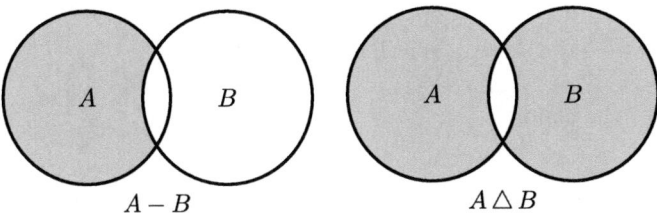

Figure 3.3: The figure on the left illustrates the set difference $A - B$, and the figure on the right illustrates the symmetric difference $A \triangle B$

The symmetric difference has the empty set as a neutral element, so

$$A \triangle \emptyset = \emptyset \triangle A = A$$

holds for all sets A. Every set A is its own inverse

$$A \triangle A = \emptyset.$$

The symmetric difference is commutative, so

$$A \triangle B = B \triangle A$$

holds for all sets A and B, since the union is a commutative operation. The symmetric difference is also an associative operation, see Proposition 3.25. The intersection distributes over the symmetric difference

$$A \cap (B \triangle C) = (A \cap B) \triangle (A \cap C).$$

The symmetric difference $A \triangle B$ consists of the elements in $A \cup B$ that do not belong to $A \cap B$. This is evident from the Venn diagram in Fig. 3.3. We will now give a formal proof that uses algebraic manipulation to derive the set identity given in the next lemma.

Lemma 3.24. *We have*

$$A \triangle B = (A \cup B) - (A \cap B).$$

Proof. Let $U = A \cup B$ be the universe. Then

$$\begin{aligned} A \triangle B &= (A - B) \cup (B - A) \\ &= (A \cap B^c) \cup (B \cap A^c) \\ &= ((A \cap B^c) \cup B) \cap ((A \cap B^c) \cup A^c) \end{aligned}$$

by the distributive law. Using the distributive law once more, it follows that

$$\begin{aligned} A \triangle B &= ((A \cup B) \cap U) \cap (U \cap (B^c \cup A^c)) \\ &= (A \cup B) \cap (A \cap B)^c \\ &= (A \cup B) - (A \cap B), \end{aligned}$$

which proves the claim. $\square$

The set difference is not an associative operation, see Exercise 3.27. The symmetric difference is an associative operation, but the proof of this fact is more involved than the other properties of the symmetric difference.

Proposition 3.25. *The symmetric difference is an associative operation, so*

$$A \triangle (B \triangle C) = (A \triangle B) \triangle C$$

holds for all sets A, B, and C.

Proof. Let $U = A \cup B \cup C$ be the universe. We have

$$A \triangle (B \triangle C) = (A \cap (B \triangle C)^c) \cup (A^c \cap (B \triangle C)).$$

Expanding the symmetric difference $B \triangle C$ in two different ways, we get

$$\begin{aligned} (B \triangle C)^c &= ((B \cup C) - (B \cap C))^c \\ &= ((B \cup C) \cap (B \cap C)^c)^c = (B^c \cap C^c) \cup (B \cap C) \end{aligned}$$

and $B \triangle C = (B - C) \cup (C - B) = (B \cap C^c) \cup (B^c \cap C)$. This yields

$$A \triangle (B \triangle C) = (A \cap B^c \cap C^c) \cup (A \cap B \cap C) \cup (A^c \cap B \cap C^c) \cup (A^c \cap B^c \cap C).$$

Since the latter expression is symmetric in A and C, we have $A \triangle (B \triangle C) = C \triangle (B \triangle A)$. By commutativity of the symmetric difference, we get

$$A \triangle (B \triangle C) = C \triangle (B \triangle A) = C \triangle (A \triangle B) = (A \triangle B) \triangle C,$$

as claimed. $\square$

3.5 Cartesian Products

EXERCISES

3.24. Prove or disprove: The set difference is a commutative operation. In other words, decide whether $A - B = B - A$ holds for all sets.

3.25. Show that $A \cap (B - C) = (A \cap B) - C$.

3.26. Show that $A \cap (B - C) = (A \cap B) - (A \cap C)$.

3.27. Show that the set difference is not an associative operation, that is, $A - (B - C)$ is in general not equal to $(A - B) - C$.

3.28. Show that for all sets A, B, and C, we have
(a) $A - (B - C) = (A - B) \cup (A \cap C)$,
(b) $(A - B) - C = A - (B \cup C)$.

3.29. Let A be a set. Show that the "complement" of A given by all $x \notin A$ cannot exist. [This explains why we used some set U relative to which we form the complement]

3.30. Show that the distributive law $A \cap (B \triangle C) = (A \cap B) \triangle (A \cap C)$ holds for all sets A, B, and C.

3.5 Cartesian Products

The pair $\{a, b\}$ contains in general two elements. Since $\{a, b\} = \{b, a\}$, this pair is not ordered. Our goal is to introduce an ordered pair (a, b) that allows us to distinguish the first coordinate a from the second coordinate b. One could introduce such ordered pairs as primitives. However, Kuratowski has shown that one can define ordered pairs in terms of unordered pairs, and this is a good example that can test your understanding of the power set.

Construction. We define an **ordered pair** (a, b) by the set

$$(a, b) = \{\{a\}, \{a, b\}\}.$$

Two ordered pairs (a, b) and (c, d) are supposed to be equal if and only if they coincide in the first coordinate as well as in the second coordinate. Let us check that Kuratowski's definition satisfies this property.

Proposition 3.26. *The equality $(a, b) = (c, d)$ holds if and only if $a = c$ and $b = d$.*

Proof. If $a = c$ and $b = d$, then

$$(a, b) = \{\{a\}, \{a, b\}\} = \{\{c\}, \{c, d\}\} = (c, d).$$

Conversely, if $(a, b) = (c, d)$, then $\{\{a\}, \{a, b\}\} = \{\{c\}, \{c, d\}\}$. We distinguish the cases (i) $a \neq b$ and (ii) $a = b$.

(i) If $a \neq b$, then $\{a\} = \{c\}$, hence $a = c$; furthermore, the equality of the sets $\{a,b\} = \{c,d\} = \{a,d\}$ implies $b = d$.

(ii) If $a = b$, then $\{\{a\},\{a,a\}\} = \{\{a\}\}$, so $\{c\} = \{a\}$ implies $a = c$, and $\{c,d\} = \{a\}$ implies $d = a = b$.

Therefore, $(a,b) = (c,d)$ implies $a = c$ and $b = d$ in both cases. □

Let A and B be sets. We would like to construct a set $A \times B$ containing all pairs (a,b) for $a \in A$ and $b \in B$. Since $\{a\}$ and $\{a,b\}$ are elements of the power set $P(A \cup B)$, the pair $(a,b) = \{\{a\},\{a,b\}\}$ must be a *subset* of $P(A \cup B)$. Therefore, $(a,b) = \{\{a\},\{a,b\}\}$ is an element of $P(P(A \cup B))$. Since each pair $(a,b) = \{\{a\},\{a,b\}\}$ can be understood as an element of $P(P(A \cup B))$, we can define the set $A \times B$ as the subset

$$A \times B = \{(a,b) \in P(P(A \cup B)) \mid a \in A \text{ and } b \in B\}.$$

We call $A \times B$ the **Cartesian product** of the sets A and B.

Usage. The Cartesian product $A \times B$ of two sets A and B is easily understood as the set of ordered pairs with the first coordinate ranging through the elements of A and the second coordinate ranging through the elements of the set B. The next example illustrates a simple case.

Example 3.27. Let $A = \{a_1, a_2, a_3\}$ and $B = \{b_1, b_2\}$. Then

$$A \times B = \{(a_1, b_1), (a_1, b_2), (a_2, b_1), (a_2, b_2), (a_3, b_1), (a_3, b_2)\}$$

The first coordinate is given by the three elements of A, and the second coordinate is given by the two elements of B. Therefore, we get a total of $6 = 3 \cdot 2$ ordered pairs in $A \times B$. An illustration of the set $A \times B$ is depicted in Fig. 3.4.

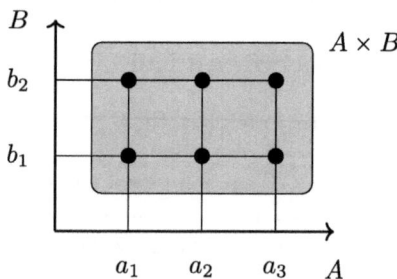

Figure 3.4: René Descartes (also known as Cartesius) introduced the Cartesian coordinate system. The first coordinate is represented by points on a horizontal line and the second by points on a vertical line. The figure illustrates the set $A \times B$ from Example 3.27. The six dots represent the elements of $A \times B$

One point of potential confusion is when one of the sets A or B is empty. How many elements are then in $A \times B$? The answer is none, as the next proposition shows.

3.6 Relations

Proposition 3.28. *The Cartesian product $A \times B = \emptyset$ if and only if $A = \emptyset$ or $B = \emptyset$.*

Proof. If $A = \emptyset$ or $B = \emptyset$, then one cannot form any pair (a,b) with $a \in A$ and $b \in B$, so $A \times B = \emptyset$.

If $A \neq \emptyset$ and $B \neq \emptyset$, then there exists elements $a \in A$ and $b \in B$ such that $(a,b) \in A \times B$; in particular, $A \times B \neq \emptyset$. It follows by contraposition that $A \times B = \emptyset$ implies $A = \emptyset$ or $B = \emptyset$. $\square$

EXERCISES

3.31. Let $A = \{a,b,c\}$ and $B = \{1,2\}$. Determine $A \times B$.

3.32. Show that the Cartesian product is in general not commutative, so
$$A \times B \neq B \times A.$$

3.33. Show that $(A \cup B) \times C = (A \times C) \cup (B \times C)$ holds for all sets A, B, and C.

3.34. Show that $(A \cap B) \times C = (A \times C) \cap (B \times C)$ holds for all sets A, B, and C.

3.35. Show that $(A - B) \times C = (A \times C) - (B \times C)$ holds for all sets A, B, and C.

3.36. Suppose that A is a finite nonempty set with m elements and B a set with n elements. Prove that $A \times B$ has mn elements.

3.6 Relations

Ordered pairs allow us to formalize relations among people such as x is a sibling of y. The entities can also belong to different sets such as a client x using a server y.

In general, a **relation** is a subset of a Cartesian product. We will focus on **binary relations** that are subsets of a Cartesian product of two sets, but one can form n-ary relations as subsets of the Cartesian product of n sets as well.

Given two sets A and B, a relation R between elements of these two sets is a subset R of $A \times B$. We will say that x is in relation R to y, and write $x R y$ to express that $(x,y) \in R$. We will call R a relation on a set A if and only if R is a subset of $A \times A$. We will often use symbols to denote relations instead of letters, since that is often more natural.

The following examples will show that relations are a familiar concept, even though you might not have seen it in our formalization.

Example 3.29. The relation on the set of positive integers $\mathbf{N}_1 = \{1, 2, 3, \ldots\}$ given by

$$\{(x, y) \mid \text{there exists positive integer } z \text{ such that } x + z = y\}$$

is the usual strict inequality and is denoted by $<$. We write $3 < 4$ to express that 3 is strictly less than 4 rather than writing that $(3, 4)$ is an element of $<$.

Example 3.30. Let S denote the set of strings over the alphabet $\{a, \ldots, z\}$. We define $x \leqslant y$ to mean that the string x comes before the string y in lexicographic order or that the two strings are equal. So aardvark $\leqslant$ aasvogel.

Example 3.31. Let A be a finite set and $S = P(A)$ its power set. We can define a relation $\sim$ on S such that $X \sim Y$ if and only if X and Y are two subsets of A that have the same number of elements.

Relations occur very frequently in applications. Therefore, it is good to know the standard terminology for some of their most important properties.

A binary relation R on a set A is called **reflexive** if and only if $x\,R\,x$ holds for all $x \in A$. For example, the relation $\sim$ in Example 3.31 is reflexive, since $X \sim X$ holds for all subsets X of A. A reflexive relation on a set A contains the equality relation on A as a subset.

We call a binary relation R on a set A **irreflexive** if and only if xRx does not hold for any $x \in A$. The strict inequality $<$ on the set of positive integers is an example of an irreflexive relation. A relation that fails to be reflexive is not necessarily irreflexive. Irreflexivity refers to the fact that no element $x \in A$ is related to itself.

It is a common mistake to think that these notions are negations of each other. However, a reflexive relation on a set A contains all diagonal pairs (x, x) with $x \in A$, whereas an irreflexive relation contains none of the diagonal pairs. You should pause for a moment and construct a relation that is neither reflexive nor irreflexive.

A binary relation R on a set A is called **asymmetric** if and only if for all x, y in A, $(x, y) \in R$ implies $(y, x) \notin R$. The strict inequality $<$ on the set of positive integers is an example of an asymmetric relation. Any relation satisfying $x\,R\,x$ for some x in A cannot be asymmetric.

Example 3.32. Let P denote the set of players in a tennis tournament. We define a relation $\prec$ on P such that $x \prec y$ if and only if x beats y in the tournament. Then $\prec$ is an asymmetric relation, since there are no ties. Thus, for all players x and y, either $x \prec y$ or $y \prec x$.

A binary relation R on a set A is called **antisymmetric** if and only if for all x, y in A, $x\,R\,y$ and $y\,R\,x$ implies $x = y$. Every asymmetric relation is antisymmetric, but not conversely. Indeed, the divisibility relation on the set of positive integers is an example of an antisymmetric relation that is not an asymmetric relation.

3.6 Relations

Proposition 3.33. *An irreflexive and antisymmetric relation is asymmetric.*

Proof. Let $<$ be an irreflexive and antisymmetric relation on a set A. Seeking a contradiction, we assume that there exist elements x and y in A such that $x < y$ and $y < x$. By antisymmetry, this implies $x = y$, contradicting the irreflexivity of $<$. Therefore, $<$ is asymmetric, as claimed. □

A binary relation R on a set A is called **symmetric** if and only if $x R y$ implies $y R x$. The relation $\sim$ from Example 3.31 is a symmetric relation.

A relation that is not symmetric does not have to be asymmetric or antisymmetric. The relation $R = \{(1,2),(2,1),(1,3),(1,1)\}$ on $A = \{1,2,3\}$ is not symmetric since it does not contain $(3,1)$. However, the relation R is not asymmetric, since it contains $(1,1)$. It is not antisymmetric either, since it contains $(2,1)$ and $(1,2)$, but $1 \neq 2$.

A binary relation R on a set A is called **transitive** if and only if for all x, y, z in A, $x R y$ and $y R z$ implies $x R z$. The strict inequality $<$ on the set of positive integers is an example of a transitive relation.

An irreflexive and symmetric relation on a set V describes the edges of an undirected graph on the set of vertices V. So undirected graphs are essentially a study of irreflexive and symmetric relations. A relation that is reflexive, symmetric, and transitive is called an equivalence relation, and such relations are fundamental in the construction of number systems and other mathematical objects, see Chap. 5. A relation that is reflexive, antisymmetric, and transitive is called a partial order. Partial orders have many applications in computer science and mathematics, see Chap. 6.

Important Terminology

A binary relation R on a set A is called

reflexive if and only if xRx for all x in A,

irreflexive if and only if xRx does not hold for any x in A,

asymmetric if and only if $(x,y) \in R$ implies $(y,x) \notin R$,

antisymmetric if and only if xRy and yRx implies $x = y$,

symmetric if and only if xRy implies yRx,

transitive if and only if xRy and yRz implies xRz.

We will now introduce some other general operations on relations. Given a relation R that is a subset of a Cartesian product $A \times B$, then the **inverse relation** is defined as

$$R^{-1} = \{w \in B \times A \mid w = (y,x) \text{ and } (x,y) \in R\}.$$

In other words, $(x, y) \in R$ if and only if $(y, x) \in R^{-1}$.

Let A, B, and C be sets. If a relation R is a subset of $A \times B$ and a relation S is a subset of $B \times C$, then the **composition** $S \circ R$, read S after R, is the relation

$$S \circ R = \{(a, c) \in A \times C \mid \text{there exists } b \in B \text{ such that } (a, b) \in R \text{ and } (b, c) \in S\}.$$

This notation follows the convention used for the composition of functions.

If R is a relation on a set A, then we define R^n to be the relation that is recursively defined by

$$R^n = \begin{cases} R & \text{if } n = 1, \\ R^{n-1} \circ R & \text{if } n > 1, \end{cases}$$

In other words, $R^2 = R \circ R$ is the composition of R with itself, $R^3 = R \circ R \circ R$, and so on.

We will study relations in more depth in subsequent chapters. The next section discusses one particularly important special case, namely functions.

EXERCISES

3.37. Determine whether the relation "is child of" on the set of all people is (a) reflexive, (b) irreflexive, (c) asymmetric, (d) antisymmetric, (e) symmetric, (f) transitive. Justify your answers.

3.38. Determine whether the relation "is sibling of" on the set of all people is (a) reflexive, (b) irreflexive, (c) asymmetric, (d) antisymmetric, (e) symmetric, (f) transitive. Justify your answers.

3.39. Determine whether the relation "has the same birthday as" on the set of all people is (a) reflexive, (b) irreflexive, (c) asymmetric, (d) antisymmetric, (e) symmetric, (f) transitive. Justify your answers.

3.40. Find a relation on a set S that is both reflexive and irreflexive. What can you say about the set S?

3.41. Find the smallest nonempty set S that allows you to construct a relation that is neither reflexive nor irreflexive. Explicitly give the relation and a valid argument that S is as small as possible.

3.42. Find a relation that is both symmetric and antisymmetric. Explain why the relation has both properties.

3.7 Functions

The concept of a function is of great importance throughout mathematics. Undoubtedly, you have encountered many examples such as the exponential, logarithmic, square-root, and trigonometric functions. One can get quite some

3.7 Functions

variety by combining such elementary functions, but the concept is too narrow. In this section, we define functions in a much broader sense as special types of relations.

A relation f is called a **function** from a set A to a set B if and only if for each $x \in A$ there exists precisely one element $y \in B$ such that $(x, y) \in f$. We follow Leonard Euler and write $f(x)$ to denote the unique value that f associates to the argument x.

We write $f \colon A \to B$ to express that f is a function from A to B. When we study such a function, there are four relevant sets:
(a) the **domain** $A = \operatorname{dom}(f) = \{x \in A \mid \text{there exists } y \in B \text{ such that } (x, y) \in f\}$ of the function,
(b) the **codomain** B of the function,
(c) the **range** $\operatorname{ran}(f) = \{y \in B \mid \text{there exists } x \in A \text{ such that } y = f(x)\}$ of the function,
(d) the set $f \subseteq A \times B$ that relates the arguments to function values.

The domain specifies the type of the arguments, the codomain specifies the type of the function values, and together $A \times B$ specifies the type of the function. The relation f determines the domain, but not the codomain. Therefore, one cannot omit to specify the type of the function (Fig. 3.5).

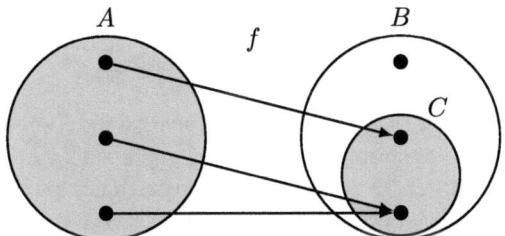

Figure 3.5: A function $f \colon A \to B$ with range $C = \operatorname{ran}(f)$. Note that not every element of the codomain is in the range of f. Since f is a function, each element of A is related to just one element of B

Example 3.34. Let A denote the set of people and D the set of dates. The function $f \colon A \to B$ that associates to each person x their birthdate $f(x)$ is indeed a function, as each person has precisely one birthdate.

Example 3.35. The function $\chi_\mathbf{Q} \colon \mathbf{R} \to \mathbf{R}$ given by

$$\chi_\mathbf{Q}(x) = \begin{cases} 1 & \text{if } x \text{ is a rational number,} \\ 0 & \text{if } x \text{ is an irrational number} \end{cases}$$

has the underlying relation

$$\{(x, 1) \mid x \in \mathbf{Q}\} \cup \{(x, 0) \mid \mathbf{R} \setminus \mathbf{Q}\}.$$

One can also express $\chi_\mathbf{Q}$ using elementary functions and limits, but that is less straightforward.

Notation. Sometimes the function $f\colon A \to B$ is denoted as

$$(F(a) \mid a \in A).$$

In some contexts, the venerable notation $f(x)$ is replaced by an index notation f_x. Then the notations

$$(f_a \mid a \in A) \quad \text{or} \quad (f_a)_{a \in A}$$

are used to specify the function f. These notations are used in calculus to denote sequences, where a sequence is typically a function $f\colon \mathbf{N}_0 \to \mathbf{R}$ or a function $f\colon \mathbf{N}_1 \to \mathbf{R}$.

Given a function $f\colon A \to B$ and a set X, then

$$f \upharpoonright X = \{(x,y) \in f \mid x \in X\}$$

is a function $f \upharpoonright X \colon A \cap X \to B$, called the **restriction** of f to X. Usually, X is chosen as a subset of the domain A. Another common notation for the restriction $f \upharpoonright X$ is $f\big|_X$.

Let $f\colon A \to B$ be a function and X a subset of B. The **preimage** $f^{-1}(X)$ of f is given by

$$f^{-1}(X) = \{x \in A \mid f(x) \in X\}.$$

The notation f^{-1} is also used to denote the inverse function that will be introduced shortly, but the meaning should always be clear from the context.

Let $f\colon A \to B$ and $g\colon B \to C$. The **composition** $g \circ f$ of the functions is given by $(g \circ f)(a) = g(f(a))$ for all $a \in A$. Note that this coincides in our case with the composition of the relations, but in the literature the definitions sometimes clash. The composition of functions is an associative operation, meaning that $h \circ (g \circ f) = (h \circ g) \circ f$.

A function $i_A\colon A \to A$ is called the **identity map** on A if and only if $i_A(a) = a$ holds for all $a \in A$. Given a function $f\colon A \to B$, we have

$$i_B \circ f = f = f \circ i_A.$$

Thus, i_B is a left identity and i_A is a right identity for the function composition.

A function $f\colon A \to B$ is called **injective** or **one-to-one** if and only if $f(x) = f(y)$ implies $x = y$. In other words, a function f is injective if and only if distinct arguments $x \neq y$ yield distinct function values $f(x) \neq f(y)$.

Example 3.36. Any function has a unique value for each function argument, but different arguments may share the same value. For instance, the function $f\colon \mathbf{R} \to \mathbf{R}$ given by $f(x) = x^2$ has a unique function value for each argument, as a function should. However, the distinct arguments -2 and 2 both share the same function value, namely $f(-2) = 4 = f(2)$. Therefore, the function f is not injective.

3.7 Functions

Example 3.37. The function $g\colon \mathbf{R}_{\geq 0} \to \mathbf{R}$ given by $g(x) = x^2$ is an injective function. Indeed, g is the restriction of the function f from the previous example to the set of nonnegative real numbers. For any two nonnegative real numbers x and y, if $f(x) = f(y)$, then $x^2 = y^2$, so we must have $x = \sqrt{x^2} = \sqrt{y^2} = y$. This shows that g is an injective function.

As the next proposition shows, injective functions have the crucial property that they possess an inverse function.

Proposition 3.38. *Let f be a function from A to B. Then the inverse relation f^{-1} is a function from $\operatorname{ran}(f)$ to A if and only if f is an injective function.*

Proof. If the inverse relation f^{-1} is a function, then $f^{-1}(f(a)) = a$ holds for all $a \in A$. For all $x, y \in A$, the equality $f(x) = f(y)$ implies $x = f^{-1}(f(x)) = f^{-1}(f(y)) = y$. It follows that the function f is injective.

Conversely, suppose that f is injective. If $(b, a_1) \in f^{-1}$ and $(b, a_2) \in f^{-1}$, then $f(a_1) = b = f(a_2)$, so $a_1 = a_2$ by the injectivity of f. Therefore, f^{-1} is a function. □

Given an injective function f, we call f^{-1} the **inverse function** of f.

Example 3.39. For instance, if b denotes a positive real number that is not equal to 1, then the inverse of the exponential function $f(x) = b^x$ is given by the logarithmic function $f^{-1}(x) = \log_b x$. Thus, $f^{-1}(f(x)) = \log_b(b^x) = x$ holds for all real numbers.

Remark. Despite the notational conflation, it is usually easy to tell apart preimages and inverse functions. Given a function $f\colon A \to B$, the inverse function f^{-1} is a function from $\operatorname{ran}(f)$ to A; in particular, it is a *function*. Given a subset X of B, the preimage $f^{-1}(X)$ denotes a *set*. It is usually clear from the context whether f^{-1} is used to specify a function or a set.

A function $f\colon A \to B$ is called **surjective** or **onto** if and only if $\operatorname{ran}(f) = B$. In other words, a function is surjective if for every element $b \in B$ there exists an element $a \in A$ such that $b = f(a)$.

Example 3.40. The function $f\colon \mathbf{R} \to \mathbf{R}_{\geq 0}$ from the set of real numbers to the set of nonnegative real numbers given by $f(x) = x^2$ is surjective. Indeed, for any nonnegative real number y in $\mathbf{R}_{\geq 0}$, the argument $\sqrt{y}$ yields the value $f(\sqrt{y}) = (\sqrt{y})^2 = y$. So the function is surjective.

A function is called **bijective** if and only if it is surjective and injective.

Proposition 3.41. *Let A and B be nonempty sets. If there exists an injective function $f\colon A \to B$, then there exists a surjective function $g\colon B \to A$ such that $g \circ f = i_A$.*

Proof. Let $f\colon A \to B$ be an injective function. Then
$$f^{-1} = \{(y,x) \in \operatorname{ran}(f) \times A \mid (x,y) \in f\}$$
is a function $f^{-1}\colon \operatorname{ran}(f) \to A$ by Proposition 3.38. Since A is not empty, we can choose a fixed element $a_1 \in A$. Define $g\colon B \to A$ by
$$g(b) = \begin{cases} a_1 & \text{if } b \in B - \operatorname{ran}(f), \\ f^{-1}(b) & \text{if } b \in \operatorname{ran}(f). \end{cases}$$
The map g is surjective, since for each $a \in A$ there exist an element $b = f(a) \in B$ such that $g(b) = g(f(a)) = f^{-1}(f(a)) = a$. □

Functions can be a useful tool to define new sets. For example, if $f\colon A \to B$ is a function, then its range
$$\operatorname{ran}(f) = \{y \in B \mid \text{there exists some } x \in A \text{ such that } y = f(x)\}$$
is again a set. We will write this set in the more descriptive form
$$\{f(x) \mid x \in A\}.$$
The next axiom asserts that replacing every element x of a set A by a function $f(x)$ is once again a set. The intuition for this axiom is that if we replace each element of A by some other set, then the resulting class is not bigger than A, so it should again be a set.

However, there is a technical snag. For example, it is reasonable to form a new set by replacing each set $x \in A$ by its power set $P(x)$. The power set is defined for every set, but we cannot form the domain "set of all sets" for the map $x \mapsto P(x)$, since such a universal set does not exist, and we cannot legally refer to the class of all sets in Zermelo–Fraenkel set theory. What we do instead is to form a property $\mathcal{F}(x,y)$, which is true if and only if "y is the power set of x".

Let $\mathcal{F}(x,y)$ be a property of two sets. We say that $\mathcal{F}$ is a **functional property** if and only if for each set x there exists precisely one set y such that $\mathcal{F}(x,y)$ holds. Roughly speaking, the purpose of a functional property is to model a function on the universe of sets. Since the universe of sets is not a set itself, we have to resort to the usage of such functional properties. A few examples will clarify what functional properties are.

Example 3.42. The property $\mathcal{F}(x,y)$ given by "y is the power set of x" is a functional property.

Example 3.43. Let z be a fixed set. The property $\mathcal{F}(x,y)$ given by "y is the intersection of x with the set z" is a functional property.

> **S8.** Let $\mathcal{F}(x,y)$ be a functional property. For each set A, there exists a set B such that $y \in B$ if and only if $\mathcal{F}(x,y)$ holds for some $x \in A$.

3.7 Functions

The functional property $\mathcal{F}$ can refer to other sets (so it can have parameters), but is not allowed to refer to the set B that we try to construct. We write the set B in the form
$$B = \{f(x) \mid x \in A\},$$
where $f(x)$ denotes the unique set y associated with x.

The axiom of replacement **S8** is a convenient tool in specifying sets.

Example 3.44. Let $S = \{1, 2, \ldots, 100\}$. Then
$$\{x^2 \mid x \in S\}$$
is the set of the first 100 perfect squares in the set of positive integers.

Given a set system S of nonempty sets, we call a function f on S a **choice function** if and only if $f(X) \in X$ holds for all sets $X \in S$. In other words, for each set X in the set system, the choice function selects an element $f(X)$ from X. The next axiom postulates the existence of choice functions.

> **S9.** There exists a choice function for each collection of nonempty sets.

The proof of the next proposition uses the Axiom of Replacement **S8** and the Axiom of Choice **S9**.

Proposition 3.45. *Let A and B be nonempty sets, and $f: A \to B$ a surjective function. Then there exists an injective function $g: B \to A$ such that $f \circ g = i_B$.*

Proof. Consider the set system
$$S = \{f^{-1}(\{y\}) \mid y \in B\}$$
of preimages of singleton sets $\{y\}$ with $y \in B$. Since f is surjective, each element of S is a nonempty set. Let s be a choice function on S. We define $g: B \to A$ by
$$g(y) = s(f^{-1}(\{y\})),$$
so the value $g(y)$ is one element of the preimage $f^{-1}(\{y\})$. Therefore, $f(g(y)) = f(s(f^{-1}(\{y\}))) = y$. If $g(y) = g(z)$, then $y = f(g(y)) = f(g(z)) = z$; hence, g is an injective function satisfying $f \circ g = i_B$. □

EXERCISES

3.43. Let A and B be nonempty sets. Prove that a function $f\colon A \to B$ is bijective if and only if there exists a function $g\colon B \to A$ such that $g \circ f = i_A$ and $f \circ g = i_B$.

3.44. Let A be a nonempty set. Let $f\colon A \to A$ be a function and denote by f^n the composition of n copies of f, so $f^n = f \circ \cdots \circ f$ with n functions f. Show that if $f^n = i_A$ for some positive integer n, then f must be bijective.

3.45. Let A, B, and C be nonempty sets. Let $f\colon A \to B$, $g\colon B \to C$, and $h\colon C \to A$ be functions satisfying $h \circ g \circ f = i_A$, $g \circ f \circ h = i_C$, and $f \circ h \circ g = i_B$. Show that all three functions must be bijective.

3.46. Let A be a finite set. Show that an injective function $f\colon A \to A$ must be surjective. [Hint: For a given $a \in A$, apply $f^n(a)$ for various n and search for repeated elements. Then exploit the injectivity of f.]

3.47. Describe the following subsets of the set $\mathbf{Z}$ of integers by functional properties (so give the sets in the form $\{f(n) \mid n \in S\}$ for some subset S of the set of integers).
(a) $A = \{1, 4, 9, 16, 25, 36, 49\}$,
(b) $B = \{3, 5, 7, 11, 13, 17, 19, 23\}$,
(c) $C = \{1, 2, 4, 8, 16, 32, 64, 128, 256\}$,
(d) $D = \{2, 6, 12, 20, 30, 42, 56, 72, 90, 110\}$.

3.8 Numbers

Set theory can serve as the foundation for many parts of mathematics. We will illustrate this by defining the nonnegative integers in terms of sets. This will allow us to characterize finite sets and pave the way to our understanding of infinite sets.

Numbers via Sets. John von Neumann gave a set-theoretic description of the nonnegative integers. He defined 0 to be the empty set, 1 to be a set with 1 element, 2 to be a set with two elements, and so forth. Specifically, his definitions of 0, 1, 2, and 3 are given by

$$
\begin{aligned}
0 &= \emptyset, \\
1 &= \{\emptyset\}, \\
2 &= \{\emptyset, \{\emptyset\}\}, \\
3 &= \{\emptyset, \{\emptyset\}, \{\emptyset, \{\emptyset\}\}\}.
\end{aligned}
$$

There is a simple principle behind this bewildering number of braces and empty sets. The definition $0 = \emptyset$ is straightforward, since there is just one empty set. Then 1 is defined to be the set $\{0\}$, 2 to be the set $\{0, 1\}$, and 3 is the set $\{0, 1, 2\}$. If you unravel the definitions, then you end up with von Neumann's definitions of 0, 1, 2, and 3.

3.8 Numbers

In general, we define $0 = \emptyset$, and $n + 1 = n \cup \{n\}$ for all positive integers. In other words, von Neumann defined a nonnegative integer n as the set that contains all previous numbers, so $m < n$ if and only if $m \in n$.

A vexing issue is that we can construct any nonnegative integer, but the axioms that we have given so far do not allow us to collect them all into a single set. In fact, there does not seem to be a way to construct an infinite set.

Let us define the **successor** of a set X to be $X \cup \{X\}$. We call a set S **inductive** if and only if $\emptyset \in S$ and for all $X \in S$, the successor $X \cup \{X\}$ is contained in S. Thus, an inductive set contains all the von Neumann numbers, but perhaps also other sets. The Axiom of Infinity asserts the existence of an inductive set.

S10. There exists an inductive set.

Let S be an inductive set. We define the set of nonnegative integers $\mathbf{N}_0$ to be the set
$$\mathbf{N}_0 = \{x \in S \mid x \in I \text{ for all inductive sets } I\}.$$
We will now show that $\mathbf{N}_0$ is inductive; hence it is the smallest inductive set.

Proposition 3.46. *The set $\mathbf{N}_0$ is inductive.*

Proof. By definition, the empty set $\emptyset$ is contained in $\mathbf{N}_0$. If $X \in \mathbf{N}_0$, then $X \cup \{X\}$ is contained in every inductive set, so $X \cup \{X\} \in \mathbf{N}_0$. We can conclude that $\mathbf{N}_0$ is an inductive set. □

Proof Principles. Since we have now defined the set of nonnegative integers, we can now establish the validity of proof by induction, an important proof technique. This proof technique is intrinsically linked to the structure of the set $\mathbf{N}_0$ of nonnegative numbers.

Theorem 3.47 (The Induction Principle). *Let $P(x)$ be a property. Suppose that*
(1) $P(0)$ holds,
(2) For all $n \in \mathbf{N}_0$, $P(n)$ implies $P(n+1)$.
Then the property $P(n)$ holds for all nonnegative integers n.

Proof. The set $S = \{n \in \mathbf{N}_0 \mid P(n)\}$ is an inductive set by the assumptions (1) and (2). Therefore, $\mathbf{N}_0 \subseteq S$, which proves the claim. □

This proof technique is so important that we will devote the entire next chapter to it. As an appetizer, we will give a closed formula for the so-called triangular numbers. For a nonnegative integer n, the triangular number T_n is defined as $T_n = 1 + 2 + \cdots + n$. Figure 3.6 illustrates why triangular numbers bear their name.

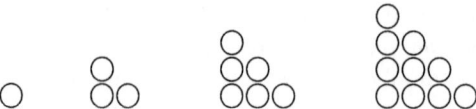

Figure 3.6: The figure illustrates the triangular numbers $T_1 = 1$, $T_2 = 1 + 2 = 3$, $T_3 = 1 + 2 + 3 = 6$ and $T_4 = 1 + 2 + 3 + 4 = 10$

Proposition 3.48. *For all nonnegative integers n, we have*
$$T_n = \frac{n(n+1)}{2}.$$

Proof. We will prove this by induction on n. Let $P(n)$ be the statement that the equality
$$T_n = n(n+1)/2$$
holds.

Base Case Since $T_0 = 0$ and $0(0+1)/2 = 0$, it follows that the statement $P(0)$ is true.

Inductive Step We need to show for all nonnegative integers n that $P(n)$ implies $P(n+1)$. In other words, we need to show that for all nonnegative n, if $P(n)$ is true, then $P(n+1)$ is true.

Suppose that $P(n)$ is true. We have
$$\begin{aligned}
T_{n+1} &= 1 + 2 + \cdots + n + (n+1) && \text{(by definition of } T_{n+1}\text{)} \\
&= T_n + n + 1 && \text{(by definition of } T_n\text{)} \\
&= \frac{n(n+1)}{2} + n + 1 && \text{(as P(n) is true).}
\end{aligned}$$

The right-hand side can be simplified to
$$T_{n+1} = \frac{n(n+1)}{2} + n + 1 = \frac{n^2 + 3n + 2}{2} = \frac{(n+1)(n+2)}{2},$$
so $P(n+1)$ holds. We can conclude by induction that $P(n)$ holds for all n, which proves the claim. □

The following variation on the induction principle is often more convenient to use.

Theorem 3.49 (The Strong Induction Principle). *Let $P(x)$ be a property. Suppose that we can establish for all nonnegative integers n the principle*

$$\text{If } P(k) \text{ holds for all } k < n, \text{ then } P(n). \tag{3.1}$$

Then the property $P(n)$ holds for all nonnegative integers n. It should be noted that (3.1) implies $P(0)$, since there are no nonnegative integers $k < 0$.

3.8 Numbers

Proof. Suppose that (3.1) is true. For all nonnegative integers, we define $Q(n)$ to be the statement: $P(k)$ holds for all $k < n$. Since there are no nonnegative integers less than 0, the statement $Q(0)$ is vacuously true.

If $Q(n)$ holds, then $P(k)$ holds for all $k < n$. So by (3.1), $P(n)$ is true. It follows that $P(k)$ holds for all $k < n+1$. In other words, the statement $Q(n+1)$ holds.

We can conclude by induction that $Q(n)$ is true for all nonnegative integers n, which implies that $P(n)$ is true for all nonnegative integers n. □

A **least element** of a set A of nonnegative integers is an element of A satisfying $a \leqslant x$ for all $x \in A$. We will use a proof by strong induction to prove that *every* nonempty subset of $\mathbf{N}_0$ has a least element. This might seem like an obvious fact. However, you might want to keep in mind that the set of integers has subsets that do not have a least element.

Theorem 3.50 (The Well-Ordering Principle). *Every nonempty subset of the set of nonnegative integers has a least element.*

Proof. Let S be a nonempty subset of $\mathbf{N}_0$. We will prove by strong induction that S has a least element a. For each $n \in \mathbf{N}_0$, we define $P(n)$ to be the statement: if $n \in S$, then S has a least element.

The statement $P(0)$ is certainly true. Indeed, if $0 \in S$, then the least element of S is of course 0.

Let us assume that $P(m)$ is true for all $m \leqslant n$. If $n+1 \notin S$, then $P(n+1)$ is vacuously true. Therefore, let us suppose that S contains $n+1$. We can now distinguish two cases:

Case 1. Suppose that there exists a nonnegative number $m \leqslant n$ such that $m \in S$. Since $P(m)$ is true, the set S has a least element.

Case 2. Suppose that there does not exist a nonnegative number $m \leqslant n$ such that $m \in S$. Then $n+1$ is the least element of S.

Therefore, we can conclude by strong induction that $P(n)$ holds for all $n \in \mathbf{N}_0$. Since S is not empty, it follows that S contains a least element. □

EXERCISES

3.48. Explicitly give von Neumann's definition of 4 and 5 in terms of sets.

A set S is called **transitive** if and only if every element of an element of S is itself an element of S. In other words, S is called transitive if and only if $A \in B$ and $B \in S$ implies $A \in S$. The notion of a transitive set should not be confused with a transitive relation.

3.49. Which of the following sets are transitive? Explain.
(a) $\emptyset$,
(b) $\{\emptyset, \{\emptyset\}\}$,
(c) $\{\emptyset, \{\{\emptyset\}\}\}$,
(d) $\{\emptyset, \{\emptyset\}, \{\emptyset, \{\emptyset\}\}\}$.

3.50. Show that
(a) S is a transitive set if and only if $\bigcup S \subseteq S$.
(b) S is a transitive set if and only if $B \in S$ implies $B \subseteq S$.
(c) S is a transitive set if and only if $S \subseteq P(S)$.

3.51. Show that if X is a transitive set, then
$$\bigcup (X \cup \{X\}) = X.$$

3.52. Show that every element of the set $\mathbf{N_0}$ of nonnegative numbers is a transitive set.

3.53. For all $n, m \in \mathbf{N_0}$, show that if the successor $n \cup \{n\}$ of n is equal to the successor $m \cup \{m\}$ of m, then $n = m$.

3.54. Let $Q(x)$ be a property. Suppose that k is nonnegative integer and
(1) $Q(k)$ holds,
(2) For all nonnegative integers $n \geq k$, $Q(n)$ implies $Q(n+1)$.
Show that the property $Q(n)$ holds for all nonnegative integers $n \geq k$. [This simple variation of the induction principle gives some flexibility.]

3.55. Give three distinct subsets of the set of integers that do not have a least element.

3.56. Does the set $\{x \in \mathbf{R} \mid 0 < x\}$ have a least element? Either determine the least element or prove that this set does not have a least element.

3.9 Cardinality

We frequently count the number of elements in a set or compare the size of sets. We have a very good intuition how to do this when the sets are finite. Our intuition fails us when we start to compare the size of infinite sets. Cantor showed that infinite sets can differ substantially in "size." The technical term for the size of a set is its cardinality.

Two sets A and B are said to have the **same cardinality** or are **equipotent** if and only if there exists a bijective function from A onto B. The cardinality of A is **less than or equal to** the cardinality of B if and only if there exists an injective function from A to B.

If there exists an injective map from A to B, then this essentially means that a copy of A can be embedded into B by means of the injective map, so B is "at least as big as" the set A. If there exists a bijective function from A onto B, then this means, roughly speaking, that the set B can be obtained by "renaming" the elements of A using the bijective map.

For sets A and B, we write $|A| \leq |B|$ if and only if the cardinality of A is less than or equal to the cardinality of B. We write $|A| = |B|$ if and only if A and B have the same cardinality.

Let us first check that the comparison of cardinalities behaves as one might expect from any inequality.

3.9 Cardinality

Proposition 3.51. *Let A, B, and C be sets. Then*
(a) $|A| = |A|$,
(b) *if $|A| \leq |B|$ and $|B| \leq |A|$, then $|A| = |B|$,*
(c) *if $|A| \leq |B|$ and $|B| \leq |C|$, then $|A| \leq |C|$.*

Proof. (a) The identity map i_A on A is a bijective map, hence $|A| = |A|$.

(b) We can reformulate the second claim as follows. If $f\colon A \to B$ and $g\colon B \to A$ are injective maps, then there exists a bijective map $h\colon A \to B$. This is a major theorem known as the Schröder-Bernstein theorem. We could prove it now, but an elementary proof is tedious, see [36]. We will prove it in Theorem 6.46, since the methods of Chap. 6 allow us to give a shorter and more elegant proof.

(c) By assumption, there exist injective maps $f\colon A \to B$ and $g\colon B \to C$. Consider the function composition $g \circ f$. If $(g \circ f)(a_1) = (g \circ f)(a_2)$, then $g(f(a_1)) = g(f(a_2))$. Since g is injective, it follows that $f(a_1) = f(a_2)$. As f is injective, it follows that $a_1 = a_2$, which proves that $g \circ f$ is injective. Therefore, we can conclude that $|A| \leq |B|$ and $|B| \leq |C|$ implies $|A| \leq |C|$. □

A set A is called **finite** if and only if there exists a nonnegative integer n such that $|n| = |A|$. If $|n| = |A|$, then we say that A has n elements and write the cardinality in the more pleasing form $n = |A|$.

The next proposition implies that there cannot be two different cardinalities of a finite set.

Proposition 3.52 (Pigeonhole Principle). *There is no injective mapping from $n = \{0, 1, \ldots, n-1\}$ onto a proper subset of n.*

A more whimsical formulation of this proposition is that it is impossible to place n pigeons into fewer than n pigeonholes without occupying at least one pigeonhole twice. For instance, it is impossible to place five pigeons

into four pigeonholes

such that each pigeonhole contains at most one pigeon. This seems self-evident, but giving a rigorous proof is not quite that simple.

Proof. We will prove the claim by induction on n. Let $P(n)$ denote the statement that there is no injective mapping from n onto a subset of n.

Base Case $P(0)$ is true, since there are no proper subsets of $0 = \emptyset$.

Inductive Step We will show that for all nonnegative integers n, the statement $P(n)$ implies $P(n+1)$. Suppose that $P(n)$ is true. Let f be an injective map from $\{0, 1, \ldots, n\}$ onto a proper subset S. We can distinguish the cases (a) $n \in S$ and (b) $n \notin S$.

(a) If $n \in S$, then there exists some nonnegative integer $m \leq n$ such that $n = f(m)$. We define a new function

$$g(i) = \begin{cases} f(i) & \text{for all } i \in \{0, \ldots, n-1\} \text{ such that } i \neq m, \\ f(n) & \text{if } i = m \text{ and } m \neq n. \end{cases}$$

Then $g \colon \{0, \ldots, n-1\} \to S - \{n\}$ is an injective map, contradicting $P(n)$.

(b) If $n \notin S$, then the restriction $f \upharpoonright n$ maps $n = \{0, \ldots, n-1\}$ onto the proper subset $S - \{f(n)\}$, contradicting $P(n)$.

Therefore, $P(n+1)$ is true. We can conclude by induction that the claim holds for all nonnegative integers n. □

There are many applications of the pigeonhole principle, especially in combinatorics. We will discuss numerous applications of this important proof principle in Chaps. 11 and 14.

Remark 3.53. Dedekind realized that for any infinite set A, one can find an injective map onto a proper subset of A. For example, the map $s \colon \mathbf{N}_0 \to \mathbf{N}_0 \backslash \{0\}$ given by $s(n) = n + 1$ is injective and maps $\mathbf{N}_0$ to the proper subset of positive integers. It follows from the pigeonhole principle and Dedekind's observation that a set S is infinite if and only if one can find an injective map from S onto a proper subset.

A set A satisfying $|A| \leq |\mathbf{N}_0|$ is called a **countable set**. In other words, it suffices to find an injective function from A into $\mathbf{N}_0$ to prove that A is countable. Alternatively, we can prove that A is countable if we can find a surjection from $\mathbf{N}_0$ onto A, see Proposition 3.45. A set A that is not countable is called **uncountable**.

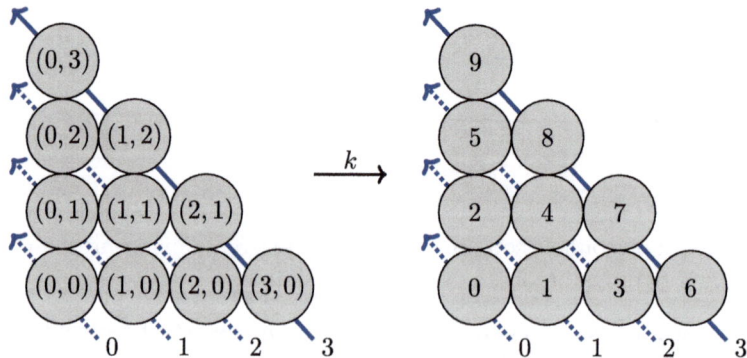

Figure 3.7: A pair of nonnegative integers (x, y) lying on the nth diagonal, where $n = x + y$, is mapped to the value $T_n + y$. For example, $(2, 1)$ lies on the 3rd diagonal. The number of elements on the previous three diagonals is $T_3 = 1 + 2 + 3$. So $(2, 1)$ is mapped to the value $T_3 + 1 = 6 + 1 = 7$

3.9 Cardinality

Proposition 3.54. *Suppose that A and B are countable sets. Then $A \times B$ is a countable set.*

Proof. By assumption, there exist injective maps $f \colon A \to \mathbf{N}_0$ and $g \colon B \to \mathbf{N}_0$. Therefore, the function $h \colon A \times B \to \mathbf{N}_0 \times \mathbf{N}_0$ given by $h(x,y) = (f(x), g(y))$ is an injective function. Therefore, it suffices to show that the set $\mathbf{N}_0 \times \mathbf{N}_0$ is countable.

Let us follow Cantor and define a function $k \colon \mathbf{N}_0 \times \mathbf{N}_0 \to \mathbf{N}_0$ by

$$k(x,y) = \frac{1}{2}(x+y+1)(x+y) + y. \tag{3.2}$$

The first few values of $k(x,y)$ are illustrated in Fig. 3.7.

Note that $k(x,y) = T_n + y$, where $n = x+y$ designates the diagonal to which (x,y) belongs and $T_n = \sum_{k=1}^{n} k$ is the n-th triangular number, which counts the number of elements on all previous diagonals. The function $k(x,y)$ is bounded by the triangular numbers,

$$T_{x+y} \leqslant k(x,y) = T_{x+y} + y < T_{x+y} + x + y + 1 = T_{x+y+1}.$$

We will now show that k is an injective function. Suppose that $k(x,y) = k(a,b)$. Seeking a contradiction, let us assume that $x+y \neq a+b$. Without loss of generality, we may assume that $x+y < a+b$. By the above bound for $k(x,y)$, we get

$$k(x,y) = T_{x+y} + y < T_{a+b} \leqslant k(a,b),$$

contradicting the equality $k(x,y) = k(a,b)$. Therefore, $x+y = a+b$.

It follows that $k(x,y) = T_{x+y} + y = T_{x+y} + b = k(a,b)$, so $y = b$, whence $x = a$. Therefore, $(x,y) = (a,b)$ and it follows that k is injective. □

We encourage you to explore further properties of Cantor's function k, see Exercise 3.62.

Proposition 3.55. *Let $S = \{A_i \mid i \in I\}$ be a family of countable sets A_i such that I is a nonempty countable set. Then $\bigcup S$ is a countable set.*

Proof. The claim holds if all sets A_i are empty, so we may assume that not all sets A_i are empty. Let $I_0 = \{i \in I \mid A_i \neq \varnothing\}$ be the indices of all nonempty A_i. Since I is countable, the subset I_0 is countable as well. Therefore, there exists a surjection $s \colon \mathbf{N}_0 \to I_0$, so we can form the set family $T = \{A_{s(n)} \mid n \in \mathbf{N}_0\}$, which satisfies $\bigcup T = \bigcup S$.

Since $A_{s(n)}$ is a nonempty set, there exists a surjective map $f_n \colon \mathbf{N}_0 \to A_{s(n)}$. Therefore, the function $f \colon \mathbf{N}_0 \times \mathbf{N}_0 \to \bigcup S$ given by $f(m,n) = f_n(m)$ is a surjective map onto $\bigcup S = \bigcup T$. If we compose this map with the inverse of Cantors map $k^{-1} \colon \mathbf{N}_0 \to \mathbf{N}_0 \times \mathbf{N}_0$, we get a surjective map $f \circ k^{-1} \colon \mathbf{N}_0 \to \bigcup S$. Therefore, $\bigcup S$ is countable. □

Let A and B be sets. We write $|A| < |B|$ if and only if there exists an injective function from A into B, but no bijective function from A onto B.

Proposition 3.56 (Cantor). *Let A be a set. There does not exist any surjection from A onto $P(A)$.*

Proof. Given any function $f \colon A \to P(A)$, we can construct the set
$$S = \{x \in A \mid x \notin f(x)\}.$$
Seeking a contradiction, let us assume that there exists an element $a \in A$ such that $f(a) = S$. Then we have $a \in f(a)$ if and only if $a \notin f(a)$, which is a contradiction. Thus, f is not surjective. $\square$

Corollary 3.57. *Let A be a set. Then $|A| < |P(A)|$.*

Proof. The function $f \colon A \to P(A)$ given by $f(x) = \{x\}$ is injective, so $|A| \leq |P(A)|$. By the previous proposition, there cannot exist any surjective map from A onto $P(A)$. Therefore, $|A| < |P(A)|$. $\square$

A remarkable consequence of the previous corollary is that one can construct sets of larger and larger cardinality.

Let us denote by $2^{\mathbf{N}_0}$ the set of all functions from the nonnegative integers to the set $\{0, 1\}$ with two elements.

Proposition 3.58. *The power set $P(\mathbf{N}_0)$ and the set $2^{\mathbf{N}_0}$ are equipotent,*
$$|2^{\mathbf{N}_0}| = |P(\mathbf{N}_0)|.$$

Proof. Define a function $f \colon P(\mathbf{N}_0) \to 2^{\mathbf{N}_0}$ by
$$f(A) = \chi_A$$
where χ_A is the characteristic function of A defined by
$$\chi_A(n) = \begin{cases} 1 & n \in A \\ 0 & n \notin A \end{cases}$$
Then f is a bijection. Indeed, if two characteristic functions χ_A and χ_B are the same, then $A = B$; thus, f is injective. The function f is surjective, since every function in $2^{\mathbf{N}_0}$ is a characteristic function of its set of arguments yielding the value 1. $\square$

Consider functions that takes a nonnegative integer as an argument and have a single bit as a value. It might seem that one can compute any such function with a suitable computer program given enough time. However, nothing could be further from the truth. We will now sketch the proof that for most functions in $2^{\mathbf{N}_0}$ one cannot write a program that will compute it.

Choose a programming language L of your preference that is expressive enough to compute functions. For instance, the `while` language and the λ-calculus are primitive choices. We call a function f in $2^{\mathbf{N}_0}$ computable in L if and only if there exists a program that for each input $n \in \mathbf{N}_0$ will compute $f(n)$.

3.9 Cardinality

Theorem 3.59. *There exist functions in $2^{\mathbf{N}_0}$ that cannot be computed in L.*

Proof. Let A_k denote the set of all programs in L that have k characters. Then A_k is a finite set. The set of all programs that can be expressed in L is given $\bigcup S$, where $S = \{A_k \mid k \in \mathbf{N}_0\}$. By Proposition 3.55, $\bigcup S$ is a countable set.

By contrast, the set $2^{\mathbf{N}_0}$ of characteristic functions is not countable, since $|2^{\mathbf{N}_0}| = |P(\mathbf{N}_0)| > |\mathbf{N}_0|$ by Corollary 3.57. This proves the claim. $\square$

One should note how crude our argument was. We simply listed all possible programs that can be written in the programming language L. Since this set is countable, the subset of programs computing a function with a nonnegative integer as the input and a single bit as the output is countable as well. Since the set of characteristic functions is not countable, there are many functions that we cannot compute. Courses on computability will explore many examples of interesting functions that cannot be computed.

EXERCISES

3.57. Show that the sets $A = \{x \in \mathbf{R} \mid 1 \leqslant x \leqslant 4\}$ and $B = \{x \in \mathbf{R} \mid 2 \leqslant x \leqslant 7\}$ have the same cardinality.

3.58. Let (a) $A_0 = \{x \in \mathbf{R} \mid 0 < x < 1\}$, (b) $A_1 = \{x \in \mathbf{R} \mid 0 \leqslant x < 1\}$, (c) $A_2 = \{x \in \mathbf{R} \mid 0 < x \leqslant 1\}$, (d) $A_3 = \{x \in \mathbf{R} \mid 0 \leqslant x \leqslant 1\}$. Show that all four sets have the same cardinality using Proposition 3.51 (b).

3.59. Let $A = \{x \in \mathbf{R} \mid 0 < x < 1\}$ and $B = \mathbf{R}$. Show that $|A| = |B|$ by giving an explicit bijection from A onto B.

3.60. Show that $|\mathbf{N}_0| = |\mathbf{Z}|$ by explicitly giving a bijective function from the set $\mathbf{N}_0$ of nonnegative integers onto the set $\mathbf{Z}$ of integers.

3.61. Suppose that A and B are sets such that $|A| = |B|$. Show that $|P(A)| = |P(B)|$.

3.62. Show that Cantor's function $k \colon \mathbf{N}_0 \times \mathbf{N}_0 \to \mathbf{N}_0$ defined in Eq. (3.2) is surjective. This means that k is a bijective function.

3.63. Show that $|P(\mathbf{N}_0)| = |P(\mathbf{N}_0 \times \mathbf{N}_0)|$.

3.64. Show that the following sets are countable:
(a) The set of nonnegative even integers.
(b) The set of prime numbers.
(c) The set of (positive and negative) integers.

3.65. Show that a subset of a countable set is countable.

3.10 Notes

Set theory was conceived by Georg Cantor [11] in a series of seminal papers. A brief informal treatment of set theory can be found in a delightful book by Kaplansky [45].

We have given a brief overview of Zermelo-Fraenkel set theory. The two books by Halmos [33] and Hrbacek and Jech [36] are excellent introductions to Zermelo–Fraenkel set theory that are recommended for further reading. Our debt to the books by Halmos and Hrbacek and Jech should be obvious. We also consulted the books by Deiser [19], Ebbinghaus [21], Enderton [24], and Schindler [66] in the preparation of this chapter.

Another axiomatic set theory was conceived by von Neumann, Bernays, and Gödel. This theory contains proper classes in addition to sets, and this can be convenient at times. We recommend the book by Smullyan and Fitting [70] for further information.

For a more advanced treatment of set theory including a thorough discussion of forcing, see Jech [41] and Kunen [54].

Chapter 4

Proofs by Induction

> *It is impossible not to fall in love at second sight with mathematical induction.*
>
> — Michael Berg, in an MAA Book Review, 2017

The principle of induction and the related principle of strong induction have been introduced in the previous chapter. However, it takes a bit of practice to understand how to formulate such proofs. In this chapter, we will illustrate both methods with several examples. Furthermore, we discuss a far-reaching generalization of these methods called well-founded induction and illustrate this method of proof by investigating properties of the Ackermann function. The chapter concludes with a discussion of recursion, recursively defined sets, and structural induction.

4.1 Perfect Squares

The perfect squares are given by

$$1^2 = 1, \quad 2^2 = 4, \quad 3^2 = 9, \quad 4^2 = 16, \ldots$$

Figure 4.1 depicts the perfect squares graphically.

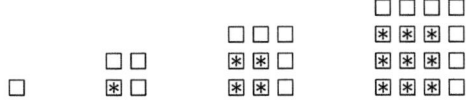

Figure 4.1: The first four perfect squares are 1, 4, 9, and 16. The illustration explains why these are also known as square numbers. The number of squares without an asterisk in the subfigures are 1, 3, 5, and 7

© The Author(s), under exclusive license to Springer Nature Switzerland AG 2025
A. Klappenecker, H. Lee, *Discrete Structures*, Undergraduate Texts in Mathematics, https://doi.org/10.1007/978-3-031-73434-2_4

The figure makes it apparent that going from the perfect square n^2 to the next, we have to add n squares on top, n squares on the side, and 1 for the corner; hence,
$$(n+1)^2 = n^2 + 2n + 1,$$
a fact that we could have just as easily obtained by algebra. However, the figure moreover suggests that
$$1 + 3 + 5 + 7 = 4^2.$$
By looking at such small examples, it is natural to suspect that in general the sum of the first n odd positive integers yields n^2, namely
$$1 + 3 + 5 + \ldots + 2n - 1 = n^2.$$
The same fact can be expressed in summation notation as
$$n^2 = \sum_{k=1}^{n} (2k-1).$$
We now want to give a compelling argument that the previous formula holds for all integers $n \geqslant 1$. It is not sufficient to consider a few examples. We can take advantage of the principle of induction to prove this fact. Let us first recall the principle of induction for positive integers.

Principle of Induction. To prove that a property $P(n)$ holds for all positive integers n, it suffices to prove the base case and the inductive step.

Base case. The property $P(1)$ holds.

Inductive step. For all positive integers n, $P(n)$ implies $P(n+1)$.

The property $P(n)$ is in our case the equality
$$n^2 = \sum_{k=1}^{n} (2k-1).$$
So let us prove that $P(n)$ holds for all integers $n \geqslant 1$.

Proposition 4.1. *The sum of the first n positive odd integers equals*
$$n^2 = \sum_{k=1}^{n} (2k-1)$$
for all $n \geqslant 1$.

4.1 Perfect Squares

Proof. *Base Case* The claim $P(n)$ holds for $n = 1$, since

$$1^2 = 1 = \sum_{k=1}^{1}(2k-1).$$

Inductive Step In the inductive step, our goal is to show that the implication $P(n) \to P(n+1)$ holds for all $n \geqslant 1$. In words, our goal is to show that the implication

$$n^2 = \sum_{k=1}^{n}(2k-1) \quad \text{implies} \quad (n+1)^2 = \sum_{k=1}^{n+1}(2k-1) \quad (4.1)$$

holds for all $n \geqslant 1$. If the hypothesis is false, then the implication is vacuously true. So let us assume that the hypothesis

$$n^2 = \sum_{k=1}^{n}(2k-1) \quad (4.2)$$

is true. Since

$$(n+1)^2 = n^2 + (2n+1),$$

substituting the right-hand side of the hypothesis (4.2) for n^2 yields

$$(n+1)^2 = \sum_{k=1}^{n}(2k-1) + (2n+1) = \sum_{k=1}^{n+1}(2k-1),$$

which proves that the implication (4.1) is true.

Therefore, we can conclude by induction that $P(n)$ holds for all $n \geqslant 1$, which proves the claim. □

> **Induction Hypothesis** In the induction step, you have to show that the implications $P(n) \to P(n+1)$ hold for all $n \geqslant 1$. Here $P(n)$ is called the **induction hypothesis**. Since the implication is trivially true if the induction hypothesis is false, the usual proof pattern is:
> Assume that the induction hypothesis $P(n)$ is true.
> Then deduce $P(n+1)$ by logical deduction.
> Note that this is the proof principle for $\forall n(P(n) \to P(n+1))$.

One of the most common confusions about the induction principle is that inductive step seems to be similar to the claim. Far from it! The claim is to prove that $P(n)$ holds for all $n \geqslant 1$, which we can also express as

$$\forall n P(n).$$

On the other hand, the inductive step seeks to show

$$\forall n (P(n) \to P(n+1)).$$

What is the difference? Suppose that $P(n)$ is false for all $n \geq 1$. Then $\forall n(P(n) \to P(n+1))$ is true, since each implication is vacuously true. Therefore, the inductive step is true. However, the claim $\forall n P(n)$ is false, since each $P(n)$ is false. This also serves as a reminder why anchoring the induction with the base case is important.

EXERCISES

4.1. Prove or disprove: $n^2 + n + 41$ is a prime for all $n \geq 1$.

4.2. Find a proof of Proposition 4.1 that Gauss could have already found in elementary school if his teacher would have dared to ask.

4.3. Prove by induction that the sum of the first n squares is given by

$$\sum_{k=1}^{n} k^2 = 1^2 + 2^2 + \cdots + n^2 = \frac{n(n+1)(2n+1)}{6} \tag{4.3}$$

for all $n \geq 1$.

4.4. Prove by induction that the sum of the first n cubes is given by

$$\sum_{k=1}^{n} k^3 = 1^3 + 2^3 + \cdots + n^3 = (1 + 2 + \cdots + n)^2 = \frac{n^2(n+1)^2}{4} \tag{4.4}$$

for all $n \geq 1$.

4.5. Prove by induction that the sum of the squares of the first n odd positive integers is given by

$$\sum_{k=1}^{n}(2k-1)^2 = 1^2 + 3^2 + 5^2 + \cdots + (2n-1)^2 = \frac{1}{3}(4n^3 - n) \tag{4.5}$$

for all positive integers n.

4.6. Prove by induction that for all integers $n \geq 1$, the integer $2^{2n} - 1$ is divisible by 3.

4.7. Prove by induction that for all positive integers n, the number 2^n divides $3^{2^n} - 1$.

4.8. Suppose that $A_1, A_2, \ldots, A_n$ are sets satisfying $A_1 \supseteq A_2 \supseteq \cdots \supseteq A_n$.
(a) Show that $A_1 - A_2 - \cdots - A_n = (A_1 - A_2) \cup (A_3 - A_4) \cup \cdots \cup (A_{n-1} - A_n)$ if n is even.

(b) Show that $A_1 - A_2 - \cdots - A_n = (A_1 - A_2) \cup (A_3 - A_4) \cup \cdots \cup (A_{n-2} - A_{n-1}) \cup A_n$ if n is odd.

Here we assume that the set difference is right associative, so $A - B - C$ is interpreted as $(A - (B - C))$. Hint: Use Exercise 3.28 and prove the claims by induction.

4.9. (Requires calculus) Show that

$$\int_0^\infty x^n e^{-x} dx = n! \tag{4.6}$$

holds for all integers $n \geq 0$.

4.2 Bernoulli's Inequality

Inequalities are frequently proved by induction. We illustrate this for an inequality by Jacob Bernoulli. This inequality is frequently useful, especially when dealing with probabilities.

What we want to establish is that the inequality

$$(1+x)^n \geq 1 + nx \tag{4.7}$$

holds for all real numbers $x \geq -1$ and all nonnegative integers n. This is clear when x is large, but is perhaps not so apparent for very small real numbers x.

Let us denote by $P(n)$ the property that the inequality (4.7) holds for all real numbers x such that $x \geq -1$. Our goal is to establish that

$$P(0), \quad P(1), \quad P(2), \ldots$$

hold, meaning that we seek to establish the validity of the inequalities

$$(1+x)^0 \geq 1 + 0x, \quad (1+x)^1 \geq 1 + 1x, \quad (1+x)^2 \geq 1 + 2x, \ldots$$

Establishing that $P(n)$ holds for all integers $n \geq 0$ suggests that we use a proof by induction. We need to be a bit careful, though. If we were to prove the base case $P(1)$ and then establish $P(n)$ for all integers $n \geq 1$ by the induction step, we would have established $P(n)$ for all positive integers n, but the proof would not cover the case $n = 0$. Of course, there is an easy fix, as we can establish $P(0)$ as our base case and then prove that $P(n) \to P(n+1)$ holds for all integers $n \geq 0$. Exercise 3.54 asserts that this is a valid variation of the induction proof principle.

Proposition 4.2 (Bernoulli's Inequality)**.** *Let n denote a nonnegative integer and x a real number such that $x \geq -1$. Then*

$$(1+x)^n \geq 1 + nx. \tag{4.8}$$

Proof. Let $P(n)$ denote the inequality (4.8). We will prove the claim by induction on n.

Induction Basis We have $P(0)$, since $(1+x)^0 = 1 = 1 + 0x$.

Inductive Step We will show that $P(n) \to P(n+1)$ holds for all $n \geq 0$. Suppose that $P(n)$ holds, that is, we assume that $(1+x)^n \geq 1 + nx$ holds. Multiplying both sides by the nonnegative factor $(1+x)$ yields

$$\begin{aligned}(1+x)^{n+1} &= (1+x)(1+x)^n \\ &\geq (1+x)(1+nx) \quad \text{(by induction hypthesis)} \\ &\geq 1+(n+1)x \quad \text{(since } nx^2 \text{ is nonnegative)}\end{aligned}$$

which proves $P(n+1)$. Therefore, the claim follows by induction on n. □

The proof of this inequality is simple, but we will use it to illustrate a technique that you can use to gain a better understanding of a proof by induction. We call it the modus ponens omnibus.

> **The Modus Ponens Omnibus** It is educational (though not entirely rigorous) to view induction as a repeated application of modus ponens. We establish in the base case the property $P(1)$, and in the inductive step the implications
>
> $$P(1) \to P(2), \quad P(2) \to P(3), \quad P(3) \to P(4), \ldots$$
>
> Now let's repeatedly use modus ponens:
> $P(1)$ and $P(1) \to P(2)$ allow us to deduce $P(2)$,
> $P(2)$ and $P(2) \to P(3)$ allow us to deduce $P(3), \ldots$
> You will gain a better understanding by unraveling a proof by induction in this way. Of course, we cannot replace induction by repeated modus ponens, since proofs must be of finite length!

Let's apply the modus ponens omnibus to the proof of Bernoulli's inequality. The base case $P(0)$ established that

$$(1+x)^0 \geq 1 + 0x \qquad (4.9)$$

holds. The implication $P(0) \to P(1)$ is established by multiplying both sides of this inequality with $(1+x)$, which led to

$$(1+x)^1 \geq 1 + 1 \cdot x. \qquad (4.10)$$

Since the inequality (4.9) was true and $x \geq -1$, we can rest assured that the inequality (4.10) holds, so $P(1)$ is true. The implication $P(1) \to P(2)$ is established by multiplying both sides of the inequality (4.10) by $(1+x) \geq 0$ and dropping the nonnegative term x^2, which yields

$$(1+x)^2 \geq 1 + 2x. \qquad (4.11)$$

4.3 Fibonacci Numbers

Since $P(1)$ was true and $P(1) \to P(2)$ is true, this means that we have established $P(2)$. At this point, you should be able to continue this pattern.

So what could go wrong in a proof by induction? The base case is often so simple that people do not pay much attention to it and may even omit it. This is dangerous, since the entire proof falls apart when the base case is wrong. In the inductive step, we established $P(n) \to P(n+1)$ for all n. For Bernoulli's inequality, it was crucial that we restricted ourselves to $x \geq -1$. Indeed, if $x < -1$, then $1 + x$ is negative and multiplication with the term would reverse the inequality. So be careful about each step in a proof!

EXERCISES

4.10. Deduce from Bernoulli's inequality that $\sqrt[n]{n} < 2$ holds for all positive integers n.

4.11. Determine when equality holds in Bernoulli's inequality.

4.12. Let $x_1, x_2, \ldots, x_n \geq -1$ be real numbers that all have the same sign. Prove by induction that

$$(1 + x_1)(1 + x_2) \cdots (1 + x_n) \geq 1 + x_1 + x_2 + \cdots + x_n.$$

4.13. Let $b > 1$ be a real number. Deduce from Bernoulli's inequality that there exists a positive constant L such that

$$Ln < b^n$$

holds for all $n \geq 1$. [In the language of Chap. 10, this establishes $b^n \in \Omega(n)$ and $n \in O(b^n)$.]

4.14. Let $b > 1$ be a real number and k a positive integer. Deduce from Bernoulli's inequality that there exists a positive integer n_0 such that

$$n^k \leq b^n$$

holds for all $n \geq n_0$. [In the language of Chap. 10, this establishes $b^n \in \Omega(n^k)$ and $n^k \in O(b^n)$.]

4.3 Fibonacci Numbers

The **Fibonacci numbers** are defined as the sequence given by

$$f_1 = 1, \quad f_2 = 1, \quad f_3 = 2, \quad f_4 = 3, \quad f_5 = 5, \quad f_6 = 8, \quad f_7 = 13, \ldots$$

The law governing this sequence is given by $f_1 = 1$ and $f_2 = 1$ and from then on the next number is simply obtained as the sum of the previous two Fibonacci numbers

$$f_n = f_{n-1} + f_{n-2}$$

for all integers $n \geq 3$. We define $f_0 = 0$ to ease the formulation of some results.

There are many interesting identities that can be derived for these numbers. For example, let us rewrite the elements in the form

$$\begin{aligned} f_3 - f_2 &= f_1 \\ f_4 - f_3 &= f_2 \\ &\vdots \\ f_{n+2} - f_{n+1} &= f_n \end{aligned}$$

If we sum the elements of the right-hand side, then we obtain the sum of the first n Fibonacci numbers. If we sum the elements of the left-hand side, then we obtain a sum in which most elements cancel, namely

$$(f_{n+2} - f_{n+1}) + (f_{n+1} - f_n) + \cdots + (f_3 - f_2) = f_{n+2} - f_2 = f_{n+2} - 1.$$

A sum in which subsequent terms cancel such that just an initial and a final term remain is called a **telescoping sum**.

Proposition 4.3. *The Fibonacci numbers satisfy*

$$f_1 + f_2 + \cdots + f_n = f_{n+2} - 1. \tag{4.12}$$

for all $n \geq 1$.

Proof. We use a proof by induction to verify the claim. Let us denote by $F(n)$ the Eq. (4.12).

Base Case The claim $F(n)$ holds for $n = 1$, since

$$f_1 = 1 = 2 - 1 = f_{1+2} - 1.$$

Inductive Step We claim that the implication $F(n) \to F(n+1)$ holds for all $n \geq 1$. Indeed, suppose that $F(n)$ is true. Then

$$\begin{aligned} \underbrace{f_1 + f_2 + \cdots + f_n}_{} + f_{n+1} &= f_{n+2} - 1 + f_{n+1} & \text{by } F(n) \\ &= f_{n+3} - 1 & \text{by definition of } f_{n+3}, \end{aligned}$$

so $F(n+1)$ holds. Therefore, the implication $F(n) \to F(n+1)$ holds for all $n \geq 1$.

It follows by induction that $F(n)$ holds for all $n \geq 1$. □

EXERCISES

4.15. Prove by induction that the sum of the first n terms of the Fibonacci sequence that have even index is given by

$$\sum_{k=1}^{n} f_{2k} = f_2 + f_4 + \cdots + f_{2n} = f_{2n+1} - 1. \tag{4.13}$$

4.4 Geometric Series

4.16. Show that the number of sequences of 1s and 2s that sum to a total of $n-1$ is given by the Fibonacci number f_n.

4.17. Prove by induction that the Fibonacci numbers satisfy the Simpson identity
$$f_{n+1}f_{n-1} - f_n^2 = (-1)^n \tag{4.14}$$
for all $n \geq 2$. [Hint: Watch out for the base case.]

4.18. Prove by induction that
$$\sum_{k=1}^{n} f_k^2 = f_n f_{n+1}. \tag{4.15}$$
holds for all $n \geq 1$.

4.4 Geometric Series

A sequence of numbers that have a fixed common ratio x is called a **geometric series**. For example, the sequence

$$2, \quad 6, \quad 18, \quad 54, \quad 162, \quad 486, \quad 1458, \ldots$$

has the property that a term divided by its previous term has the ratio $x = 3$. The next proposition shows how to sum a finite geometric series.

Proposition 4.4. *Let x be a real number such that $x \neq 1$, and c a real number. If n is a nonnegative integer, then*

$$\sum_{k=0}^{n} cx^k = c\left(\frac{1 - x^{n+1}}{1 - x}\right). \tag{4.16}$$

Proof. Let us denote the Eq. (4.16) by $P(n)$. We prove the claim by induction on n.

Base Case The claim $P(n)$ is true for $n = 0$, since
$$\sum_{k=0}^{0} cx^k = c = c(1 - x^{0+1})/(1 - x).$$

Inductive Step Our goal is to show that the implication $P(n) \to P(n+1)$ holds for all $n \geq 0$. Suppose that $P(n)$ is true. Then

$$\sum_{k=0}^{n+1} cx^k = \sum_{k=0}^{n} cx^k + cx^{n+1} = c\left(\frac{1 - x^{n+1}}{1 - x}\right) + cx^{n+1}$$

by induction hypothesis. Bringing to a common denominator and simplifying yields

$$\sum_{k=0}^{n+1} cx^k = c\left(\frac{1 - x^{n+1} + x^{n+1} - x^{n+2}}{1 - x}\right) = c\left(\frac{1 - x^{n+2}}{1 - x}\right),$$

so $P(n+1)$ is true.

Therefore, we can conclude by induction that $P(n)$ holds for all $n \geq 0$. □

Corollary 4.5. *Suppose that x is a real number such that $|x| < 1$ and c a real number. Then*
$$\sum_{k=0}^{\infty} cx^k = \frac{c}{1-x}$$

Proof. By definition,
$$\sum_{k=0}^{\infty} cx^k = \lim_{n \to \infty} \sum_{k=0}^{n} cx^k = \lim_{n \to \infty} c\left(\frac{1 - x^{n+1}}{1 - x}\right) = \frac{c}{1-x},$$
where the second equality follows from the previous proposition and the last equality follows from the fact that $\lim_{n \to \infty} x^{n+1} = 0$ when $|x| < 1$. □

EXERCISES

4.19. Evaluate the infinite geometric series $\sum_{k=0}^{\infty} x^k$ for
(a) $x = 1/2$,
(b) $x = 1/3$,
(c) $x = 2$,
(d) $x = 1$.
Explain your results.

4.20. Evaluate the geometric series
$$\sum_{k=1}^{\infty} \frac{9}{10^k}$$
and explain its significance for the decimal number system.

4.21. Let us define some geometric figures as follows. The figure K_0 is defined as an equilateral triangle with side length 1 (shown on the left). For all $n \geq 1$, the figure K_n is obtained from the figure K_{n-1} by subdividing each line segment into three equal parts and replacing the middle segment by two segments of the same length pointing outward. So K_1 is a 12 sided polygon. The figures K_0, K_1, K_2, and K_3 are shown here:

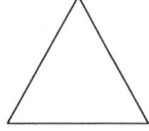  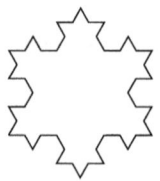 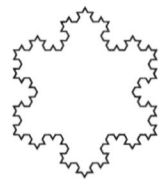

(a) Determine the number N_n of sides of the polygon K_n.
(b) Determine the length L_n of a line segment in the polygon K_n, assuming that the line segments in K_0 have unit length, $L_0 = 1$.
(c) Determine the perimeter P_n of K_n.
(d) Determine the area A_n of K_n.
(e) Determine the area and the perimeter of the Koch snowflake $\lim_{n \to \infty} K_n$.

4.5 Binomial Theorem

Given two variables x and y, the **binomial theorem** explicitly describes the expansion of the term $(x + y)^n$ for any nonnegative integer n. For example, the distributive law yields

$$(x + y)^2 = (x + y)(x + y) = x^2 + 2xy + y^2.$$

Multiplying with another $(x + y)$ term yields

$$(x + y)^3 = x^3 + 3x^2y + 3xy^2 + y^3.$$

For the general case, we need to recall some notation. For a positive integer n, the **factorial function** $n!$ is defined as the product of the first n positive integers,

$$n! = n(n-1)\cdots 2 \cdot 1,$$

and $0! = 1$. We define the **binomial coefficient** $\binom{n}{k}$ as

$$\binom{n}{k} = \frac{n!}{k!(n-k)!}.$$

It follows from this definition that

$$\binom{n}{0} = 1, \quad \binom{n}{n} = 1.$$

A direct computation yields Pascal's identity

$$\binom{n}{k} = \binom{n-1}{k-1} + \binom{n-1}{k}$$

for $1 \leq k \leq n$.

Proposition 4.6. *Let x and y be variables. Then*

$$(x + y)^n = \sum_{k=0}^{n} \binom{n}{k} x^k y^{n-k} \qquad (4.17)$$

holds for all nonnegative integers n.

Proof. Let $P(n)$ denote the Eq. (4.17). We prove the claim by induction on n.
Base Case For $n = 0$, we have $(x + y)^0 = 1$, which is equal to $\binom{0}{0}x^0y^0 = 1$. Therefore, the equation $P(0)$ is true.
Induction Step We are going to show that the implication $P(n) \to P(n + 1)$ holds for all $n \geq 0$. Suppose that $P(n)$ is true, so $(x + y)^n = \sum_{k=0}^{n} \binom{n}{k}x^k y^{n-k}$ holds. By multiplying with x and with y, we respectively get

$$y(x+y)^n = \binom{n}{0}x^0 y^{n+1} + \binom{n}{1}x^1 y^n + \binom{n}{2}x^2 y^{n-1} + \cdots + \binom{n}{n}x^n y,$$
$$x(x+y)^n = \phantom{\binom{n}{0}x^0 y^{n+1} + {}} \binom{n}{0}x^1 y^n + \binom{n}{1}x^2 y^{n-1} + \cdots + \binom{n}{n-1}x^n y + \binom{n}{n}x^{n+1}.$$

Adding these two sums yields

$$(x+y)^{n+1} = \binom{n+1}{0}x^0 y^{n+1} + \binom{n+1}{1}x^1 y^n + \cdots + \binom{n+1}{n}x^n y + \binom{n+1}{n+1}x^{n+1},$$

where we used Pascal's identity $\binom{n}{k} + \binom{n}{k-1} = \binom{n+1}{k}$ to simplify the coefficients of the terms $x^k y^{n+1-k}$ for all k in the range $1 \leq k \leq n$, and the fact that $\binom{n}{0} = \binom{n+1}{0} = 1$ and $\binom{n}{n} = \binom{n+1}{n+1} = 1$ holds. In other words, $P(n+1)$ holds. We can conclude by induction that $P(n)$ holds for all $n \geq 0$. □

EXERCISES

The first few exercises rely on the definition of binomial coefficients. You learn how to evaluate them and prove some simple properties.

4.22. Numerically evaluate the binomial coefficients

$$\binom{5}{0}, \binom{5}{1}, \binom{5}{2}, \binom{5}{3}, \binom{5}{4}, \binom{5}{5}.$$

4.23. Show that

$$\binom{n}{k} = \binom{n}{n-k}$$

holds for all nonnegative integers n and integers k in the range $0 \leq k \leq n$.

The next few exercises are simple applications of the binomial theorem.

4.24. Use the binomial theorem to expand

$$(2x + 3y)^4.$$

4.25. Use the binomial theorem to expand

$$\left(1 - \frac{1}{x}\right)^3.$$

4.26. Show that

(a) $\sum_{k=0}^{n} \binom{n}{k} = 2^n.$

(b) $\sum_{k=0}^{n} \binom{n}{k}(-1)^k = 0.$

(c) $\sum_{k=0}^{n} \binom{n}{k} 2^k = 3^n.$

4.27. Evaluate the sum
$$\sum_{k=0}^{499} (-1)^k \binom{999}{2k}$$
and simplify the result to a power of an integer. [Hint: The path to the real solution is literally complex. Just don't get fooled into believing it is difficult. Have a look at the powers of $(1+i)$ and $(1-i)$ for $i = \sqrt{-1}$.]

4.28. Let p be a prime number. Use induction to show that for all positive integers n, the prime p divides $n^p - n$. This enormously useful result is known as **Fermat's Little Theorem**. [In the notation of Chap. 5, we can state this divisibility result as the congruence $n^p \equiv n \pmod{p}$.]

4.6 Strong Induction

In the principle of strong induction, we still have to prove the base case, but we grant ourselves a much more generous induction hypothesis. We assume the induction hypothesis $P(1) \wedge P(2) \wedge \cdots \wedge P(n)$ and show that it implies $P(n+1)$.

> **Principle of Strong Induction.** To prove that a property $P(n)$ holds for all positive integers n, it suffices to prove the base case and the inductive step.
>
> **Base case.** The property $P(1)$ holds.
>
> **Inductive step.** If $P(k)$ holds for all k in the range $1 \leq k \leq n$, then $P(n+1)$ holds.

The Fibonacci sequence
$$f_1 = 1, \quad f_2 = 1, \quad f_3 = 2, \quad f_4 = 3, \quad f_5 = 5, \quad f_6 = 8, \quad \ldots$$
does not contain every positive integer. However, it has a curious property. We claim that each positive integer can be expressed as a sum of terms
$$f_{a_k} + f_{a_{k-1}} + \ldots + f_{a_1}$$
where the indices $a_k > a_{k-1} > \cdots > a_1$ are pairwise distinct positive integers. For example, the numbers 4, 6, 7, and 9 are missing from the Fibonacci sequence, but we can express them as sums of Fibonacci numbers with distinct

indices,
$$\begin{aligned} 4 &= f_4 + f_1 = 3 + 1 \\ 6 &= f_5 + f_1 = 5 + 1 \\ 7 &= f_5 + f_2 + f_1 = 5 + 1 + 1 \\ 9 &= f_6 + f_1 = 8 + 1 \end{aligned}$$

It is not obvious that every positive integer should have such a representation. However, it does seem to work for larger numbers as well. For instance, the fairy tale number 1001 can be expressed as

$$1001 = f_{16} + f_7 + f_1 = 987 + 13 + 1.$$

Proposition 4.7. *Every positive integer can be expressed as a sum of terms of the Fibonacci sequence with pairwise distinct indices.*

Proof. We prove this by strong induction. Let $P(n)$ denote the property that n is a sum of terms of the Fibonacci sequence with pairwise distinct indices.
Base Case Since $f_1 = 1$, the property $P(1)$ holds.
Induction Step We claim that $P(1) \wedge \cdots \wedge P(n)$ implies $P(n+1)$. Indeed, suppose that all positive integers k in the range $1 \leqslant k \leqslant n$ can be written as a sum of terms of the Fibonacci sequence with distinct indices.

If $n+1$ is a Fibonacci number, then $P(n+1)$ holds.

It remains to show that if $n+1$ is not a Fibonacci number, then $P(n+1)$ holds as well. Let m be a positive integer such that f_m is the largest Fibonacci number less than $n+1$; hence, $f_m < n+1 < f_{m+1}$. Such an integer m exists, since the Fibonacci sequence is monotonically increasing and $n+1 > f_2 = 1$. Then

$$n + 1 - f_m < f_{m+1} - f_m = f_{m-1} < n + 1$$

is a positive integer such that $P(n+1-f_m)$ holds by induction hypothesis, so there exist integers $a_k > \cdots > a_1 > 0$ such that

$$n + 1 - f_m = f_{a_k} + f_{a_{k-1}} + \cdots + f_{a_1}.$$

Since $n + 1 - f_m < f_{m-1}$, we have $f_m > f_{m-1} > f_{a_k}$. Therefore,

$$n + 1 = f_m + f_{a_k} + f_{a_{k-1}} + \cdots + f_{a_1}$$

is a representation of $n+1$ by a sum of terms of the Fibonacci sequence with pairwise distinct indices. Therefore, $P(n+1)$ holds.

We can conclude by strong induction that the claim holds for all positive integers. □

A more common way to represent integers is given in the next proposition.

Proposition 4.8. *Every positive integer n can be written in the form*

$$n = \sum_{k=0}^{m} c_k 2^k \qquad (4.18)$$

for some nonnegative integer m and coefficients c_k being 0 or 1.

4.6 Strong Induction

Proof. We prove the claim by strong induction. Let $P(n)$ denote the property that n can be written in the form (4.18).
Base Case The number $n = 1$ can be represented by $c_0 = 1$, so $P(1)$ holds.
Inductive Step We claim that $P(1) \wedge P(2) \wedge \cdots \wedge P(n)$ implies $P(n+1)$. Suppose that $P(k)$ holds for all k in the range $1 \leq k \leq n$. We are going to distinguish between the two cases (a) $n+1$ is even and (b) $n+1$ is odd. Our goal is to show that $P(n+1)$ holds in both cases.

(a) If $n+1$ is even, then $(n+1)/2$ is a positive integer and $P((n+1)/2)$ holds. Thus, there exists a representation

$$\frac{n+1}{2} = \sum_{k=0}^{m} c_k 2^k.$$

It follows that

$$n+1 = 2 \sum_{k=0}^{m} c_k 2^k = \sum_{k=0}^{m} c_k 2^{k+1},$$

so $P(n+1)$ holds.

(b) If $n+1 > 1$ is an odd integer, then n is an even integer satisfying $P(n)$, so

$$n = \sum_{k=0}^{m} c_k 2^k.$$

We necessarily have $c_0 = 0$, since n is an even integer. It follows that

$$n+1 = \sum_{k=0}^{m} c_k 2^k + 1 = \sum_{k=0}^{m} d_k 2^k$$

with $d_0 = 1$ and $d_k = c_k$ for k in the range $1 \leq k \leq n$. Therefore, $P(n+1)$ holds.

We can conclude by the principle of strong induction that $P(n)$ holds for all positive integers n. □

Proposition 4.9. *For all nonnegative integers a and b with $b > 0$, there exist nonnegative numbers q and r such that*

$$a = qb + r \quad \text{with} \quad 0 \leq r < b.$$

The numbers q and r are respectively called the **quotient** and **remainder** when a is divided by b.

Proof. We are going to prove the claim by strong induction on a. Let $P(a)$ denote that there exist nonnegative integers q and r such that $a = qb + r$ with $0 \leq r < b$.
Base Case For $a = 0$, the numbers $q = 0$ and $r = 0$ yield $a = 0 = 0 \times b + 0$, so $P(0)$ holds.

Inductive Step For all positive integers a, we are going to show that $P(0) \wedge P(1) \wedge \cdots \wedge P(a-1)$ implies $P(a)$. We distinguish the cases (a) $a < b$ and (b) $a \geqslant b$.

(a) If $a < b$, then $a = qb + r$ with $q = 0$ and $r = a < b$. So $P(a)$ holds.

(b) If $a \geqslant b$, then $a' = a - b$ satisfies $P(a')$. Thus, there exist nonnegative integers q' and r' such that $a' = q'b + r'$ with $0 \leqslant r' < b$. Then $a = a' + b = (q' + 1)b + r'$, so a divided by b has quotient $q = q' + 1$ and remainder $r = r'$. Therefore, $P(a)$ holds.

We can conclude by strong induction that $P(a)$ holds for all $a \geqslant 0$. □

Recall that an integer $p > 1$ is called a **prime** if and only if it does not have any positive divisors apart from 1 and p.

Proposition 4.10. *Every integer $n > 1$ can be written as a product*

$$n = p_1 p_2 \cdots p_m$$

of $m \geqslant 1$ prime numbers.

Proof. We prove the claim by strong induction. Let $P(n)$ denote the property that n can be written as a product of prime numbers.
Base Case Since $n = 2$ is a prime number, the property $P(2)$ holds.
Inductive Step Suppose that $P(k)$ holds for all k in the range $2 \leqslant k < n$. Our goal is to show that $P(n)$ is true. We can distinguish two cases.
(a) If n is prime, then $P(n)$ holds.
(b) If n is composite, then there exist integers $a, b > 1$ such that $n = ab$. Since $P(a)$ and $P(b)$ hold, there exist prime numbers $a_1, \ldots, a_k$ such that $a = a_1 \cdots a_k$ and prime numbers $b_1, \ldots, b_\ell$ such that $b = b_1 \cdots b_\ell$. Then

$$n = a_1 \cdots a_k b_1 \cdots b_\ell$$

is a product of primes, so $P(n)$ holds.
The claim follows by the principle of strong induction. □

Sometimes it is convenient to establish several base cases to ease the formulation of the inductive step. We illustrate this in the proof of the next proposition.

Proposition 4.11. *For all positive integers n, the number $n^4 - n^2$ is divisible by 12.*

Proof. We prove the claim by strong induction. Let $P(n)$ denote the predicate that $n^4 - n^2$ is divisible by 12.
Base cases. We notice that

$$\begin{array}{llll} 1^4 - 1^2 = 0 \times 12, & 2^4 - 2^2 = 1 \times 12, & 3^4 - 3^2 = 6 \times 12, \\ 4^4 - 4^2 = 20 \times 12, & 5^4 - 5^2 = 50 \times 12, & 6^4 - 6^2 = 105 \times 12. \end{array}$$

4.6 Strong Induction

Therefore, $P(1)$, $P(2)$, $P(3)$, $P(4)$, $P(5)$, and $P(6)$ hold.
Inductive step. Let $n \geq 6$, Suppose that $P(k)$ holds for all k in the range $1 \leq k \leq n$. We are going to show that $P(n+1)$ must hold.

Let $k = n + 1 - 6$. By induction hypothesis, $P(k)$ holds, so $k^4 - k^2$ is a multiple of 12. Thus,

$$(n+1)^4 - (n+1)^2 = (k+6)^4 - (k+6)^2$$
$$= (k^4 + 24k^3 + 216k^2 + 864k + 1296) - (k^2 + 12k + 36).$$

If we rearrange terms and factor out 12, we get

$$(n+1)^4 - (n+1)^2 = (k^4 - k^2) + (24k^3 + 216k^2 + 852k + 1260)$$
$$= (k^4 - k^2) + 12(2k^3 + 18k^2 + 71k + 105).$$

Since $k^4 - k^2$ is a multiple of 12, it follows that $(n+1)^4 - (n+1)^2$ is a multiple of 12 as well. Therefore, $P(n+1)$ holds.

The claim follows by the principle of strong induction. □

Base Cases The number of base cases depends on the formulation of the inductive step. In the previous proposition, we essentially used $P(k) \to P(k+6)$. Therefore, we needed six base cases to anchor the induction for numbers that respectively have remainders 1, 2, 3, 4, 5, and 0 when divided by 6. We recommend to first formulate the inductive step and then determine all bases cases that are needed.

EXERCISES

4.29. A positive integer n can be expressed as a sum of distinct terms of the Fibonacci sequence. Either show that the representation is unique or give a counter example.

4.30. Let $g_0 = 1$, $g_1 = 2$, $g_2 = 3$, and $g_n = g_{n-1} + g_{n-2} + g_{n-3}$ for all integers $n \geq 3$. Prove by strong induction that $g_n \leq 2^n$ holds for all nonnegative integers n.

4.31. Let f_n be a sequence of nonnegative integers satisfying the recurrence relation $f_n = (n^3 - 3n^2 + 2n)f_{n-3}$, as well as $f_1 = 1$, $f_2 = 2$, and $f_3 = 6$. Prove by strong induction that $f_n = n!$ holds for all integers $n \geq 1$.

4.32. Let ϕ denote the golden ratio

$$\phi = \frac{1 + \sqrt{5}}{2}.$$

The Fibonacci numbers f_n given by $f_1 = 1$, $f_2 = 1$, and $f_n = f_{n-1} + f_{n-2}$ satisfy the lower bound
$$f_n \geq \phi^{n-2}$$
for all $n \geq 2$.

4.33. Let z be a real number such that $z + 1/z$ is an integer. Prove by strong induction that $z^n + 1/z^n$ is an integer for all positive integer exponents n.

*4.7 Well-founded Induction

In this section, we give a generalization of the induction principle that is applicable in a wider variety of situations. It encompasses the principles of induction and strong induction as special cases. However, well-founded induction is a much more powerful tool.

Let S be an arbitrary set. Recall that a relation $\triangleleft$ on S is a subset of the set $S \times S$ of ordered pairs with elements from S. We will use the infix notation $s \triangleleft s'$ to denote that a pair (s, s') is an element of the relation $\triangleleft$.

Let T be a nonempty subset of S. We say that an element m in T is $\triangleleft$-**minimal** in T if and only if there does not exist an element t in T such that $t \triangleleft m$. Note that there might exist an element s in S such that $s \triangleleft m$, but s cannot belong to T.

A relation $\triangleleft$ over S is called **well-founded** if and only if every nonempty subset T of S has a $\triangleleft$-minimal element.

Example 4.12. The successor relation $\triangleleft = \{(n, n+1) \mid n \in \mathbf{N}_1\}$ on the set $\mathbf{N}_1$ of positive integers is a well-founded relation. Indeed, the smallest element of a nonempty subset T of $\mathbf{N}_1$ is a $\triangleleft$-minimal element. For example, 2 is a $\triangleleft$-minimal element of $\{2, 3\}$. A subset might have even more than one $\triangleleft$-minimal element. For instance, 2 and 4 are $\triangleleft$-minimal elements of $\{2, 4, 5\}$.

Example 4.13. The strictly-less-than relation $\triangleleft = \{(m, n) \mid m < n\}$ on the set $\mathbf{N}_1$ of positive integers is a well-founded relation. The smallest element of a nonempty subset T of the set of positive integers is a $\triangleleft$-minimal element.

Example 4.14. The strict lexicographic order $\triangleleft$ on the set $\mathbf{N}_0 \times \mathbf{N}_0$ of pairs of nonnegative integers is given by $(a, b) \triangleleft (a', b')$ if and only if $a < a'$, or $a = a'$ and $b < b'$. The relation $\triangleleft$ is wellfounded. Indeed, given a nonempty subset T of $\mathbf{N}_0 \times \mathbf{N}_0$, we can choose the element (a_0, b_0) such that $a_0 = \min\{a \in \mathbf{N}_0 \mid (a, b) \in T\}$ and $b_0 = \min\{b \in \mathbf{N}_0 \mid (a_0, b) \in T\}$. Then (a_0, b_0) is a $\triangleleft$-minimal element of T.

We write $y \triangleright x$ if and only if $x \triangleleft y$. We say that a relation $\triangleleft$ has an infinite descending chain if and only if there exist elements a_k with $k \in \mathbf{N}_1$ such that
$$a_1 \triangleright a_2 \triangleright a_3 \triangleright \cdots$$

The next proposition gives an alternative characterization of well-founded relations in terms of the absence of infinite descending chains.

4.7 Well-founded Induction

Proposition 4.15. *A relation $\triangleleft$ on a set S is well-founded if and only if there does not exist an infinite sequence $(a_1, a_2, a_3, \ldots)$ of elements in S such that $a_{k+1} \triangleleft a_k$ holds for all integers $k \geq 1$. In other words, $\triangleleft$ is well-founded if and only if there does not exist an infinite descending chain*

$$a_1 \triangleright a_2 \triangleright a_3 \triangleright \cdots$$

of elements in S.

Proof. Let $\triangleleft$ be a well-founded relation on a set S. Seeking a contradiction, suppose that there exists an infinite sequence $(a_1, a_2, a_3, \ldots)$ of elements in S such that $a_{k+1} \triangleleft a_k$ holds for all integers $k \geq 1$. Then $T = \{a_k \mid k \in \mathbf{N}_1\}$ would be a nonempty subset of S without a $\triangleleft$-minimal element, contradicting the well-foundedness of $\triangleleft$.

Conversely, suppose that there does not exist an infinite sequence

$$(a_1, a_2, a_3, \ldots)$$

of elements in S such that $a_{k+1} \triangleleft a_k$ holds for all integers $k \geq 1$. Seeking a contradiction, suppose that T is a nonempty subset of S without a $\triangleleft$-minimal element, meaning that for every element t in T, there exist an element s in T such that $s \triangleleft t$. Choose an arbitrary element a_1 in T. Thus, there exists an element a_2 in T such that $a_2 \triangleleft a_1$. Given any finite sequence $(a_1, a_2, \ldots, a_n)$ of elements in T such that $a_n \triangleleft a_{n-1} \triangleleft \cdots \triangleleft a_2 \triangleleft a_1$, we can find an element a_{n+1} such that $a_{n+1} \triangleleft a_n$. Inductively, we obtain an infinite sequence $(a_1, a_2, a_3, \ldots)$ of elements in $T \subseteq S$ such that $a_{k+1} \triangleleft a_k$, contradicting our assumption that such sequences cannot exist. Therefore, T must contain a $\triangleleft$-minimal element. □

Well-founded relations are for example used in termination proofs of programs. We can also use it to show that some games finish in a finite number of steps, as the next example shows.

Example 4.16. Ernie has a bag full of red and blue chips. He plays a game all morning. Bert gets curious and asks him about the rules of the game. Ernie explains, "You always draw two random chips from the bag. Your goal is to get the bag empty." Bert asks, "What if there is only one chip left?" "You take that chip and the game is finished," Ernie replies. He continues, "If one of the two chips is blue, you discard a blue chip and put the other chip back into the bag." Bert mumbles, "Seems boring." Ernie adds, "If both chips are red, then you discard one red chip and put the other red chip and five blue chips into the bag." Bert sneers, "Ernie, that game might never finish!" "Not at all! If we have r red balls and b blue balls, then we go from (r, b) to

$$(r, b-1) \triangleleft (r, b)$$

or

$$(r-1, b+5) \triangleleft (r, b)$$

in the strict lexicographic order. Since this is a wellfounded relation, we will always finish in a finite number of steps." Ernie replies and chuckles.

The next theorem justifies the principle of wellfounded induction (also known as the principle of Noetherian induction).

Theorem 4.17 (Well-Founded Induction). *Let $\triangleleft$ be a well-founded relation on a set S. In order to show that a property $P(x)$ holds for all elements x in S, it suffices to show that*

(a) **Induction Basis.** $P(m)$ *holds for all $\triangleleft$-minimal elements m of S.*

(b) **Inductive Step.** *If x in S is not a $\triangleleft$-minimal element of S, and $P(a)$ holds for all elements $a \in S$ such that $a \triangleleft x$, then $P(x)$ holds.*

Proof. Seeking a contradiction, suppose that the subset

$$T = \{x \in S \mid P(x) \text{ is false}\}$$

is not empty. Since $\triangleleft$ is a well-founded relation on S, the set T must contain a $\triangleleft$-minimal element m. By the induction basis, the element m cannot be a $\triangleleft$-minimal element of S. The elements a in S such that $a \triangleleft m$ do not belong to T, so the property $P(a)$ holds for all of them. Therefore, the inductive step guarantees that $P(m)$ must hold, contradicting the fact that m is an element of T. Therefore, T must be empty and thus $P(x)$ holds for all $x \in S$. □

In other words, in the inductive step, we show that $P(x)$ holds under the assumption that $P(a)$ holds for all a satisfying $a \triangleleft x$.

Boolean Formulas. Recall that the set of Boolean formulas over a set $\mathcal{B}$ of Boolean variables is the smallest set of strings over the alphabet

$$\mathcal{B} \cup \{\neg, \vee, \wedge, \rightarrow, \leftrightarrow\}$$

such that
(a) a variable from $\mathcal{B}$ is a Boolean formula,
(b) if A is a Boolean formula, then $\neg A$ is a Boolean formula,
(c) if A and B are Boolean formulas, then $(A \vee B)$, $(A \wedge B)$, $(A \rightarrow B)$, and $(A \leftrightarrow B)$.

Put differently, a Boolean formula can be obtained by a finite number of applications of the rules (a), (b), and (c).

A Boolean formula is a highly structured string of symbols. Apparently, each Boolean formula has the same number of opening as closing parentheses. We will now show how to prove this fact by well-founded induction.

We first need to introduce a well-founded relation $\triangleleft$ on the set of Boolean formulas by letting immediate subformulas preceed a formula. In other words,

4.7 Well-founded Induction

if A and B are Boolean formulas, then

$$A \triangleleft \neg A,$$
$$A, B \triangleleft (A \vee B),$$
$$A, B \triangleleft (A \wedge B),$$
$$A, B \triangleleft (A \to B),$$
$$A, B \triangleleft (A \leftrightarrow B),$$

and no other relations hold. For instance,

$$A \triangleleft (A \to B) \triangleleft ((A \to B) \wedge C),$$

but $A \triangleleft ((A \to B) \wedge C)$ does not hold.

Proposition 4.18. *In a Boolean formula, the number of opening parentheses is equal to the number of closing parentheses.*

Proof. Induction Basis The set of $\triangleleft$-minimal Boolean formulas coincides with the set $\mathcal{B}$ of Boolean variables. Since Boolean variables do not involve any parentheses, the claim certainly holds for them.

Inductive Step The set of Boolean formulas that are not $\triangleleft$-minimal are of the form $\neg A$, $(A \vee B)$, $(A \wedge B)$, $(A \to B)$, or $(A \leftrightarrow B)$.

Suppose that the number of opening and closing parentheses is equal for the subformulas A and B. Since $A \triangleleft \neg A$, and $\neg A$ does not introduce additional parentheses, the claim holds for $\neg A$. Since $A, B \triangleleft (A \star B)$, where

$$\star \in \{\vee, \wedge, \to, \leftrightarrow\},$$

and each composite formula introduces one additional opening and one additional closing parenthesis, the number of opening and closing parentheses remains balanced.

Therefore, we can conclude by well-founded induction that all Boolean formulas have the same number of opening parentheses as the number of closing parentheses. $\square$

We could have proven this proposition by induction or strong induction, but it would have required us to introduce an auxiliary depth function that assigns a nonnegative integer to each Boolean formula. The argument using wellfounded induction is more direct, so it is often the preferred method of proof.

Our next example shows that well-founded induction can be applied in situations where it is difficult or impossible to apply mathematical induction or strong induction.

Ackermann Function. We will illustrate well-founded induction by looking at some properties of the following recursively defined function:

$$A(x, y) = \begin{cases} y + 1 & \text{if } x = 0, \\ A(x - 1, 1) & \text{if } x \neq 0 \text{ and } y = 0, \\ A(x - 1, A(x, y - 1)) & \text{if } x \neq 0 \text{ and } y \neq 0. \end{cases}$$

This function is known as Ackermann's function. The function $A(x, y)$ is defined for nonnegative integer arguments x and y.

The definition of $A(x, y)$ might appear a bit enigmatic. It is instructive to see this recursion in action. For instance, evaluating $A(1, 2)$ yields

$$\begin{aligned} A(1, 2) &= A(0, A(1, 1)) &&\text{as } x = 1 \neq 0 \text{ and } y = 2 \neq 0 \\ &= A(0, A(0, A(1, 0))) &&\text{as } x = 1 \neq 0 \text{ and } y = 1 \neq 0 \\ &= A(0, A(0, A(0, 1))) &&\text{as } x = 1 \neq 0 \text{ and } y = 0 \\ &= A(0, A(0, 2)) &&\text{as } A(0, 1) = 2 \\ &= A(0, 3) &&\text{as } A(0, 2) = 3 \\ &= 4. \end{aligned}$$

Unfortunately, most arguments lead to prohibitively long evaluation chains and large results. For example, one can show that

$$A(4, 0) = 2^{2^2} - 3 = 13,$$

$$A(4, 1) = 2^{2^{2^2}} - 3 = 65533,$$

$$A(4, 2) = 2^{2^{2^{2^2}}} - 3 = 2^{65,536} - 3.$$

The number of atoms in the observable universe is estimated to be around 10^{82}, so the value of $A(4, 2)$ vastly exceeds this number.

In general, the nested levels of recursion are so high that not even a computer is of much use in experimentation with $A(x, y)$, especially when the arguments get larger.

We can demystify the function by taking advantage of the well-founded relation from Example 4.14. We notice that when A is used on the right-hand side of the definition, then the arguments of $A(x-1, 1)$, $A(x, y-1)$, and $A(x - 1, A(x, y - 1))$ satisfy

$$(x - 1, 1) \lhd (x, y),$$
$$(x, y - 1) \lhd (x, y)$$
$$(x - 1, A(x, y - 1)) \lhd (x, y).$$

As $\lhd$ is a well-founded relation, this means that there are just a finite number of recursive calls possible, for otherwise we would obtain an infinite descending chain of arguments.

Let us now use well-founded induction to formally prove that Ackermann's function is well-defined and never goes into an infinite recursion for all nonnegative integer arguments.

4.7 Well-founded Induction

Proposition 4.19. *Ackermann's function $A(x,y)$ is a well-defined function that yields for each input $(x,y) \in \mathbf{N}_0 \times \mathbf{N}_0$ a nonnegative integer value.*

Proof. Let $\lhd$ denote the strict lexicographic order on $\mathbf{N}_0 \times \mathbf{N}_0$, that is,

$$(m_1, m_2) \lhd (n_1, n_2) \text{ if and only if } \begin{cases} m_1 < n_1, \\ m_1 = n_1 \text{ and } m_2 < n_2. \end{cases}$$

Then $\lhd$ is a well-founded relation on $\mathbf{N}_0 \times \mathbf{N}_0$. The element $(0,0)$ is the only $\lhd$-minimal element of $\mathbf{N}_0 \times \mathbf{N}_0$.
Induction Basis The value $A(0,0)$ is defined and equal to 1.
Inductive Step Let us assume that $A(m', n')$ is defined for all $(m', n') \lhd (m, n)$. Then we have the following three cases for $A(m,n)$:
(a) If $m = 0$, then $A(0, n)$ is defined and equal to $n + 1$, since $A(0, y) = y + 1$.
(b) If $m \neq 0$ and $n = 0$, then $(m-1, 1) \lhd (m, 0)$, so $A(m-1, 1)$ is defined by induction hypothesis; hence $A(m, 0)$ is defined and equal to $A(m-1, 1)$.
(c) If $m \neq 0$ and $n \neq 0$, then $(m, n-1) \lhd (m, n)$, so $A(m, n-1)$ is defined; furthermore, $(m-1, y) \lhd (m, n)$ for all y in $\mathbf{N}_0$, so by induction hypothesis $A(m-1, A(m, n-1))$ is defined. However, this is precisely $A(m, n)$, so $A(m, n)$ is defined as well.

Therefore, the Ackermann function yields a nonnegative integer value for all inputs from $\mathbf{N}_0 \times \mathbf{N}_0$. □

The next proposition gives some simple lower bound for the value of the Ackermann function $A(x, y)$ in terms of the right argument y. This inequality is tight for $x = 0$, but can be loose for larger values of x.

Proposition 4.20. *The Ackermann function $A(x, y)$ satisfies for all nonnegative integers x and y the inequality*

$$A(x, y) > y.$$

Proof. We use well-founded induction over the strict lexicographic order from Example 4.14.
Induction Basis Since $A(0,0) = 1 > 0$, the property holds for the $\lhd$-minimal element $(0, 0)$.
Inductive Step Let us assume that $A(m', n') > n'$ holds for all $(m', n') \lhd (n, m)$. Then we have the following three cases for $A(m, n)$:
(a) If $m = 0$ and $n \neq 0$, then $A(0, n) = n + 1 > n$.
(b) If $m \neq 0$ and $n = 0$, then $(m-1, 1) \lhd (m, 0)$, so $A(m-1, 1) > 1$; hence

$$A(m, 0) = A(m-1, 1) > 1 > 0.$$

(c) If $m \neq 0$ and $n \neq 0$, then $(m, n-1) \lhd (m, n)$, so $A(m, n-1) > n-1$; furthermore, $(m-1, y) \lhd (m, n)$ for all y in $\mathbf{N}_0$, so by induction hypothesis $A(m-1, A(m, n-1)) > A(m, n-1)$. It follows that

$$A(m, n) = A(m-1, A(m, n-1)) > A(m, n-1) > n-1.$$

Thus, $A(m-1, n-1)$ is at least n, allowing us to deduce that
$$A(m, n) = A(m-1, A(m, n-1)) > n.$$
Therefore, the claim follows by well-founded induction on $\triangleleft$. □

The impenetrable recursive definition of the Ackermann function obscures the fact that its function values can be described without too much difficulty. Most values of the Ackermann function are simply of the form
$$2^z - 3,$$
for some positive integer z. Alas, the common mathematical notations are not well-suited to describe the exponent z, since it is simply too enormous!

We get ready for the task by introducing some notation to concisely express huge numbers. Knuth's up-arrow notation for iterated powers is convenient for this purpose. For an integer x and a nonnegative integer n, the notation expresses x^n in the form $x \uparrow^1 n$, which reminds us of the notation x^n that is used in many programming languages for exponentiation. In other words, $x \uparrow^1 n$ is the product
$$x \uparrow^1 n = x \cdot x \cdot \ldots \cdot x,$$
of n terms.

The next big notational leap is to introduce iterated powers $\uparrow^m$ for integers $m > 1$, which will allow us to describe extremely large numbers. We recursively define $\uparrow^m$ to be of the form
$$x \uparrow^m n = \begin{cases} x & \text{if } n = 1, \\ x \uparrow^{m-1} (x \uparrow^m (n-1)) & \text{if } n > 1. \end{cases}$$

So the operators $\uparrow^m$ are defined in terms of the $\uparrow^{m-1}$ operators, and so forth. For instance, repeatedly applying this definition yields
$$\begin{aligned} x \uparrow^3 4 &= x \uparrow^2 (x \uparrow^3 3) \\ &= x \uparrow^2 (x \uparrow^2 (x \uparrow^3 2)) \\ &= x \uparrow^2 (x \uparrow^2 (x \uparrow^2 (x \uparrow^3 1))) \\ &= x \uparrow^2 (x \uparrow^2 (x \uparrow^2 x)). \end{aligned}$$

In other words, the notation $x \uparrow^m n$ yields
$$x \uparrow^m n = x \uparrow^{m-1} (x \uparrow^{m-1} (x \uparrow^{m-1} (\cdots (x \uparrow^{m-1} x) \cdots))),$$
where the right-hand side contains n times the term x with the up-arrow operator $\uparrow^{m-1}$ between them. We can then further express the operators $\uparrow^{m-1}$ in terms of $\uparrow^{m-2}$ operators, and so on.

We give a few small examples to illustrate how the iterated powers can be reduced to ordinary powers.

4.7 Well-founded Induction

Example 4.21. The iterated powers allow one to quickly form large numbers. For instance,

$$2 \uparrow^1 3 = 2^3 = 8,$$
$$2 \uparrow^2 3 = 2 \uparrow^1 (2 \uparrow^1 2) = 2^{2^2} = 16,$$
$$2 \uparrow^3 3 = 2 \uparrow^2 (2 \uparrow^2 2) = 2^{2^{2^2}} = 65{,}536,$$
$$2 \uparrow^4 3 = 2 \uparrow^3 (2 \uparrow^3 2) = 2 \uparrow^3 4,$$

where the last number is the unfathomably large iterated power

$$2^{2^{\cdot^{\cdot^{\cdot^2}}}}$$

consisting of 65,536 terms of 2s. For larger m, the repeated expansion of $\uparrow^m$ yields humongous numbers.

It will be convenient to extend the notation $\uparrow^m$ to exponents less than 1. We define the up-arrow $\uparrow^m$ for $m \in \{-2, -1, 0\}$ as
(a) $x \uparrow^0 n = xn$ (multiplication),
(b) $x \uparrow^{-1} n = x + n$ (addition),
(c) $x \uparrow^{-2} n = 1 + n$ (increment).

Lemma 4.22. *For all integers m and n such that $m \geq -1$ and $n \geq 2$, the up-arrow operator satisfies*

$$2 \uparrow^m n = 2 \uparrow^{m-1} (2 \uparrow^m (n-1)).$$

Proof. For $m \geq 2$, the claim follows directly from the definition of the up-arrow operator $\uparrow^m$.

For $m = 1, 0$, and -1, it follows from the definitions that

$$2 \uparrow^1 n = 2^n = 2 \cdot 2^{n-1} = 2 \uparrow^0 (2 \uparrow^1 (n-1)),$$
$$2 \uparrow^0 n = 2 \cdot n = 2 + 2(n-1) = 2 \uparrow^{-1} (2 \uparrow^0 (n-1)),$$
$$2 \uparrow^{-1} n = 2 + n = 1 + (2 + (n-1)) = 2 \uparrow^{-2} (2 \uparrow^{-1} (n-1)),$$

which proves the claim. □

Proposition 4.23. *The Ackermann function satisfies*

$$A(x, y) = 2 \uparrow^{x-2} (y+3) - 3.$$

for all nonnegative integer arguments x and y.

Proof. We will prove the claim by well-founded induction on the strict lexicographic order relation $\lhd$ given in Example 4.14.
Induction Basis For $x = 0$ and all nonnegative integers y, the Ackermann function satisfies
$$A(0, y) = y + 1 = 2 \uparrow^{-2} (y+3) - 3.$$

Inductive Step We will now show that $A(x,y) = 2 \uparrow^{x-2} (y+3) - 3$ holds for $x > 0$, assuming that the claim holds for all $A(x', y')$ with $(x', y') \lhd (x, y)$.

If $x > 0$ and $y = 0$, then by definition of the Ackermann function and the Induction Hypothesis,

$$A(x, 0) = A(x-1, 1) = 2 \uparrow^{x-3} 4 - 3.$$

Since $(2 \uparrow^m 2) = 4$ for all $m \geq -1$, we can rewrite $2 \uparrow^{x-3} 4$ as

$$2 \uparrow^{x-3} 4 = 2 \uparrow^{x-3} (2 \uparrow^{x-2} 2) = 2 \uparrow^{x-2} 3.$$

This allows us to conclude that

$$A(x, 0) = 2 \uparrow^{x-2} 3 - 3,$$

so the claim holds in this case as well.

If $x > 0$ and $y > 0$, then

$$\begin{aligned} A(x,y) &= A(x-1, A(x, y-1)) \quad \text{by definition} \\ &= 2 \uparrow^{x-3} (A(x, y-1) + 3) - 3 \quad \text{by Induction Hyp.} \\ &= 2 \uparrow^{x-3} (2 \uparrow^{x-2} (y+2)) - 3 \quad \text{by Induction Hyp.} \\ &= 2 \uparrow^{x-2} (y+3) - 3 \quad \text{by Lemma 4.22,} \end{aligned}$$

so the claim holds in this case as well.

By well-founded induction, we can conclude that

$$A(x, y) = 2 \uparrow^{x-2} (y+3) - 3$$

holds for all $(x, y) \in \mathbf{N}_0 \times \mathbf{N}_0$. □

EXERCISES

4.34. Let $\lhd$ be a well-founded relation on a set S. Show that
(a) $\lhd$ must be irreflexive, meaning that there cannot exist an element a in S such that $a \lhd a$.
(b) $\lhd$ must be asymmetric, meaning that $a \lhd b$ implies that $b \lhd a$ cannot hold.
(c) there cannot exist elements $a_1, a_2, \ldots a_n$ in S such that $a_k \lhd a_{k+1}$ for all k in the range $1 \leq k < n$ and $a_n \lhd a_1$.

4.35. Let S be an arbitrary nonempty set. Show that the "is an element of" relation $\in$ on the set S is a well-founded relation.

4.36. Prove or disprove: The relation $<$ on the set of integers is a well-founded relation.

4.37. Prove or disprove: The strict lexicographic ordering $\lhd$ on the set of finite strings over the alphabet $\{a, b\}$ is a well-founded relation. The strict lexicographic ordering is the usual alphabetical dictionary ordering. So we

4.7 Well-founded Induction

have $a \triangleleft b$. If the strings are of unequal length, we append blank spaces $\textvisiblespace$ to the shorter string, where we assume that $\textvisiblespace \triangleleft a$ and $\textvisiblespace \triangleleft b$. Given the strings $s_1 s_2 \cdots s_m$ and $t_1 t_2 \cdots t_n$, we have

$$s_1 s_2 \cdots s_m \triangleleft t_1 t_2 \cdots t_m$$

if and only if $s_1 = t_1, \ldots s_{k-1} = t_{k-1}$, and $s_k \triangleleft t_k$ for some integer k in the range $1 \leq k \leq m$. For example, $ab \triangleleft b$ and $b \triangleleft ba$, but $\neg(bb \triangleleft ba)$ and $\neg(bb \triangleleft bb)$.

4.38. Let $\triangleleft_1$ be a well-founded relation on a set A, and $\triangleleft_2$ a well-founded relation on a set B. Let the product relation $\triangleleft$ on $A \times B$ be given by $(a, b) \triangleleft (a', b')$ if and only if $a \triangleleft_1 a'$ and $b \triangleleft_2 b'$. Show that $\triangleleft$ is a well-founded relation on $A \times B$.

4.39. A relation $\ll$ on S is called transitive if and only if $a \ll b$ and $b \ll c$ implies $a \ll c$. Let $\triangleleft$ be a well-founded relation on a set S. Let $\triangleleft^+$ denote the smallest transitive relation on S containing $\triangleleft$. Show that $\triangleleft^+$ is a wellfounded relation on S as well. [Remark: The relation $\triangleleft^+$ is irreflexive and transitive, and such a relation is called a strict order. We will study strict orders in Sect. 6.2.]

4.40. Let $\triangleleft$ denote the successor relation $\triangleleft = \{(n, n+1) \mid n \in \mathbf{N}_1\}$ on the set of positive integers. Show that the smallest transitive relation $\triangleleft^+$ containing $\triangleleft$ is given by the usual strict inequality relation $<$ on the set of positive integers.

4.41. Let S and S' be sets and let $f: S \to S'$ be a function from S to S'. Suppose that $\triangleleft'$ is a well-founded relation on S'. Define a relation $\triangleleft$ on S by

$$a \triangleleft b \quad \text{if and only if} \quad f(a) \triangleleft' f(b).$$

Show that $\triangleleft$ is a well-founded relation on S.

4.42. Show that it is possible to formulate the principle of well-founded induction using a single condition (replacing the induction basis and inductive step):

For all x in S, show that $P(x)$ holds under the assumption that $P(a)$ holds for all a in S such that $a \triangleleft x$.

Then $P(x)$ holds for all $x \in S$.

Explain why this form of well-founded induction is equivalent to the well-founded induction principle stated in this section.

4.43. Let S be a set and $\triangleleft$ a well-founded relation on S. For an element s in S, let us call $\text{pred}(s) = \{t \in S \mid t \triangleleft s\}$ the set of predecessors of s.

Show that if T is a subset of S that contains an element s in S whenever it contains its set $\text{pred}(s)$ of predecessors, then T cannot be a proper subset of S, meaning that $T = S$.

4.44. Show that well-founded induction over the set of positive integers using the successor relation $\triangleleft$ from Example 4.12 is the familiar induction principle over the positive integers.

4.45. Show that well-founded induction over the set of positive integers using the strictly-less-than relation ◁ from Example 4.13 is the familiar strong induction principle over the positive integers.

4.46. Prove by induction that the Ackermann function satisfies
(a) $A(1, y) = y + 2$ for all $y \in \mathbf{N_0}$.
(b) $A(2, y) = 2(y + 3) - 3$ for all $y \in \mathbf{N_0}$.
(c) $A(3, y) = 2^{y+3} - 3$ for all $y \in \mathbf{N_0}$.
These are the formulas for the "small" values of the Ackermann function.

4.47. This exercise allows you to get familiar with Knuth's up-arrow notation.
(a) Determine $2 \uparrow^2 5$ and $5 \uparrow^2 2$.
(b) How many decimal digits are needed to express $3 \uparrow^2 3$?
(c) Which number is bigger $3 \uparrow^2 4$ or $4 \uparrow^2 3$?
(d) Show that $2 \uparrow^m 2 = 4$ holds for all integers m such that $m \geq -1$.

4.48. Use well-founded induction to show that the recursively defined function $f\colon \mathbf{N_0} \times \mathbf{N_0} \to \mathbf{N_0}$ given by

$$f(x, y) = \begin{cases} x & \text{if } y = 0, \\ y & \text{if } x = 0, \\ f(x-1, y-1) & \text{if } x > 0 \text{ and } y > 0. \end{cases}$$

satisfies

$$f(x, y) = |x - y|$$

for all nonnegative integers x and y. Use the following approach:
(a) Let ◁ be the relation in the set $\mathbf{N_0} \times \mathbf{N_0}$ given by

$$(a, b) \triangleleft (a+1, b+1)$$

for all nonnegative integers a and b. Show that ◁ is a well-founded relation.
(b) Determine the ◁-minimal elements of $\mathbf{N_0} \times \mathbf{N_0}$.
(c) Use well-founded induction with respect to ◁ to prove that $f(x, y) = |x-y|$ holds for all nonnegative integers x and y.

4.49. Show using well-founded induction that the function $f\colon \mathbf{Z} \to \mathbf{Z}$ given by

$$f(x) = \begin{cases} x - 10 & \text{if } x > 100, \\ f(f(x+11)) & \text{otherwise,} \end{cases}$$

and the function $g\colon \mathbf{Z} \to \mathbf{Z}$ given by

$$g(x) = \begin{cases} x - 10 & \text{if } x > 100, \\ 91 & \text{otherwise,} \end{cases}$$

satisfy

$$f(x) = g(x)$$

for all integer arguments x. Hint: Use the relation ◁ on the set $\mathbf{Z}$ of integers given by $a \triangleleft b$ if and only if $b < a \leq 101$.

4.50. Suppose that x and y are two nonnegative integers such that $x \geq y$ and that are not both equal to 0. Then the greatest common divisor of x and y can be calculated by the recursively defined function

$$\gcd(x,y) = \begin{cases} x & \text{if } y = 0, \\ \gcd(y, x \bmod y) & \text{if } y > 0. \end{cases}$$

Let $S = \{(x,y) \in \mathbf{N}_0 \times \mathbf{N}_0 \mid x \geq y \text{ and } (x,y) \neq (0,0)\}$.
(a) Show that the relation $\triangleleft$ on S given by

$$(x', y') \triangleleft (x, y) \quad \text{if and only if} \quad y' < y$$

is a wellfounded relation.
(b) Show that $\gcd(x,y)$ terminates in a finite number of steps using wellfounded induction.

*4.8 Recursion

It might be a good time to pause and reflect on some of the objects that we reasoned about in this chapter. We began this chapter by studying the square function that associates the square n^2 to a positive integer argument n. We realized that we need to add $2n - 1$ to the square $(n-1)^2$ to go to the next square n^2, as illustrated in Fig. 4.2.

Figure 4.2: Suppose that we are given an $(n-1) \times (n-1)$ square such as the 3×3 square shown above. We need to add $(n-1) + 1 + (n-1) = 2n - 1$ empty squares $(3 + 1 + 3 = 7$ in the example shown) to obtain an $n \times n$ square

So another definition of the square function s on the set of positive integers is given by

$$s(n) = \begin{cases} 1 & \text{if } n = 1, \\ s(n-1) + 2n - 1 & \text{if } n > 1. \end{cases}$$

The function $s(n)$ is recursively defined in terms of a function g that depends on n and previously defined terms (here $s(n-1)$), namely

$$g(n, s(n-1)) = s(n-1) + 2n - 1.$$

The function g shows how to compute $s(n)$ from the previous term $s(n-1)$.
The reason why we expect that the recursive structure of $s(n)$ defines a well-defined function is that it is defined in terms of "previous" terms with smaller

arguments. Applying the recurrence $s(n) = s(n-1) + 2n - 1$ repeatedly yields

$$s(n) = s(n-1) + 2n - 1$$
$$= s(n-2) + 2n - 3 + 2n - 1,$$

and so forth. Eventually, it reaches the base case $n = 1$ that directly defines $s(1)$, without relying on other terms.

As a caveat, notice that we did *not* define the recursion as

$$w(n) = w(n+1) - (2n+1),$$

with base case $w(1) = 1$. Even though $n^2 = (n+1)^2 - (2n+1)$, we need to formulate the recursive step so that the recursion terminates in a finite number of steps. Evidently, $w(2) = w(3) - 5 = w(4) - 5 - 7$. Further repeated applications of the recursive step yields larger and larger arguments. The problem with this approach is that a base case is not reached in a finite number of steps when $w(n)$ is called with an argument $n > 1$.

So how can we establish that s is defined in a sensible way? If someone is unaware of the motivation that gave rise to the recursive definition of $s(n)$, then the equality $s(n) = n^2$ might come as a surprise. Yet, verifying that this equality holds is not difficult. The functions $s(n)$ and n^2 are both equal to 1 for the base case $n = 1$. For $n > 1$, they both satisfy the same recursive relation,

$$s(n) = s(n-1) + 2n - 1,$$

and

$$n^2 = (n-1)^2 + 2n - 1.$$

The equality $s(n) = n^2$ can be established using induction. Better yet, the next theorem obviates the need for such a proof. It asserts that two functions satisfying the same base case and the same recursive relation must be the same, as long as the recursive step merely uses the previous term.

Theorem 4.24 (Recursion Theorem). *Let S be a nonempty set, s an element of S, and $g\colon \mathbf{N}_0 \times S \to \mathbf{N}_0$ a function. Then there exists a unique function $f\colon \mathbf{N}_0 \to S$ satisfying*
(a) $f(0) = s$ *(the base case)*,
(b) $f(n) = g(n, f(n-1))$ *(the recursive step)*.

We recommend that you solve Exercises 4.53–4.55, so that you get familiar with applying the Recursion Theorem. We will forgo the proof of the Recursion Theorem, since it is a special case of a more general result that we will prove below.

The aforementioned Recursion Theorem is somewhat limited in scope, as $f(n)$ can only depend on n and a single previous function value, namely $f(n-1)$. So we cannot apply it to the Fibonacci numbers that depend on two previous

4.8 Recursion

values or to the Ackermann function that depends on two parameters. We will prove a General Recursion Theorem that relies on a well-founded relation to ensure that the recursion terminates in a finite number of steps.

Let $\triangleleft$ be a well-founded relation on a set S. The **initial segment** $\operatorname{seg} x$ of an element x from S is defined as

$$\operatorname{seg} x = \{s \in S \mid s \triangleleft x\}.$$

The initial segment comprises all elements that precede x. The initial segment $\operatorname{seg} x$ is a finite set.

In well-founded recursion, the function value $f(x)$ depends on x and the finite number of function values $f \upharpoonright \operatorname{seg} x = \{f(s) \mid s \in \operatorname{seg} x\}$.

Let $\operatorname{seq} S = \bigcup_{n \in \mathbf{N}_0} S^n$ denote the set of finite sequences over S.

Theorem 4.25 (The General Recursion Theorem). *Let $\triangleleft$ be a well-founded relation on the set S and $g \colon S \times \operatorname{seq} S \to S$. Then there is a unique function f on S such that*

$$f(x) = g(x, f \upharpoonright \operatorname{seg} x),$$

for all $x \in S$.

Proof. The main idea of the proof is to approximate f using functions with potentially smaller domains than f.

We say that a function h is *admissible* if and only if $\operatorname{dom} h \subseteq S$, the domain of h is closed in the sense that if $x \in \operatorname{dom} h$, then $\operatorname{seg} x \subseteq \operatorname{dom} h$, and

$$h(x) = g(x, h \upharpoonright \operatorname{seg} x)$$

holds for all x in $\operatorname{dom} h$. Thus, an admissible function follows the same recursive step and has the same base case values as f, as long as x is in the domain of h.

Let $\mathcal{A}$ denote the set of all admissible functions.

Suppose that h_1 and h_2 are two admissible functions in $\mathcal{A}$. We will now show that these two functions agree for all arguments in their shared domain $\operatorname{dom} h_1 \cap \operatorname{dom} h_2$. Seeking a contradiction, let us suppose that the set

$$C = \{x \in \operatorname{dom} h_1 \cap \operatorname{dom} h_2 \mid h_1(x) \neq h_2(x)\}$$

of contradictory values is not empty. Let m be a $\triangleleft$-minimal element of C. Then $s \triangleleft m$ implies that s is in the domain of both h_1 and h_2, as both functions are admissible. Furthermore, $h_1(s) = h_2(s)$, as s does not belong to C. Put differently, the values of h_1 and h_2 coincide on $\operatorname{seg} x$. Therefore, we can conclude that

$$h_1(x) = g(x, h_1 \upharpoonright \operatorname{seg} x) = g(x, h_2 \upharpoonright \operatorname{seg} x) = h_2(x),$$

which contradicts that x is an element of C.

Viewing the elements of $\mathcal{A}$ as sets of argument/value pairs, we can form the union

$$f = \bigcup \mathcal{A}.$$

The resulting relation is a function with domain

$$\operatorname{dom} f = \bigcup_{h \in \mathcal{A}} \operatorname{dom} h.$$

If x is in the domain of f, then it must be in the domain of some admissible function h in $\mathcal{A}$. Therefore,

$$f(x) = h(x) = g(x, h \upharpoonright \operatorname{seg} x) = g(x, f \upharpoonright \operatorname{seg} x).$$

It remains to show that $\operatorname{dom} f = S$. Seeking a contradiction, suppose that $\operatorname{dom} f$ is a proper subset of S. Let m be a $\triangleleft$-minimal element of $S - \operatorname{dom} f$. So $\operatorname{seg} m \subseteq \operatorname{dom} f$. Then $D = \operatorname{dom} f \cup \{m\}$ is closed, meaning that the initial segment $\operatorname{seg} x \subseteq D$ for all $x \in D$. We can define a function $h \colon D \to S$ by

$$h(x) = \begin{cases} f(x) & \text{if } x \in \operatorname{dom} f, \\ g(m, f \upharpoonright \operatorname{seg} m) & \text{if } x = m. \end{cases}$$

Evidently, $h \in \mathcal{A}$, so $h \subseteq f$, and $m \in \operatorname{dom} f$, which contradicts the choice of m. Therefore, we can conclude that $\operatorname{dom} f = S$.

One can show the uniqueness of f using the same argument that we used to show that h_1 and h_2 take on the same values. $\square$

Since two different elements of S may have the same initial segment, we cannot omit the argument x from the function g, as it is sometimes erroneously done in the literature. Apart from x, the function g can depend on the value of *all* terms $f(s)$ with s in $\operatorname{seg} x$.

EXERCISES

4.51. Repeatedly use the recursive definition of the square function $s(n)$ to compute $s(6)$. Confirm by direct calculation that $s(6) = 6^2$.

4.52. The function $f \colon \mathbf{N}_1 \to \mathbf{N}_1$ is recursively defined by

$$f(n) = \begin{cases} 1 & \text{if } n = 1, \\ 1 - f(n-1) & \text{if } n > 1. \end{cases}$$

Describe the function f without recursion.

4.53. The function $f \colon \mathbf{N}_1 \to \mathbf{N}_1$ is recursively defined by

$$f(n) = \begin{cases} 1 & \text{if } n = 1, \\ nf(n-1) & \text{if } n > 1. \end{cases}$$

Describe the function f without recursion.

4.8 Recursion

4.54. The function $f\colon \mathbf{N}_0 \to \mathbf{N}_0$ is recursively defined by

$$f(n) = \begin{cases} 1 & \text{if } n = 0, \\ 2f(n-1) & \text{if } n > 0. \end{cases}$$

Describe the function f without recursion and prove equality using the Recursion Theorem.

4.55. Show that the function $f\colon \mathbf{N}_0 \to \mathbf{N}_0$ given by

$$f(n) = \begin{cases} 0 & \text{if } n = 0, \\ f(n-1) + 2n & \text{if } n > 0. \end{cases}$$

defines the oblong numbers $f(n) = n^2 + n$ for all nonnegative integers n. Prove your result using the Recursion Theorem.

4.56. Give a self-contained proof of the Recursion Theorem (Theorem 4.24) without appealing to Theorem 4.25.

4.57. The function $f\colon \mathbf{N}_0 \to \mathbf{N}_0$ is recursively defined by

$$f(n) = \begin{cases} 1 & \text{if } n = 0, \\ 2 & \text{if } n = 1, \\ f(n-1) + 2f(n-2) & \text{if } n > 1. \end{cases}$$

Describe the function f without recursion and prove equality using the General Recursion Theorem.

4.58. The function $f\colon \mathbf{N}_0 \to \mathbf{Z}$ is recursively defined by

$$f(n) = \begin{cases} -4 & \text{if } n = 0 \text{ or } 1, \\ -3f(n-1) + 4f(n-2) & \text{if } n > 1. \end{cases}$$

Describe the function f without recursion and prove equality using the General Recursion Theorem.

4.59. Show that the function $f\colon \mathbf{N}_0 \to \mathbf{N}_0$ given by

$$f(n) = \begin{cases} 0 & \text{if } n = 0, \\ 1 & \text{if } n = 1, \\ -1 & \text{if } n = 2, \\ -f(n-1) + f(n-2) + f(n-3) & \text{if } n \geqslant 3. \end{cases}$$

Describe the function f without recursion and prove your result using the General Recursion Theorem.

*4.9 Recursively Defined Sets

In this section, we will detail how to define a set using recursion. This approach to defining sets is widely used in computer science. We will also show how to prove properties of such recursively defined sets using a proof principle that is known as structural induction.

Motivation. Let U be a set that we will call the universe. We can specify a subset S of the universe U using enumeration or set builder notation. Enumeration confines us to the definition of finite sets and set builder notation sometimes requires intricate predicates to specify the subset. Alternatively, we have the option to define a subset S of the universe U by recursion. In this case, we specify
(a) one or more elements of S using a base set,
(b) and one or more rules to construct new elements of S given some known elements.

The construction of S is anchored by the explicitly given elements that belong to the base base. The remaining elements of S are constructed by repeatedly applying the recursive rules mentioned in part (b). Evidently, we need to make the recursive part (b) a bit more precise.

We have seen this way of construction already when defining the set of nonnegative integers. Simplifying a bit, we defined the element 0 as belonging to the base set, and the successor function that allowed us to construct from a known element n a new element $f(n) = n + 1$.

Our first example is a variation of this construction.

Example 4.26. Let us consider as the universe U the set $\mathbf{N}_0$ of nonnegative integers. We can define a subset S of $\mathbf{N}_0$ by asserting that
(a) the number 3 is an element of S (so $B = \{3\}$ is our base set),
(b) If n is an element of S, then $f(n) = n + 3$ is an element of S. This construction rule is the recursive part of the definition, and the function $f(n) = n + 3$ is called a constructor of the set S.

Which set S is described by these rules? We know that 3 is an element of S by part (a). Since $n = 3$ is an element of S, we can deduce that $n + 3 = 6$ is an element of S as well due to the rule (b). By repeated applications of the rule (b), we can infer that $9, 12, 15$, and so on are also elements of S. In fact, we claim that the subset S described by these rules[1] is given by the set of positive integers that are multiples of 3, so $S = \{3m \mid m \in \mathbf{N}_1\}$.

In this example, the constructor $f(n) = n + 3$ depends on a single argument. In the general case, a constructor can depend on more than one argument, but always on at most a finite number of arguments. So another constructor could be the function $f(m, n) = m + 2n$ that "constructs" a new element $m + 2n$ of the set given two known elements m and n.

---
[1] At least when properly interpreted as the smallest possible set satisfying these rules.

4.9 Recursively Defined Sets

Recursively Defined Sets. A recursively defined set S is defined as a subset of a universe U using explicitly given elements of the set S that belong to a nonempty subset B of the universe U that is known as the **base set**. The remaining elements of S are defined by repeatedly applying constructors from a finite set C of constructors.

A **constructor** f in C is a function $f\colon U^n \to U$ that relates n known elements to one or more new elements in the set, where the parameter n is a positive integer that depends on the function f. We call n the **arity** of the constructor f.

The subset S of the universe U **recursively defined** (or **inductively defined**) by the base set B and the set C of constructors is the *smallest subset* of the universe U such that

R1. the base set B is a subset of S,

R2. If $x_1, x_2, \ldots, x_n$ are elements in S and f is an n-ary constructor in C, then $f(x_1, x_2, \ldots, x_n) \in S$.

The rule **R2** can also be expressed more compactly as $f(S^n) \subseteq S$.

Let us now rigorously show the existence of the recursively defined set S. There are many subsets S' of the universe U that satisfy the conditions **R1** and **R2** if we ignore the stipulation that S' is supposed to be the smallest such set. For instance, the universe U itself satisfies the rules **R1** and **R2**. Consider the family

$$\mathcal{F} = \{S' \in P(U) \mid S' \text{ satisfies } \mathbf{R1} \text{ and } \mathbf{R2}\}$$

of all subsets S' of the universe U satisfying the conditions **R1** and **R2**. Then their intersection

$$S = \bigcap \mathcal{F}$$

is a set satisfying **R1** and **R2**, so the set S itself belongs to the family $\mathcal{F}$. Since S is a subset of every set in the family $\mathcal{F}$, it is clearly the smallest possible set satisfying **R1** and **R2**. In particular, this means that S does not contain any extraneous elements that are not implied by the defining rules **R1** and **R2**.

There is also a more constructive way to obtain the recursively defined set S. We define level sets L_k for nonnegative integers k as follows. Let

$$L_0 = B$$

be the base set. For any nonnegative integer k, let

$$L_{k+1} = L_k \cup \bigcup_{f \in C} f((L_k)^n).$$

The set L_{k+1} contains all elements that are obtainable from the base set by $k+1$ or fewer applications of constructors. We claim that

$$S = \bigcup_{k \geq 0} L_k.$$

Indeed, the righthand side contains B, so it satisfies **R1**. Since $f(L_k) \subseteq L_{k+1}$ holds for all constructors f in C, we can deduce that the right-hand side satisfies **R2** as well. So the right-hand side belongs to the family $\mathcal{F}$. As each set in the family $\mathcal{F}$ contains L_k for all nonnegative k, we can conclude that the righthand side must be contained in each set of the family $\mathcal{F}$. Therefore, the set $\bigcup_{k \geq 0} L_k$ must be equal to S.

Examples. We will now give a few more examples of recursively defined sets. One of the strengths of recursive definitions is that it is possible to describe sets that are otherwise more difficult to specify. The next example illustrates this point.

Example 4.27. If the universe U is given by the set $\mathbf{Z}$ of integers, the base set $B = \{2\}$ contains the single element 2, and the set $C = \{f, g\}$ of constructors contains the functions $f(x) = 2x$ and $g(x) = x^3 - 1$. The set S defined by the base set B and the set C of constructors contains, for instance, the elements

- 2, as it is an element of $B \subseteq S$,
- 4, since $f(2) = 4$,
- 7, since $g(2) = 2^3 - 1 = 7$,
- 8, since $f(f(2)) = f(4) = 8$,
- 16, since $f(f(f(2))) = f(8) = 16$,
- 32, since $f(16) = 32$,
- 63, since $g(f(2)) = g(4) = 4^3 - 1 = 63$,
- 64, since $f(32) = 64$.

Evidently, S contains all integers of the form 2^n with $n \geq 2$, as repeated application of the constructor f shows. When both constructors are used, the results are not straightforward to predict. In fact, we are not aware of any simple nonrecursive description of the set S.

There are many ways to recursively define a set. The next example recursively defines a set, but arguably not in the most straightforward way.

Example 4.28. If the universe U is given by the set $\mathbf{Z}$ of integers, the base set $B = \{0\}$, and the set $C = \{f_4, f_{-6}\}$ of constructors contains the functions $f_4(n) = n + 4$ and $f_{-6}(n) = n - 6$. Since 0 is in the base set B, it is contained in S. As 0 is contained in S, it follows that $f_4(0) = 0 + 4 = 4$ is in S. Applying the constructor f_4 to 4 shows that $8 \in S$. Since $f_{-6}(8) = 2$, we can conclude that $2 \in S$. We claim that the smallest subset S of the universe $\mathbf{Z}$ defined by the base set B and the set $C = \{f_4, f_{-6}\}$ of constructors is given by the set of even integers, $S = \{2n \mid n \in \mathbf{Z}\}$.

Example 4.29. Let $U = \{\mathsf{a}, \mathsf{b}\}^*$ denote the universe consisting of the set of strings of finite length over the alphabet $\{\mathsf{a}, \mathsf{b}\}$. So U contains strings such as a, b, ab, ba, aab, and so on. The empty string is also contained in the universe U. We will denote the empty string[2] by λ. We denote the concatenation of

---

[2] The empty string is commonly denoted by ε or λ in the literature. In programming languages, strings are often enclosed in double quotes, so "" denotes the empty string.

4.9 Recursively Defined Sets

strings by juxtaposition. So ab concatenated with aab yields abaab. For any string $x \in U$, concatenation with the empty string yields $x\lambda = x = \lambda x$.

We can recursively define a subset S of the universe $U = \{a, b\}^*$ by the base set $B = \{\lambda\}$ containing the empty string, and set $C = \{f, g, h\}$ containing the three constructors

- $f(x) = axb$,
- $g(x) = bxa$,
- $h(x, y) = xy$,

where the constructors f and g concatenate the given string x with a and b in the stated way. The constructor $h(x, y)$ concatenates the two given strings x and y.

Since the empty string λ is contained in the base set $B \subseteq S$, and $f(\lambda) = a\lambda b = ab$), it follows that $ab \in S$. Similarly, g applied to λ shows that ba is contained in S. Applying h to $x = $ ba and $y = $ ba shows that $h(x, y) = $ baba is contained in S. Applying h to $x = $ ab and $y = $ ba shows that $h(x, y) = $ abba $\in S$. One can show that the set S consists of all strings in the universe U that have the same number of a's and b's.

Structural Induction. Proving that a property $P(x)$ holds for all elements x of a recursively defined set S can be done using weak or strong induction. An alternative way to prove such properties is by structural induction.

Theorem 4.30 (The Structural Induction Principle). *Let U be a set called the universe. Let S be a subset of U that is recursively defined by a base set B and a set C of constructors. Suppose that*
(a) (Induction Basis) $P(b)$ holds for all elements b in the base set B,
(b) (Induction Step) for all constructors $f \in C$, if the property P holds for $x_1, x_2, \ldots, x_n$ in S, and $y = f(x_1, x_2, \ldots, x_n)$, then $P(y)$ holds as well.
Then $P(x)$ holds for all elements x in S.

Proof. Let $M = \{x \in S \mid P(x) \text{ holds }\}$ be the subset of elements of S for which $P(x)$ holds. By the induction basis, $P(b)$ holds for all $b \in B$, so M satisfies **R1**. By the induction step, M contains the value $f(x_1, x_2, \ldots, x_n)$ of any constructor f, given that it contains the arguments $x_1, x_2, \ldots, x_n$. Therefore, M satisfies **R2**. It follows that M must contain S. We can conclude that S and M must be equal, which shows that $P(x)$ holds for all $x \in S$. □

The proof strategy of structural induction is to show that every element in the base set satisfies the property P. In the induction step, one must show that every constructor preserves the truth in the sense that if all arguments satisfy the property P, then the element created by the constructor must satisfy P as well. It should be stressed that the induction step must make the argument for *every* constructor in the set C of constructors.

The next example illustrates the structural induction proof principle.

Example 4.31. Let $U = \mathbf{Z} \times \mathbf{Z}$ be the set consisting of pairs of integers. We define S to be the smallest subset of the universe U with base set $B = $

$\{(3, 14), (6, 11), (7, 10)\}$ and set $C = \{f, g\}$ of constructors given by $f((a, b)) = (a + 4, b + 13)$ and $g((a, b)) = (a + 22, b + 12)$. We claim that all pairs $(a, b) \in S$ have a sum $a + b$ that is divisible by 17.

We prove this claim by structural induction.

Induction Basis The pairs $(3, 14)$, $(6, 11)$, and $(7, 10)$ all sum to 17, as $3 + 14 = 6 + 11 = 7 + 10 = 17$. Therefore, the claim holds for all element in the base set B.

Induction Step For the induction hypothesis, let (a, b) be a pair in S such that $a + b$ is divisible by 17, say $a + b = 17m$ for some integer m. Then

- the pair $f((a, b)) = (a + 4, b + 13)$ has the sum $(a + 4) + (b + 13) = a + b + 17 = 17(m + 1)$, which is divisible by 17,

- the pair $g((a, b)) = (a + 22, b + 12)$ has the sum $(a + 22) + (b + 12) = (a + b) + 34 = 17(m + 2)$, which is divisible by 17.

We can conclude by structural induction that all pairs (a, b) in S have a sum $a + b$ that is divisible by 17.

Example 4.32. Let S be the smallest set of strings recursively defined by the base set $B = \mathbf{N_0} \cup \{x\}$ and set $C = \{f, g\}$ of constructors given by $f(y, z) = (y+z)$ and $g(y, z) = (yz)$. We claim that all formulas in S have the same number of left and right parentheses.

We prove this claim by structural induction.

Induction Basis Since the base set B contains only numbers and a variable, there are no parentheses. So each element has 0 left parentheses and 0 right parentheses.

Induction Step Suppose that y is an element of S that contains m left and m right parentheses, and z an element of S that contains n left and n right parentheses. Then

- $f(y, z) = (y+z)$ contains $m + n + 1$ left and $m + n + 1$ right parentheses,

- $g(y, z) = (yz)$ contains $m + n + 1$ left and $m + n + 1$ right parentheses.

So each constructor preserves the balance of left and right parentheses.

We can conclude by structural induction that all formulas in S contain an equal number of left and right parentheses.

Freely Generated Sets. Our next goal is to define functions on a recursively defined set. Before we deal with that topic, it might be worthwhile to point out that the elements of a recursively defined set are not necessarily created in a unique way, as the next example shows.

Example 4.33. In Example 4.28, we defined the set of even integers using a base set $B = \{0\}$ and constructors $f_4(x) = x + 4$ and $f_{-6}(x) = x - 6$. Applying the constructor f_4 three times and then the constructor f_{-6} twice yields

$$f_{-6}(f_{-6}(f_4(f_4(f_4(x))))) = x.$$

4.9 Recursively Defined Sets

So any element in the set S is produced not just once, but actually infinitely often during the recursive construction! For instance, 0 is already contained in the base set B, but it is also created by the following sequence of constructors

$$f_{-6}(f_{-6}(f_4(f_4(f_4(0))))) = 0.$$

In fact, any sequence consisting of $2m$ constructors f_{-6} and $3m$ constructors f_4 will create 0.

We can create the set of even integers more economically. The next example shows that only a single constructor is needed. More crucially, every element in the set of even integers is created in a unique way in this example.

Example 4.34. Let $U = \mathbf{Z}$ be the universe of integers. Let S be recursively defined by the base set $B = \{0\}$ and the set $C = \{h\}$ consisting of the single constructor $h\colon \mathbf{Z} \to \mathbf{Z}$ given by

$$h(n) = \begin{cases} -n+2 & \text{if } n \leqslant 0, \\ -n & \text{if } n > 0. \end{cases}$$

In this case, the set S is created by repeatedly applying the constructor h, so

$$\{0, h(0), h(h(0)), h(h(h(0))), h(h(h(h(0)))), \ldots\} = \{0, 2, -2, 4, -4, \ldots\}.$$

It is not difficult to see that S is the set of even integers, as $h^{(2m)}(0) = -2m$ and $h^{(2m+1)}(0) = 2m+2$, where m is a nonnegative integer and $h^{(k)}$ means that h is composed with itself k times.

Let S be a set that is recursively defined by a base set B and a set C of constructors. The recursive definition of S is called **free** if and only if for every element x in S, either x belongs to the base set B or there is just one constructor f and one n-tuple of arguments $x_1, x_2, \ldots, x_n$ such that $x = f(x_1, x_2, \ldots, x_n)$.

Equivalently, a recursive definition of S is free if and only if the base set B and the sets $f(S^n)$ for all constructors f in C are pairwise disjoint, and the restriction $f \upharpoonright S^n$ of each constructor to arguments in S is an injective function.

The key feature of a free recursive definition is that each element is constructed in an essentially unique way. If an element x of S belongs to the base set, then this element cannot be produced by a constructor. If x is the value of a constructor f, then no other constructor can produce it, and there is just one tuple of arguments of f that yields x.

Example 4.35. In the previous example, the constructor $h \upharpoonright 2\mathbf{Z}$ is an injective function, and $h(2\mathbf{Z}) = 2\mathbf{Z}\setminus\{0\}$ has an image that is disjoint from $B = \{0\}$. Therefore, the previous example gives a free recursive definition of the set of even integers.

Functions. If we are given a recursively defined set S, then it is tempting to also recursively define functions on S. Naturally, one would like to exploit the recursive structure of the domain S in the recursive definition of the function as well. It turns out that some care is needed to ensure the success of this approach.

Let U denote the universe of argument values. In addition, let V denote the universe of function values. Let S be a subset of the universe U that is recursively defined by a base set B and a set C of constructors. Suppose that we want to recursively define a function $g \colon S \to V$.

We can proceed as follows. We map each element x in the base set B to an element $F(x)$ of V, and each n-ary constructor $f \colon U^n \to U$ to a function $F(f) \colon V^n \to V$. Then g is defined by

T1. $g(y) = F(y)$ for all y in the base set B

T2. If f is an n-ary constructor in C, and $x_1, x_2, \ldots, x_n$ are elements in S with value $y = f(x_1, x_2, \ldots, x_n)$, then $g(y) = F(f)(g(x_1), g(x_2), \ldots, g(x_n))$.

The rules **T1** and **T2** guarantee that for each element y in the domain S at least one value $g(y)$ is defined. However, in general, these two rules do not guarantee that g is a well-defined function, meaning that the same value is assigned for every possible way to construct y. Proving that a function is well-defined on S can be a nuisance. We do not run into this issue when the recursive definition of S is free, as the next proposition shows.

Proposition 4.36. *If the recursive definition of a set S with base set B and constructor set C is free, then defining $g \colon S \to V$ through the rules **T1** and **T2** yields a well-defined function.*

Proof. This can be shown by structural induction, see Exercise 4.72. This is not a difficult exercise, but it is instructive. □

Example 4.37. Let $U = A^*$ be the universe of strings over the alphabet $A = \{0, 1, \mathrm{x}, (,), +, *\}$. Let S be the subset of U that is recursively defined by the base set $B = \{0, 1, \mathrm{x}\}$ and the set of constructors $C = \{a, m\}$ given by

$$a(e_1, e_2) = (e_1 + e_2),$$

and

$$m(e_1, e_2) = (e_1 * e_2).$$

Therefore, S consists of strings that contain fully parenthesized arithmetic expressions such as $((1 + \mathrm{x}) * \mathrm{x})$.

When restricted to S, the constructors a and m are injective functions, and their range are disjoint sets of strings. As each string in the image of the constructors a and m contain parentheses, they are also disjoint from the base set. It follows that S is freely generated by the base set B and this set C of constructors.

4.9 Recursively Defined Sets

The strings in S do not have a value. However, we can recursively define a function $g: S \to \mathbf{N}_0$ that evaluates these expressions for instance by

$$F(0) = 0, \quad F(1) = 1, \quad F(\mathsf{x}) = 7,$$

and assigning to the constructor a the usual addition function on $\mathbf{N}_0$, and to the constructor m the usual multiplication function on $\mathbf{N}_0$. Then the function g defined by the rules **T1** and **T2** yields for instance

$$g(\,((1+\mathsf{x})*\mathsf{x})\,) = g(\,(1+\mathsf{x})\,)g(\,\mathsf{x}\,) = (g(1)+g(\mathsf{x}))g(\,\mathsf{x}\,),$$

which evaluates to

$$g(\,((1+\mathsf{x})*\mathsf{x})\,) = (1+7)7 = 56.$$

This value of g is uniquely defined, as the underlying set S is given by a free recursive definition.

We have already seen recursively defined functions similar to the evaluation function g in the previous example, namely the valuation function v in proposition logic, see Sect. 2.3. Many other examples can be found in the computer science literature.

Even though it is often quite routine to verify that the recursive definition of S is free, it should not be omitted when a function g is recursively defined on S. If S is not defined by a free recursive definition, then we cannot rely on Proposition 4.36 to guarantee that g is well-defined. Unfortunately, the argument that the set S has a free recursive definition is often omitted in the literature.

EXERCISES

4.60. Using the set of real numbers as a universe, formulate a recursive definition of the set of integers.

4.61. Let U be the set of integers, and smallest subset S of U that is recursively defined by the base set $B = \{1\}$ and the set $C = \{f\}$ containing the constructor $f(x) = 2x$. (a) Derive the five smallest elements of the set S. For each element, show how it is obtained using the element in the base set and the constructor f. (b) Explicitly describe the set S. You should explain how to obtain each element in S.

4.62. Recursively define the set $S = \{n^2 \mid n \in \mathbf{N}_1\}$ as a subset of the universe $U = \mathbf{Z}$.

4.63. Recursively define the set S of palindromic strings over the alphabet $\{\mathsf{a},\mathsf{b}\}$ as a subset of the universe $U = \{\mathsf{a},\mathsf{b}\}^*$.

4.64. Recursively define the set S of strings over the alphabet $\{\mathsf{a},\mathsf{b}\}$ such that no letter b occurs before a letter a.

4.65. Let U be the universe $U = \{(,)\}^*$. Let S denote the smallest subset of U containing the base set $B = \{\lambda\}$ consisting of the empty string, and closed under the set $C = \{f\}$ of constructors, where $f(x) = (x)$. Show by structural induction that all strings in S contain an equal number of left and right parentheses.

4.66. Let $U = \{a,b\}^*$ denote the set of strings over the alphabet $\{a,b\}$. Let S denote the subset of U that is recursively defined by the base set $B = \{bb\}$ and the set $C = \{f,g\}$ of constructors $f(x) = aax$ and $g(x) = bxb$. Prove by structural induction that every string in S ends in bb.

4.67. Let S be the smallest subset of the universe $U = \{a,b\}$ that is recursively defined by the base set $B = \{\lambda\}$ containing the empty string, and the set $C = \{f,g\}$ of constructors such that $f(x) = aaxb$ and $g(x) = baxa$. Show by structural induction that each string in S contains twice as many letters a than letters b.

4.68. Let S be the smallest subset of the universe $U = \mathbf{Z} \times \mathbf{Z}$ that is recursively defined by the base set $B = \{(0,0)\}$ and the set $C = \{f,g\}$ of constructors given by $f((x,y)) = (x+1, y)$ and $g((x,y)) = (x+1, y-1)$. Prove by structural induction that every $(x,y) \in S$ satisfies $x \geq -y$.

4.69. (a) Show by structural induction that every element in the recursively defined set S in Example 4.28 is even. (b) Show that every even integer can be constructed by applying the constructors f_4 and f_{-6} a finite number of times to the element of the base set.

4.70. Bob wants to draw his ancestor tree, so he wants to start with himself, two branches to his parents, and then branches to their parents (his grandparents), and so on, as long as the ancestors are known. Show by structural induction that an ancestor tree for g generations contains at most $2^g - 1$ persons. [Hint: The ancestor tree starting at Bob's mother might not have the same number of generations than the ancestor tree starting at Bob's father. However, Bob does not include an ancestor unless he also has the information about the spouse of the ancestor.]

4.71. Show that every proof by structural induction can be replaced by a proof by strong induction. [Hint: Introduce a suitable notion of depth for the elements of the recursively defined set.]

4.72. Show by structural induction that if the recursive definition of a set S with base set B and constructor set C is free, then defining $g\colon S \to V$ through the rules **T1** and **T2** yields a well-defined function. [Hint: For the induction basis, argue that no other value than the one given by **T1** can be assigned to an argument in the base set B. For the induction step, argue that if $y = f(x_1, x_2, \ldots, x_n)$ and $g(y)$ is the value given in **T2**, then no other value can be assigned to y. The freeness of the recursive definition is the key in these arguments.]

4.10 Notes

The books by Eccles [23] and Velleman [77] contain more information about proofs by induction. This proof technique is ubiquitous and is used in almost every textbook of mathematics. Andreescu and Crişan [6] wrote an entire book dedicated to induction proofs that contains numerous challenging problems. Well-founded induction and structural induction are discussed in Loeckx and Sieber [58]. They also show how to use these methods in computer science for program verification. Manna [59] is another good source for applications of well-founded induction in computer science.

One should note that there are different definitions of structural induction in the literature, which can cause some confusion. Burstall introduced structural induction as a special case of well-founded induction. However, now it is common to use a more relaxed definition of structural induction that is different from Burstall's version.

Chapter 5

Equivalence Relations

> *Equivalence relations are so ubiquitous in everyday life that we often forget about their proactive existence.*
>
> — T. Britz, M. Mainetti, L. Pezzoli

The elements of a set may differ in many ways. Sometimes we want to single out some of their properties or attributes. An equivalence relation allows one to relate and "identify" elements that have the same properties. For instance, the relation "has the same model year" is an equivalence relation on the set of cars. In this chapter, we derive the most common properties of equivalence relations and give some applications.

5.1 Generalities

Let S be a set. An **equivalence relation** on the set S is a reflexive, symmetric, and transitive relation[1] $\sim$ on S. In other words, $\sim$ is an equivalence relation if and only if it satisfies the following three properties:

E1. For all x in S, we have $x \sim x$ (reflexivity).

E2. For all x, y in S, if $x \sim y$, then $y \sim x$ (symmetry).

E3. For all x, y, z in S, if $x \sim y$ and $y \sim z$, then $x \sim z$ (transitivity).

An equivalence relation allows us to describe what elements of a set we consider to be "essentially the same," so it broadens the concept of an equality.

Example 5.1. The identity relation $=$ on a set S is an equivalence relation. It is given by the set of pairs $\{(x,x) \mid x \in S\}$.

---

[1]You might recall that we introduced these notions in Chap. 3. In this chapter, we will study this important trio of properties in more detail.

© The Author(s), under exclusive license to Springer Nature Switzerland AG 2025
A. Klappenecker, H. Lee, *Discrete Structures*, Undergraduate Texts in Mathematics, https://doi.org/10.1007/978-3-031-73434-2_5

Example 5.2. The relation "has the same birthday as" is an equivalence relation on the set of people.

Example 5.3. The relation "has the same absolute value" on the set of real numbers is an equivalence relation.

Example 5.4. Let S be the set of strings over an alphabet A. Then the relation "has the same string length as" is an equivalence relation on the set S of strings.

Example 5.5. Let S and T be nonempty sets and $f \colon S \to T$. Then the relation $\sim$ on S defined by $x \sim y$ if and only if $f(x) = f(y)$ is an equivalence relation. All previous examples are special cases of this example.

Let us verify that $\sim$ is indeed an equivalence relation. Since $f(x) = f(x)$ holds for all x in S, we can deduce that $x \sim x$ for all x in S; thus, the relation $\sim$ is reflexive. If $x \sim y$, then $f(x) = f(y)$, so $f(y) = f(x)$, which implies $y \sim x$; therefore, the relation $\sim$ is symmetric. Finally, if $x \sim y$ and $y \sim z$, then $f(x) = f(y)$ and $f(y) = f(z)$, which implies $f(x) = f(z)$, whence $x \sim z$; thus, the relation $\sim$ is transitive.

For an element x in a set S, we can define its **equivalence class** $[x]$ under an equivalence relation $\sim$ as the set

$$[x] = \{y \in S \mid x \sim y\}.$$

In other words, $[x]$ is the set of all elements in S that are equivalent to x.

Lemma 5.6. *Two equivalence classes are either the same or are disjoint.*

Proof. Suppose that the equivalence classes $[x]$ and $[y]$ have an element z in common. Then $x \sim z$ and $y \sim z$. It follows that $z \sim y$ by symmetry. Since $x \sim z$ and $z \sim y$, we have $x \sim y$. Thus, $y \in [x]$ and the transitivity of $\sim$ allows us to conclude $[y] \subseteq [x]$. As $x \sim y$ implies by symmetry that $y \sim x$, we can conclude in the same vein that $[x] \subseteq [y]$. Therefore, if two equivalence classes of $\sim$ contain a common element, then they are the same. □

Example 5.7. We can define a relation $\sim$ on Euclidean plane $\mathbf{R}^2$ such that $(x_1, y_1) \sim (x_2, y_2)$ if and only if $\sqrt{x_1^2 + y_1^2} = \sqrt{x_2^2 + y_2^2}$. This relation is reflexive, symmetric, and transitive, so it is an equivalence relation.

Since $\sqrt{x_1^2 + y_1^2}$ denotes the distance of the point (x_1, y_1) from the origin $(0, 0)$, all points that are at the same distance from the origin belong to the same equivalence class. So an equivalence class of points forms a circle about the origin. Figure 5.1 shows three equivalence classes.

A **partition** of a set S is a family P of nonempty subsets of S such that $S = \bigcup P$ and the sets in P are pairwise disjoint. The elements of P are called **blocks**. For instance, the partition $P = \{\{1, 2\}, \{3\}\}$ of the set $S = \{1, 2, 3\}$ consists of two blocks, namely $\{1, 2\}$ and $\{3\}$.

Proposition 5.8. *Let S be a nonempty set and $\sim$ an equivalence relation on S. Then the equivalence classes of $\sim$ partition the set S.*

5.1 Generalities

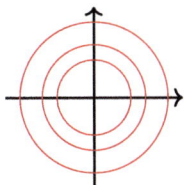

Figure 5.1: The equivalence classes $[(0, 0.5)]$, $[(0, 0.7)]$, and $[(0, 1)]$

Proof. Let $P = \{[x] \mid x \in S\}$. The sets in the family P are pairwise disjoint by Lemma 5.6. Since the relation $\sim$ is reflexive, we have $x \in [x]$, so the equivalence classes are not empty. Furthermore, it follows that $\bigcup P = S$. □

Proposition 5.9. *Let P be a partition of a nonempty set S. For x, y in S, we define $x \equiv y$ if and only if x in C and y in C for some C in P. Then $\equiv$ is an equivalence relation on S and P is its set of equivalence classes.*

Proof. It follows from the definition that the relation $\equiv$ is reflexive and symmetric. If x, y, z are elements of S such that $x \equiv y$ and $y \equiv z$, then there exist sets C and D in P such that x, y in C and y, z in D. Since C and D are elements of a partition and the element y is contained in both C and D, we must have $C = D$. Thus, $x \equiv z$, which proves that the relation is transitive. We can conclude that $\equiv$ is indeed an equivalence relation.

Let C be a set in the partition P. For all x, y in C, we have by definition $[x] = C = [y]$. This implies the second claim. □

Example 5.10. Consider the partition

$$P = \{\{1, 2\}, \{3\}, \{4, 5\}\}.$$

of the set $S = \{1, 2, 3, 4, 5\}$. Each block of the partition corresponds to elements of the set S that are identified under the equivalence relation $\equiv$. The equivalence relation $\equiv$ on S corresponding to P is given by

$$\equiv\,=\,\{(1, 1), (2, 2), (1, 2), (2, 1)\} \cup \{(3, 3)\} \cup \{(4, 4), (5, 5), (4, 5), (5, 4)\}.$$

For instance, $1 \equiv 2$ in the equivalence relation $\equiv$, since 1 and 2 belong to the same block in the partition P.

Given an equivalence relation $\sim$ on a set S, we denote by $S/\sim$ the set

$$S/\sim\,=\,\{[x] \mid x \in S\}$$

of all equivalence classes. The set $S/\sim$ is called the **quotient set** of the set S under $\sim$. The map $x \mapsto [x]$ from S to its quotient set $S/\sim$ is called a **natural map**.

Example 5.11. Consider the set

$$R = \{(x,y) \in \mathbf{R}^2 \mid 0 \leq x \leq 2, 0 \leq y \leq 1\}$$

of points that form a rectangle in the Euclidean plane. We can form an equivalence relation $\sim$ on R that identifies points on the lower and upper border of the rectangle. In other words, the equivalence classes of a point (x,y) is given by

$$[(x,y)] = \begin{cases} \{(x,y)\} & \text{if } 0 < y < 1, \\ \{(x,0),(x,1)\} & \text{if } y = 0 \text{ or } y = 1. \end{cases}$$

The equivalence relation identifies the line segment $L_0 = \{(x,0) \mid 0 \leq x \leq 2\}$ with the line segment $L_1 = \{(x,1) \mid 0 \leq x \leq 2\}$ by identifying the points $(x,0) \sim (x,1)$ for all x in the range $0 \leq x \leq 2$. We can think of the quotient space $R/\sim$ as a cylinder that is obtained from the rectangle R by gluing the line segments L_0 and L_1 together.

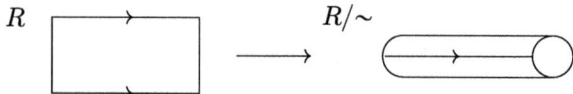

Of course, we cannot take this representation too literally, but it conveys the idea of the identification of the points on the top and bottom border of the rectangle. In topology, this identification is actually made precise.

Equivalence relations permeate all areas of computer science and mathematics. In the subsequent sections, we show how to obtain integers, rational numbers, and more using equivalence relations.

EXERCISES

5.1. Let n be a positive integer. Define on the set $\mathbf{R}$ of real numbers an approximate equality $\approx_n$ such that $x \approx_n y$ if and only if $|x-y| \leq 10^{-n}$. Either show that $\approx_n$ is an equivalence relation or give a counterexample.

5.2. Let L denote the set of all lines in the Euclidean plane $\mathbf{R}^2$. Let ℓ_1 and ℓ_2 be lines in L. We say that $\ell_1 \sim \ell_2$ if and only if the lines ℓ_1 and ℓ_2 either coincide or have no point in common. In other words, $\ell_1 \sim \ell_2$ if and only if the lines ℓ_1 and ℓ_2 are parallel. (a) Show that $\sim$ is an equivalence relation. (b) Describe the equivalence class $[\ell]$ of a line ℓ in L.

5.3. Dr. S. Marty Pants claims that any symmetric and transitive relation $\sim$ is an equivalence relation. He argues as follows: Given x, y in S, $x \sim y$ implies $y \sim x$ by symmetry. Then transitivity yields $x \sim x$, so $\sim$ is reflexive. What is the flaw in this argument?

5.1 Generalities

5.4. We define on the set $\mathbf{N}_1 = \{1, 2, 3, \cdots\}$ of positive integers a relation $\sim$ such that two positive integers x and y satisfy $x \sim y$ if and only if $x/y = 2^k$ for some integer k. Show that $\sim$ is an equivalence relation.

5.5. Consider the set $S = \mathbf{R}^2 \setminus \{(0,0)\}$ consisting of the points in the Euclidean plane without the origin. We define a relation $\sim$ on S by $(x_1, y_1) \sim (x_2, y_2)$ if and only if there exists a nonzero real number λ such that $(x_2, y_2) = (\lambda x_1, \lambda y_1)$.
(a) Show that the relation $\sim$ is an equivalence relation on S. (b) Describe the equivalence class $[(x, y)]$ of a point (x, y) in S.

5.6. The set of integers $\mathbf{Z}$ can be partitioned in the following three subsets: the set of positive integers, the singleton set $\{0\}$ containing 0, and the set of negative integers. Describe the equivalence relation $\equiv$ on $\mathbf{Z}$ that corresponds to this partition.

5.7. Show that every equivalence relation $\sim$ on a set S can be defined by a function $f \colon S \to S$ such that for all x, y in S, we have
$$x \sim y \quad \text{if and only if} \quad f(x) = f(y).$$

5.8. Let $\equiv_c$ and $\equiv_s$ denote the equivalence relations on the set of real numbers respectively induced by the cosine function and the sine function. In other words, $x \equiv_c y$ if and only if $\cos(x) = \cos(y)$, and $x \equiv_s y$ if and only if $\sin(x) = \sin(y)$. Are the equivalence relations $\equiv_c$ and $\equiv_s$ the same or different? Prove your claim.

5.9. Suppose that an equivalence relation $\equiv_f$ on the set of real numbers is induced by a function f, meaning that $x \equiv_f y$ if and only if $f(x) = f(y)$. Determine the equivalence classes of $\equiv_f$, when
(a) $f(x) = x^2$,
(b) $f(x) = x^3$,
(c) f is an injective function,
(d) f is a constant function.

5.10. Consider the set $S = \{1, 2, \ldots, 6\}$. Find the equivalence relations that correspond to the following partitions of S:
(a) $P_1 = \{\{1, 2\}, \{3\}, \{4, 5, 6\}\}$,
(b) $P_2 = \{\{1, 4\}, \{3, 5\}, \{2, 6\}\}$.

5.11. Consider the set
$$R = \{(x, y) \in \mathbf{R}^2 \mid 0 \leqslant x \leqslant 2, 0 \leqslant y \leqslant 1\}$$
of points that form a rectangle in the Euclidean plane, as in Example 5.11. Define an equivalence relation $\equiv$ on R that models gluing of the left and right border line segments, and gluing of the top and bottom line segments.

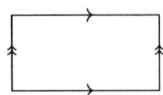

So the equivalence relation identifies the corresponding points on the left and right border line segments, and identifies points on the top and bottom border line segments. (a) Explicitly give the equivalence class $[(x, y)]$ for each point (x, y). (b) Find a geometrically inspired interpretation of the quotient space $R/\equiv$.

5.12. Determine all equivalence relations on the set $\{a, b\}$ with two distinct elements a and b.

5.13. Determine all equivalence relations on the set $\{a, b, c\}$.

5.14. Determine the number of different equivalence relations on $\{a, b, c, d\}$. You do not need to give the equivalence relations themselves, but merely count their number.

5.15. Let S be a set and I an arbitrary nonempty index set. For every $k \in I$, let R_k be an equivalence relation on S. Show that

$$M = \bigcap_{k \in I} R_k$$

is an equivalence relation on S.

5.16. Let R_1 and R_2 be equivalence relations on a nonempty set S. Is their union $R_1 \cup R_2$ an equivalence relation? Prove your claim or give a counter example.

5.17. Find the smallest equivalence relation $\equiv$ on the set $S = \{1, 2, 3, 4, 5\}$ that contains the relation $R = \{(1, 2), (1, 3), (4, 5)\}$.

5.18. Let R be a relation on a set S. Show that the intersection of all equivalence relations containing R is the uniquely determined minimal equivalence relation on S containing R.

5.19. Let R be a relation on a nonempty set S. Then the relation given by

$$M_R = \bigcup_{k=0}^{\infty} (R \cup R^{-1})^k$$

is equal to the minimal equivalence relation containing R.

Here we follow the convention that $(R \cup R^{-1})^0$ is the identity relation on S. For $k \geq 1$, the relation $(R \cup R^{-1})^k$ is the k-fold composition of $R \cup R^{-1}$ with itself.

5.20. Show that the relation $\sim$ on the set $\mathbf{Z}$ of integers given by

$$x \sim y \text{ if and only if } x + 2y \text{ is divisible by } 3$$

is an equivalence relation.

5.2 Integers

Given two nonnegative integers a and b, we can form the difference

$$a - b$$

when $a \geq b$. Since we would like to form the difference between any two nonnegative integers, we need to extend the set of nonnegative integers to the set of integers.

How do we form the set of integers? Naturally, the guiding principle is that we want to define an integer as a difference of two nonnegative integers, but we have to do this without relying on the undefined difference $a-b$ when $a < b$. We circumvent this problem by letting pairs (a, b) of nonnegative integers represent the elusive value $a - b$.

Since a pair (c, d) with the same difference $c - d = a - b$ should represent the same number, we will define an equivalence relation $\sim$ on the set of pairs of nonnegative integers

$$(a, b) \sim (c, d) \quad \text{if and only if} \quad a + d = b + c. \tag{5.1}$$

Since $a - b$ is undefined when $a < b$, we simply rewrote $c - d = a - b$ into the equivalent form $a + d = b + c$. Let us verify that the relation $\sim$ is indeed an equivalence relation.

Proposition 5.12. *The relation $\sim$ defined by (5.1) is an equivalence relation.*

Proof. The relation is reflexive, since $a + b = b + a$, hence $(a, b) \sim (a, b)$.

The relation is symmetric. Indeed, since $(a, b) \sim (c, d)$ implies $a + d = b + c$, we obtain by commutativity of addition the equation $c + b = d + a$. It follows that $(c, d) \sim (a, b)$.

Finally, we are going to prove that the relation $\sim$ is transitive. Suppose that $(a_1, b_1) \sim (a_2, b_2)$ and $(a_2, b_2) \sim (a_3, b_3)$ hold. The first relation implies $a_1 + b_2 = b_1 + a_2$. Adding b_3 to both sides yields

$$(a_1 + b_3) + b_2 = b_1 + (a_2 + b_3). \tag{5.2}$$

Since the second relation $(a_2, b_2) \sim (a_3, b_3)$ implies $a_2 + b_3 = b_2 + a_3$, this allows us to deduce from (5.2) the equation

$$(a_1 + b_3) + b_2 = b_1 + (b_2 + a_3) = (b_1 + a_3) + b_2.$$

Subtracting b_2 from both sides yields $a_1 + b_3 = b_1 + a_3$, which implies $(a_1, b_1) \sim (a_3, b_3)$. Therefore, the relation is transitive. □

We define the set **Z** of integers as the quotient set

$$\mathbf{Z} = \mathbf{N}_0 \times \mathbf{N}_0 / \sim .$$

The motivation for introducing the equivalence relation should now be clear. For example, 0 is represented by $(0, 0) \sim (1, 1) \sim (2, 2) \sim \cdots$. We defined the

value 0 in **Z** as the equivalence class that combines all the various representations of 0, namely
$$0 = \{(n,n) \mid n \in \mathbf{N}_0\}.$$
A positive integer k is represented in **Z** by the equivalence class
$$k = \{(n+k,n) \mid n \in \mathbf{N}_0\}.$$
Our goal was to obtain a representation of negative integers. We accomplished this goal by representing -2 by any pair where the second coordinate is two more than the first coordinate, so $(0,2)$ and $(5,7)$ both represent the value -2. The equivalence class of $-k$ is given by
$$-k = \{(n,n+k) \mid n \in \mathbf{N}_0\}.$$

Perhaps you find it bewildering to represent a single integer value by an infinite number of pairs of nonnegative integers. However, most people have no qualms about the fact that the rational number $2/3$ really represents an infinite set
$$\{2k/3k \mid k \in \mathbf{Z}, k \neq 0\}$$
of equivalent representations. In both cases, we use equivalence classes of numbers. Of course, the representative $2/3$ in lowest terms is in many ways preferable, but allowing non-reduced representations such as $165/935$ instead of $3/17$ can save time when conversions to the representatives are somewhat costly. Exercise 5.21 shows how to identify canonical representatives of integers.

EXERCISES

5.21. Show that each equivalence class $[(a,b)]$ of $\mathbf{Z} = \mathbf{N}_0 \times \mathbf{N}_0/\sim$ contains a unique element that has at least one zero coordinate.

5.22. Explain how to define addition and multiplication when the set of integers **Z** is given by $\mathbf{N}_0 \times \mathbf{N}_0/\sim$. The definition should work for any representative.

5.3 Modular Arithmetic

Consider the set **Z** of integers and a positive integer n. We follow Gauss and write
$$x \equiv y \pmod{n},$$
if and only if x and y are two integers such that $x - y$ is an integer multiple of n. We call this relation the **congruence modulo** n on the set of integers. The notation is a bit idiosyncratic but turns out to be convenient, since the modulus is more readable than in the alternate subscript form $\equiv_n$.

Proposition 5.13. *Let n be a positive integer. The congruence modulo n is an equivalence relation on the set of integers.*

5.3 Modular Arithmetic

Proof. The relation is reflexive, since $x - x = 0n$, so $x \equiv x \pmod{n}$ holds for all integers x. The relation is symmetric, since $x \equiv y \pmod{n}$ implies that there exist some integer k such that $x - y = kn$, but then $y - x = -kn$, which implies $y \equiv x \pmod{n}$. The relation is transitive, since $x \equiv y \pmod{n}$ and $y \equiv z \pmod{n}$ imply that there exist integers k and ℓ such that $x - y = kn$ and $y - z = \ell n$, so $x - z = x - y + y - z = (k + \ell)n$, hence $x \equiv z \pmod{n}$. Therefore, the congruence modulo n is an equivalence relation on the set of integers, as claimed. □

The equivalence class of an integer x for the congruence modulo n is given by
$$[x] = \{y \in \mathbf{Z} \mid x \equiv y \pmod{n}\}.$$
Explicitly, this equivalence class is given by the set $[x] = \{x + kn \mid k \in \mathbf{Z}\}$. The integer x can be expressed in the form $x = qn + r$ with $0 \leqslant r < n$, hence $[x] = [r]$. Therefore, the quotient set $S/\sim$ is given by
$$S/\sim \, = \{[x] \mid 0 \leqslant x \leqslant n - 1\}.$$
The set $S/\sim$ is usually denoted by $\mathbf{Z}/n\mathbf{Z}$.

The congruence modulo n is more than just an equivalence relation, since it is compatible with addition and multiplication of integers.

Proposition 5.14. *Suppose that x_1, x_2, y_1, y_2 are integers such that $x_1 \equiv x_2 \pmod{n}$ and $y_1 \equiv y_2 \pmod{n}$. Then*
$$\begin{aligned} x_1 + y_1 &\equiv x_2 + y_2 \pmod{n}, \\ x_1 y_1 &\equiv x_2 y_2 \pmod{n}. \end{aligned}$$

Proof. By assumption there exist integers k and ℓ such that $x_1 - x_2 = kn$ and $y_1 - y_2 = \ell n$. Adding these equalities, we get $(x_1 + y_1) - (x_2 + y_2) = (k + \ell)n$, which implies $x_1 + y_1 \equiv x_2 + y_2 \pmod{n}$.

For the second equation, we notice that $x_1 - x_2 = kn$ and $y_1 - y_2 = \ell n$ imply $(x_1 - x_2)y_1 = y_1 kn$ and $(y_1 - y_2)x_2 = x_2 \ell n$. Adding the latter two equations yields $(x_1 - x_2)y_1 + (y_1 - y_2)x_2 = (y_1 k + x_2 \ell)n$. Canceling terms, we get $x_1 y_1 - y_2 x_2 = (ky_1 + \ell x_2)n$, which implies $x_1 y_1 \equiv x_2 y_2 \pmod{n}$. □

As a consequence of this proposition, we can define an addition and a multiplication operation on $\mathbf{Z}/n\mathbf{Z}$ by defining for integers x and y the operations
$$\begin{aligned} [x + y] &= [x] + [y], \\ [xy] &= [x][y] \end{aligned}$$
on the equivalence classes. The previous proposition ensures that this definition does not depend on the representatives x and y of the equivalence classes. Indeed, if $[x_1] = [x_2]$ and $[y_1] = [y_2]$, then the previous proposition ensures that $[x_1 + y_1] = [x_2 + y_2]$ and $[x_1 y_1] = [x_2 y_2]$ hold.

EXERCISES

5.23. Derive the addition and multiplication tables for $\mathbf{Z}/4\mathbf{Z}$.

5.24. Find the last digit in the decimal expansion of 7^{500}.

5.25. Consider the equivalence relation $a \equiv b \pmod{9}$. Prove by induction that the equivalence classes $[10^n]$ and $[1]$ are the same for all positive integers n.

5.26. Let a and n be positive coprime integers. Show that there exists an integer b such that $ab \equiv 1 \pmod{n}$. In other words, if you want to divide by $[a]$, then you multiply by $[b]$. The element $[b]$ is the multiplicative inverse of $[a]$ in $\mathbf{Z}/n\mathbf{Z}$ and is denoted as $[a^{-1}]$.

5.27. Prove that a positive integer written in base 10 is divisible by 3 if and only if the sum of its digits is divisible by 3.

5.28. Prove that a positive integer a written in base 10 is divisible by 11 if and only if the alternating subtraction and addition of the digits of a is divisible by 11. For example, 12221 is divisible by 11, since $1 - 2 + 2 - 2 + 1 = 0$ is divisible by 11. The integer 987654321 is not divisible by 11, since $9 - 8 + 7 - 6 + 5 - 4 + 3 - 2 + 1 = 5$ is not divisible by 11.

5.29. Sow that an integer a is divisibly by 8 if and only if the number formed by the least three digits of a is divisible by 8. For example, 9,234,232,224 is divisible by 8, since 224 is divisible by 8.

5.30. Formulate a divisibility rule for divisibility by 24. Is the integer

$$111, 222, 333, 221, 112$$

divisible by 24?

5.4 Rational Numbers

For integers b and a with $b \neq 0$, it is not always possible to solve the equation

$$bx = a$$

in the integers. This suggests to use a larger domain to solve the equation. We will introduce the rational numbers a/b for this purpose.

Let us define the set of pairs

$$Q = \{(x, y) \in \mathbf{Z} \times \mathbf{Z} \mid y \neq 0\},$$

where x represents the numerator and y the denominator of the rational number, so y is assumed to be nonzero. We define a relation $\sim$ on Q such that $(a, b) \sim (c, d)$ if and only if $ad = bc$. We claim that this is an equivalence relation (which, unsurprisingly, realizes the cross-multiplication check for the equality of rational numbers).

5.4 Rational Numbers

Proposition 5.15. *The relation $\sim$ is an equivalence relation on Q.*

Proof. One easily verifies that the relation is reflexive and symmetric. The relations $(a,b) \sim (c,d)$ and $(c,d) \sim (e,f)$ imply the equations (i) $bc = ad$ and (ii) $de = cf$. Multiplying both sides of equation (i) by f yields $bcf = adf$, and substituting equation (ii) yields equation (iii) $bde = adf$. Since $d \neq 0$, we can divide both sides of equation (iii) by d and obtain $be = af$, so $(a,b) \sim (e,f)$. Therefore, $\sim$ is transitive and thus an equivalence relation. $\square$

The set $\mathbf{Q}$ of **rational numbers** is defined to be the quotient set $Q/\sim$. The equivalence class $[(a,b)]$ of a pair (a,b) is usually denoted in the fraction notation a/b. For example, the rational number $1/2$ represents the equivalence class

$$\frac{1}{2} = \{(k, 2k) \mid k \in \mathbf{Z}, k \neq 0\},$$

so the equivalence class $1/2$ contains for instance the numerator and denominator pairs $(-2,-4)$, $(-1,-2)$, $(1,2)$, and $(2,4)$, among others.

We define on the set $\mathbf{Q}$ of rational numbers two binary operations. The addition of two rational numbers is defined as

$$\frac{a}{b} + \frac{c}{d} = \frac{ad+bc}{bd}.$$

Since there are several different ways to represent the same rational number, one needs to worry that this definition depends on the particular choice of the representatives. Exercise 5.31 shows that this is not the case. The product of two rational numbers is defined as

$$\frac{a}{b} \cdot \frac{c}{d} = \frac{ac}{bd}.$$

Here one can show as well that the definition does not depend on the representatives of the equivalence classes, see Exercise 5.32.

EXERCISES

5.31. Show that the addition of rational numbers is well-defined, that is, show that if $a/b = a'/b'$ and $c/d = c'/d'$, then

$$\frac{a}{b} + \frac{c}{d} = \frac{a'}{b'} + \frac{c'}{d'}.$$

5.32. Show that the product of rational numbers is well-defined, that is, show that if $a/b = a'/b'$ and $c/d = c'/d'$, then

$$\frac{a}{b} \cdot \frac{c}{d} = \frac{a'}{b'} \cdot \frac{c'}{d'}.$$

5.5 Notes

Equivalence relations are discussed in nearly every book on algebra, combinatorics, or discrete mathematics, see for example Cohn [15] or Jacobson [39]. One can say much more about the construction of number systems, even to the level of detail of an entire book. The most famous example is Landau [56], see also the book by Little et al. [57].

Chapter 6

Partial Orders and Lattices

> *Order is the sanity of the mind, the health of the body, the peace of the city, the security of the state. As the beams to a house, as the bones to a microcosm of man, so is the order to all things.*
>
> — Robert Southey, *The Doctor etc.*

Many daily tasks require a little bit of organization. For example, you would not want your coffee poured before you got your cup. Some tasks simply need to take precedence before other tasks, otherwise you might get into a mess. This becomes particularly apparent if you try to teach a robot how to help with such tasks. In this chapter, we investigate partial orders and lattices, which allow one to abstract the precedence relation of tasks. Partial orders are absolutely fundamental in mathematical arguments, and they have many applications in computer science.

6.1 Partial Orders

Let S be a set. A relation[1] $\leqslant$ on S is called a **partial order** if and only if it satisfies the following three properties:

O1. For all x in S, we have $x \leqslant x$ (reflexivity).

O2. For all x and y in S, $x \leqslant y$ and $y \leqslant x$ imply $x = y$ (antisymmetry).

O3. For all x, y, z in S, $x \leqslant y$ and $y \leqslant z$ imply $x \leqslant z$ (transitivity).

If a relation $\leqslant$ on S satisfies the properties **O1**–**O3**, then $(S, \leqslant)$ is called a **partially ordered set**.

On occasion, we will encounter relations that are reflexive and transitive, but not necessarily antisymmetric. Such a relation is called a **preorder**.

---
[1] You might recall that we introduced relations in Chap. 3. Relations satisfying this trio of properties are particularly important in practice.

Notation. A relation is often denoted by a capital letter such as R, but for partial orders, it is customary to use symbols that resemble the inequality sign. Aside from $\leqslant$, common symbols to denote partial orders are $\leq$, $\subseteq$, and $\sqsubseteq$.

A prototypical example of a partial order is given by the power set ordered by set inclusion.

Example 6.1. Let A be a set and $S = P(A)$ its power set. The set inclusion $\subseteq$ defines a partial order on S. Indeed, the set inclusion is reflexive, since every set is a subset of itself. For two sets X and Y in S, if $X \subseteq Y$ and $Y \subseteq X$ holds, then the sets X and Y must be the same, so $\subseteq$ is antisymmetric. For all sets X, Y, Z in S, the inclusions $X \subseteq Y$ and $Y \subseteq Z$ imply that $X \subseteq Z$, so the relation is transitive.

Two elements x and y of a partially ordered set $(S, \leqslant)$ are called **comparable** if and only if $x \leqslant y$ or $x \geqslant y$. In a partial order, there might exist elements that are not comparable, as the next example illustrates.

Example 6.2. Let $S = P(A)$ be the power set of a set A ordered by inclusion. Then two elements X and Y of S are comparable if and only if X is a subset of Y or Y is a subset of X. For example, if $A = \{1, 2, 3\}$, then the subsets $X = \{1, 2\}$ and $Y = \{2, 3\}$ are not comparable, as none is a subset of the other.

A subset T of a partially ordered set S is called a **chain** if and only if every two elements of T are comparable. If S is finite, then the **height** of S is defined to be the maximum cardinality of any chain in S.

If the entire set S is a chain, then it is also called a **total order** or a **linear order**.

Example 6.3. The set $\mathbf{R}$ of real numbers with its natural order $\leqslant$ is a total order.

Example 6.4. The set $\{1, 2, \ldots, n\}$ of the first n positive integers is a total ordered set under the natural order $1 \leqslant 2 \leqslant \cdots \leqslant n$.

If two elements x and y of a partially ordered set $(S, \leqslant)$ are not comparable, then they are called **incomparable**. A subset A of a partially ordered set S is called an **antichain** if and only if any two elements of A are incomparable. The next example shows the rather extreme case that a partially ordered set S itself is an antichain; this is also known as the **discrete partial order** on S.

Example 6.5. Let S be a set. The identity relation $=$ on S is a partial order called the discrete partial order on S. This relation is explicitly given by $\{(x, x) | x \in S\}$. It relates every element x of S to itself, but distinct elements are incomparable.

Example 6.6. Tom likes to eat fruits for dessert. Among the choices

$$F = \{\text{grapefruit}, \text{apple}, \text{orange}, \text{strawberry}\},$$

his preference relation $\leqslant$ is

$$\text{grapefruit} \leqslant \text{apple}, \ \text{grapefruit} \leqslant \text{orange},$$
$$\text{apple} \leqslant \text{strawberry}, \ \text{orange} \leqslant \text{strawberry},$$

6.2 Strict Order

and grapefruit $\leqslant$ strawberry, as well as $f \leqslant f$ for all fruits $f \in F$. This is a partial order. Apples and oranges are incomparable, as Tom has no preference of one over the other.

EXERCISES

6.1. Let $\mathbf{N}_1 = \{1, 2, 3, \ldots\}$ denote the set of positive integers. For positive integers x and y, we define $x \leqslant y$ if and only if there exists a positive integer q such that $y = xq$. Show that $(\mathbf{N}_1, \leqslant)$ is a partially ordered set. The divisibility relation $x \leqslant y$ is often denoted as $x \mid y$.

6.2. Let m and n be positive integers. Consider the set $P = \{1, 2, \ldots, m\} \times \{1, 2, \ldots, n\}$ and a relation $\leqslant$ on P given by $(a_1, a_2) \leqslant (b_1, b_2)$ if and only if $a_1 \leqslant b_1$ and $a_2 \leqslant b_2$.
(a) Show that $\leqslant$ is partial order.
(b) Which elements in P are incomparable?
(c) Determine the height of $(P, \leqslant)$.
(d) Determine the width of $(P, \leqslant)$. The width is defined in Sect. 6.4.

6.3. Let $(A, \leqslant_A)$ and $(B, \leqslant_B)$ be two partially ordered sets. One can define on the Cartesian product $A \times B$ a relation $\leqslant$ by setting $(a_1, b_1) \leqslant (a_2, b_2)$ if and only if $a_1 \leqslant_A a_2$ and $b_1 \leqslant_B b_2$. (a) Show that $(A \times B, \leqslant)$ is a partially ordered set known as the product order. (b) When is the product order a total order?

6.4. Let $(A, \leqslant_A)$ and $(B, \leqslant_B)$ be two partially ordered sets. One can define on the Cartesian product $A \times B$ a relation $\leqslant$ by setting $(a_1, b_1) \leqslant (a_2, b_2)$ if and only if $a_1 <_A a_2$ or $(a_1 = a_2$ and $b_1 \leqslant_B b_2)$. (a) Show that $(A \times B, \leqslant)$ is a partially ordered set known as the lexicographic order. (b) When is the lexicographic order a total order?

6.5. Let $(P, \leqslant_P)$ be a partially ordered set and R a subset of P. Show that $(R, \leqslant_R)$ with the relation $\leqslant_R$ given by the restriction of $\leqslant_P$ to pairs in $R \times R$ is a partially ordered set.

6.2 Strict Order

Let S be a set. A relation $<$ on S is called a **strict order** if and only if it satisfies the following two properties

SO1. the relation is irreflexive, meaning that $x < x$ does not hold for any x in S,

SO2. the relation is transitive, that is, $x < y$ and $y < z$ imply that $x < z$ holds for all x, y, and z in S.

If a relation $<$ on S satisfies the properties **SO1** and **SO2**, then $(S, <)$ is called a **strictly ordered set**. Strict orders cannot be partial orders, but the two concepts are closely related as we will see. We offer a few typical examples of strict orders.

Example 6.7. The set **N** of positive integers with the strict inequality $<$ is a strictly ordered set.

Example 6.8. Let A be a set and $S = P(A)$ its power set. Then the proper subset relation $\subsetneq$ on S is a strict order.

Example 6.9. In distributed computing, one often considers a collection of processes that communicate by sending messages. A process is understood to be a totally ordered sequence of events (such as machine instructions, the sending or receiving of messages). Lamport defined the **happened-before** relation $\to$ on the events in the distributed system as follows:
(a) Suppose that a and b are events that belong to the same process. Then $a \to b$ if and only if the event a happened before the event b.
(b) Let a be the event of sending a message and b the event of receiving the same message. Then $a \to b$.
(c) The relation $a \to a$ does not hold for any event a.
(d) For all events a, b, c, if $a \to b$ and $b \to c$, then $a \to c$.
The happened-before relation $\to$ is irreflexive by property (c) and transitive by property (d); hence, it is a strict order. Events that are not comparable are potentially concurrent events.

The next two propositions show that partial orders and strict orders always occur in pairs. It is a matter of personal preference whether one first specifies a partial order and derives the associated strict order or the other way round.

Proposition 6.10. *Let $\leqslant$ be a partial order on a set S. We can define a relation $<$ on S by setting $x < y$ if and only if $x \leqslant y$ and $x \neq y$. Then the relation $<$ is a strict order on S.*

Proof. The relation $<$ is by construction irreflexive. Suppose that $x < y$ and $y < z$ holds. This means that $x \leqslant y$ and $y \leqslant z$, as well as $x \neq y$ and $y \neq z$ hold. It follows from the transitivity of $\leqslant$ that $x \leqslant z$ holds. Seeking a contradiction, let us assume that $x = z$; this would mean that $x \leqslant y$ and $y \leqslant x$ hold, so by the antisymmetry of the partial order $\leqslant$, we must have $x = y$, contradicting the fact that $x \neq y$. Thus, $x < y$ and $y < z$ imply that $x \leqslant z$ and $x \neq z$, so $x < z$. Therefore, $<$ is transitive. This shows that $<$ is a strict order, as claimed. □

We record a simple observation before proving our next proposition.

Lemma 6.11. *A strict partial order $<$ on a set S is an asymmetric relation, meaning that $x < y$ implies that $y < x$ cannot hold.*

Proof. Seeking a contradiction, let us suppose that there are two elements x and y in S such that $x < y$ and $y < x$ hold. By transitivity, this implies that $x < x$ must hold. However, this contradicts the irreflexivity of the relation. □

6.3 Cover Relations and Hasse Diagrams

Proposition 6.12. *Let $<$ be a strict order on a set S. Define a relation $\leqslant$ on S by setting $x \leqslant y$ if and only if $x < y$ or $x = y$. Then $\leqslant$ is a partial order relation.*

Proof. By definition, $\leqslant$ is reflexive.

The relation $\leqslant$ is antisymmetric, since $x \leqslant y$ and $y \leqslant x$ implies $x = y$. This is evident when $x = y$. If x and y are distinct elements of S, then this follows from the asymmetry of the strict order relation $<$.

Suppose that $x \leqslant y$ and $y \leqslant z$. If $x = y$ or $y = z$, then it immediately follows that $x \leqslant z$. If $x \neq y$ and $y \neq z$, then $x < y$ and $y < z$ holds, which implies $x < z$ by transitivity of the strict order; hence, $x \leqslant z$. Therefore, $\leqslant$ is a transitive relation. □

The previous two propositions show that one can associate with each strict order a partial order by taking the union with the identity relation. Furthermore, one can obtain a strict order from a partial order by intersection with the relation of distinctness.

It is customary to denote the strict order associated with $\leqslant, \leqslant, \subseteq,$ and $\sqsubseteq$ respectively by $<, <, \subset,$ and $\sqsubset$. Sometimes it is convenient to reverse the symbols. For instance, we write $x \geqslant y$ for $y \leqslant x$, and $x > y$ for $y < x$.

EXERCISES

6.6. Show that the set of people with the relation *"is ancestor of"* is a strictly ordered set.

6.7. Show that an asymmetric and transitive relation is a strict order.

6.8. Is an irreflexive and asymmetric relation also transitive? Prove it or give a counter example.

6.9. Construct the smallest strict order.

6.10. Let $<$ be a strict order on a finite set with n elements. Give a tight lower bound on the number of pairs in the strict order $<$.

6.11. Show that the set $\mathbf{N}_1$ of positive integers with the relation $\|$ is a strictly ordered set, where $a \| b$ if and only if $a \neq b$ and a divides b.

6.3 Cover Relations and Hasse Diagrams

Let $(S, \leqslant)$ be a partially ordered set. We say that an element z in S **covers** an element x in S if and only if $x < z$, and there does not exist an element y in S such that $x < y < z$. We write $x \lessdot z$ to denote that z covers x.

Example 6.13. Let $P = \{1, 2, 3, 6\}$ be a set partially ordered by the relation

$$\leqslant \; = \{(1,1), (1,2), (1,3), (1,6), (2,2), (2,6), (3,3), (3,6), (6,6)\}.$$

The cover relation is given by

$$\lessdot \; = \{(1,2), (1,3), (2,6), (3,6)\}.$$

For finite sets, the partial order $\leqslant$ can be recovered from the cover relation $\lessdot$ by adding relations that are implied by reflexivity and transitivity. Indeed, reflexivity implies here the relations $(1,1)$, $(2,2)$, $(3,3)$, and $(6,6)$ and transitivity implies the relation $(1,6)$. So the main benefit of the cover relation is that it is a concise representation of the partial order.

Example 6.14. As a cautionary tale, let us consider an infinite partially ordered set, namely the set $\mathbf{Q}$ of rational numbers in their natural order $\leqslant$. If a and b are rational numbers with $a < b$, then there always exists a rational number that is strictly between the two, namely $a < (a+b)/2 < b$. Hence, the cover relation is the empty set. In this case, the partial order cannot be recovered from the cover relation.

We will now show that a *finite* partially ordered set can be recovered from its cover relation. We need the concept of the reflexive and transitive closure for this purpose.

Let R be a relation on a set S. This relation does not need to be reflexive or transitive. By adding pairs (x, y) of elements to R, we can obtain a relation that is reflexive and transitive. The **reflexive and transitive closure** of R is the *smallest* reflexive and transitive relation on S containing R.

Lemma 6.15. *The reflexive and transitive closure of a relation R on a set S always exists.*

Proof. Let $\mathcal{F}$ be the family of reflexive and transitive relations on S that contain the relation R. This family is not empty, since it contains $S \times S$. The intersection $T := \bigcap \mathcal{F}$ of these relations is again a reflexive and transitive relation. Evidently, T is the smallest reflexive and transitive relation containing R, so it is the reflexive and transitive closure of R. □

For a finite set S, the cover relation completely determines the partial order.

Proposition 6.16. *Let $\leqslant$ be a partial order on a finite set S with associated cover relation $\lessdot$. Then the reflexive and transitive closure of the cover relation $\lessdot$ is precisely the partial order $\leqslant$.*

Proof. By definition, the cover relation $\lessdot$ is a subset of the reflexive and transitive partial order relation $\leqslant$. Therefore, the reflexive and transitive closure T of the cover relation $\lessdot$ is contained in the partial order $\leqslant$.

Seeking a contradiction, let us suppose that x and y are elements in S such that $x \leqslant y$, but (x, y) is not in T. Then x cannot be equal to y and

6.3 Cover Relations and Hasse Diagrams

$x < y$ must hold. Choose a chain of maximal length in $(S, \leqslant)$ such that $x = z_1 < z_2 < \cdots < z_n = y$. Then $z_k \lessdot z_{k+1}$ holds for all k in the range $1 \leqslant k < n$. It follows that $(x,y) = (z_1, z_n)$ is in the reflexive and transitive closure T of the cover relation, which is a contradiction. □

The **cover graph** of the partial order $(S, \leqslant)$ is given by $(S, \lessdot)$. The edges of this directed graph are the cover relations. The **Hasse diagram** of the partial order $(S, \leqslant)$ is the cover graph drawn in the Euclidean plane such that each pair $a \lessdot b$ of the cover relation is represented by a line segment and the y-coordinate of b is larger than the y-coordinate of the covered element a. The line segments in a Hasse diagram are arranged such that there are no more than two vertices of the cover graph on any line segment representing an edge.

The Hasse diagram is a simple concept that is helpful in visualizing small examples. The next two examples illustrate the concept.

Example 6.17. Let $S = \{1, 2, 3, 4, 5, 6, 7, 8, 9, 10, 12\}$ be a set of integers partially ordered by divisibility. The Hasse diagrams for $(S, |)$ are shown below.

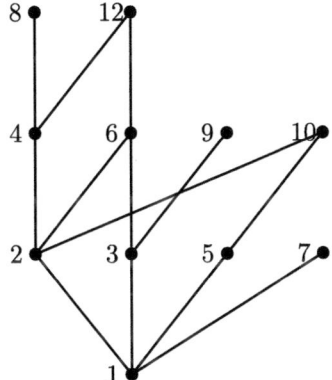

There is an edge between 3 and 6, since 3 is a proper divisor of 6, and these two numbers differ by a single prime factor. So $(3, 6)$ belongs to the cover relation of the divisibility relation $|$. There is no edge between 3 and 12, since that is implied by the divisibility relation $3 \mid 6$ and $6 \mid 12$, thus $(3, 12)$ does not belong to the cover relation of $(S, |)$. The cover relation consists of pairs (a, b) such that both are elements of S and b can be obtained from a by multiplication with a single prime factor.

Example 6.18. The power set $P(\{1, 2, 3\})$ is partially ordered by set inclusion. A set covers a subset if and only if precisely one element is omitted. The Hasse diagram is shown as follows.

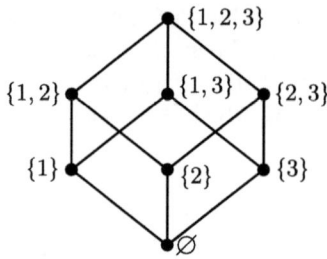

Since the direction of an edge is determined by the upward direction, arrows are not needed. This is advantageous for large lecture rooms, where a Hasse diagram is easier to read than a cover graph with directed arrows.

EXERCISES

6.12. Determine the cover relation of the set of positive integers in its natural order.

6.13. Determine the cover relation of the set of positive integers ordered by divisibility. In the divisibility relation $a \leqslant b$ if and only if there exists a positive integer q such that $b = aq$. The divisibility relation is usually written as $a \mid b$.

6.14. Determine the cover relation of $(P(\{1,2,3\}, \subseteq)$.

6.15. Draw the Hasse diagram for the divisibility relation on the set

$$\{1, 2, \ldots, 10\}$$

of positive integers from 1 to 10.

6.16. The set $\{1, 2, \ldots, 10\}$ of positive integers from 1 to 10 is partially ordered by the divisibility relation.
(a) Find an antichain of maximal cardinality and argue that this is a maximal antichain.
(b) Find a partition of $\{1, 2, \ldots, 10\}$ into a minimal number of chains.

6.4   Dilworth's Theorem

Arguably, the simplest partially ordered set is a chain. If a partially ordered set is not a chain, then we can still ask how many chains are needed to cover all elements. Evidently, if not all elements are comparable, we need more than one chain. Dilworth gave a surprisingly simple characterization of the fewest number of chains that cover a finite partially ordered set S. He showed that this figure coincides with the cardinality of the largest antichain in S.

We begin with a simple example that illustrates this fact.

6.4 Dilworth's Theorem

Example 6.19. The power set $P(\{1,2,3\})$ is partially ordered by set inclusion. The Hasse diagram on the left shows an antichain of cardinality 3. This is the largest cardinality of any antichain in $P(\{1,2,3\})$.

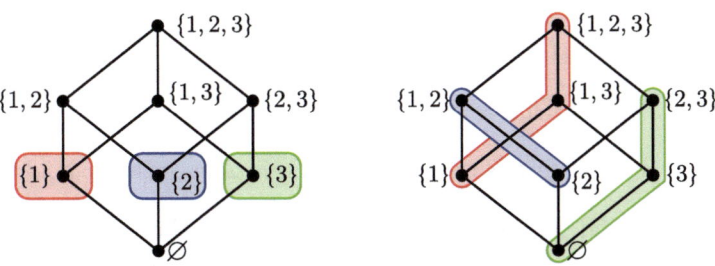

The Hasse diagram on the right shows that $P(\{1,2,3\})$ can be partitioned into the three chains (1) $\{1\} \subseteq \{1,3\} \subseteq \{1,2,3\}$, (2) $\{2\} \subseteq \{1,2\}$, and (3) $\emptyset \subseteq \{3\} \subseteq \{2,3\}$. We cannot partition $P(\{1,2,3\})$ into fewer chains, since $\{1\}$, $\{2\}$, and $\{3\}$ each must lie in a different chain. Indeed, none is a subset of the other two sets. Dilworth's theorem shows that the largest cardinality of an antichain in a finite partially ordered set is always equal to the fewest number of chains that partition this partially ordered set.

Let S be a partially ordered set. The maximal cardinality of an antichain in S is called the **width** of S. If S has width m, then any partition of S into pairwise disjoint chains $C_1, \ldots, C_n$ satisfies $n \geqslant m$, see Exercise 6.17. Dilworth showed the remarkable result that there always exist as few as $n = m$ pairwise disjoint chains that partition S. This is an example of a so-called min-max theorem, as it shows that the *minimal* number of chains that partition S is equal to the *maximal* number of elements in an antichain of S.

Theorem 6.20 (Dilworth's Theorem). *Let $(S, \preccurlyeq)$ be a finite partially ordered set. Then the minimal number n of chains that partition S is equal to the maximal cardinality m of an antichain in S.*

Proof. By Exercise 6.17, the width m of S satisfies $m \leqslant n$. We will show that $m \geqslant n$ using a proof by induction.

If $|S| = 0$, then no chains are needed to cover S and the maximal cardinality of an antichain is 0, so the claim holds in this case.

Let us assume that the claim holds for all partially ordered sets of cardinality $|S| < k$. Consider a partially ordered set of cardinality $|S| = k$, and let C be a maximal chain of S.

Case 1. If the chain C meets every maximal antichain in S of cardinality m, then the cardinality of a maximal antichain in $S \backslash C$ is $m-1$. By induction hypothesis, $S \backslash C$ can be partitioned into $m-1$ chains. So adding the chain C, we can conclude that S can be partitioned into $n = m$ chains.

Case 2. If the chain C does not meet some maximal antichain $A = \{a_1, \ldots, a_m\}$ in S, then we need to proceed in a different way to show that S can be partitioned into $n = m$ chains.

Let us define the downward closure A_d of A by

$$A_d = \{x \in S \mid \text{there exists an } a \text{ in } A \text{ such that } x \leqslant a\},$$

and its upward closure A_u by

$$A_u = \{x \in S \mid \text{there exists an } a \text{ in } A \text{ such that } x \geqslant a\}.$$

Since A_u and A_d both include the antichain A, they still have width m.

We will show that A_d and A_u are proper subsets of S such that $A_d \cup A_u = S$ and $A_d \cap A_u = A$. By induction hypothesis, we will be able to cover A_d as well as A_u both with m chains, and these chains can be stitched together to obtain m chains that cover S.

We will first show that $A_d \cup A_u = S$. Seeking a contradiction, let us suppose that $A_d \cup A_u$ is a proper subset of S. Thus, there exists an element x in $S \backslash (A_d \cup A_u)$. It follows that x is incomparable with every element in A, but this would imply that $A \cup \{x\}$ is an antichain, contradicting the maximality of the antichain A. Therefore, $S = A_d \cup A_u$.

We will show next that $A_d \cap A_u = A$. Indeed, if the intersection $A_d \cap A_u$ would contain an element x that does not belong to A, then this would imply that there exist elements a_d and a_u in A such that $a_u < x$ and $x < a_d$. However, it would follow that $a_u < a_d$, so a_d and a_u are distinct elements in A that are comparable, but this contradicts that A is an antichain. Therefore, $A_d \cap A_u = A$.

We will now show that A_d and A_u are both proper subsets of A. Indeed, the greatest element of C cannot belong to A_d, since it would be less than some element of A, contradicting the maximality of the chain C. Thus, A_d is a proper subset of S. Similarly, the least element of C cannot belong to A_u, since it would be greater than some element of A, contradicting the maximality of the chain C. Thus, A_u is a proper subset of S as well.

By induction hypothesis, the proper subset A_d of S can be partitioned into m chains C_a with $a \in A$, where the index a is chosen such that $a \in C_a$. Similarly, A_u is a proper subset of S that can be partitioned by induction hypothesis into m chains D_a with $a \in A$ such that $a \in D_a$. Then $C_a \cap D_a = \{a\}$ holds for each a in A. The element a must be the greatest element of C_a and the least element of D_a, so $C_a \cup D_a$ is a chain in S. Therefore, S can be partitioned into the m chains $C_a \cup D_a$ with a in A, as claimed.

In summary, we can conclude that S can always be covered with m chains. $\square$

The next example illustrates the second case of the previous proof.

6.4 Dilworth's Theorem

Example 6.21. Let $S = \{a, b, c, d, e, f\}$ be a partially ordered set given by the Hasse diagram Fig. 6.1. In other words, $(S, \leq)$ is the reflexive and transitive closure of the order relations $a < b < c$, $d < e < f$, and $e < b$. Consider the maximal chain $C = \{d, e, b, c\}$. This chain does not meet the antichain $A = \{a, f\}$. So this is an example where the second case in the proof of Dilworth's theorem applies. The upward closure A_u of the antichain A is given by $A_u = \{a, b, c, f\}$, and the downward closure A_d of the antichain A is given by $A_d = \{a, d, e, f\}$.

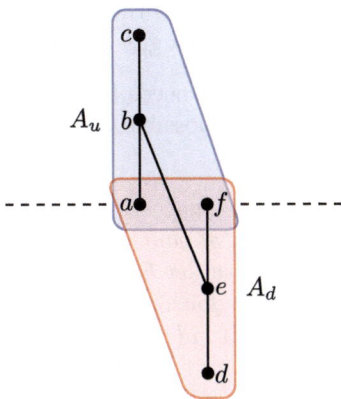

Figure 6.1: Hasse diagram of a partially ordered set with six elements. The set S is divided along the antichain $A = \{a, f\}$ into the two parts $A_u = \{a, b, c, f\}$ and $A_d = \{a, d, e, f\}$

Since A_d and A_u are proper subsets of S, they can be partitioned into chains by the induction hypothesis. The downward closure A_d of the antichain A is partitioned into the chains $C_a = \{a\}$ and $C_f = \{d, e, f\}$. The upward closure A_u of A is partitioned into the chains $D_a = \{a, b, c\}$ and $D_f = \{f\}$. Therefore, the partially ordered set S can be partitioned into the two chains $C_a \cup D_a = \{a, b, c\}$ and $C_f \cup D_f = \{d, e, f\}$.

Corollary 6.22. *Let m and n be positive integers. Any partially ordered set $(S, \leqslant)$ with $mn+1$ or more elements has a chain of length $m+1$ or an antichain with $n + 1$ elements.*

Proof. Seeking a contradiction, let us suppose that each chain in S has at most m elements and each antichain has at most n elements. Therefore, the width of S is at most n, so S is covered by n or fewer chains. Since each chain has at most m elements, the total number of elements in the chain cover is at most mn, contradicting the fact that S has $mn + 1$ or more elements. □

Dilworth's theorem has many consequences in combinatorics and computer science. We give one playful application.

Suppose that you have m different gifts $g_1, g_2, \ldots, g_m$. You want to give each of your n friends $f_1, f_2, \ldots, f_n$ one gift that they like. Suppose that your friend f_k likes a subset L_k of the set of gifts $\{g_1, g_2, \ldots, g_m\}$. Is it possible to get everyone a gift that they like? Frustratingly, this is not always possible, as the next example shows.

Example 6.23. Suppose that you have five gifts $G = \{g_1, g_2, \ldots, g_5\}$. Your friends f_1, f_2, f_3, and f_4 respectively like the gifts

$$L_1 = \{g_1\}, \quad L_2 = \{g_1, g_3, g_5\}, \quad L_3 = \{g_2\}, \quad L_4 = \{g_1, g_2\}.$$

Since the three friends f_1, f_3, f_4 only like the first two gifts g_1 and g_2, it is not possible to present each of them with a gift that they like.

Evidently, any subset of x friends needs to like at least x gifts. Amazingly, if this condition is satisfied for all possible number of friends, then everyone can receive a gift that they like.

Theorem 6.24 (Hall's Theorem). *Suppose that $G = \{g_1, g_2, \ldots, g_m\}$ is a finite set of gifts. You want to give each of your n friends $f_1, f_2, \ldots, f_n$ one gift that they like. The friend f_k likes the subset L_k of the set G of gifts, where k is in the range $1 \leqslant k \leqslant n$. Suppose that for any subset I of the set $\{1, 2, \ldots, n\}$, the number of gifts liked by the friends indexed by I is at least*

$$\left| \bigcup_{k \in I} L_k \right| \geqslant |I|.$$

Then it is possible to give each of your n friends one gift that they like.

Proof. We define a partial order $\leqslant$ on the set

$$M = \{g_1, g_2, \ldots, g_m, f_1, f_2, \ldots, f_n\}$$

by the reflexive relation that satisfies $g_k \leqslant f_\ell$ if and only if $g_k \in L_\ell$. Distinct gifts are not comparable and distinct friends are not comparable either. A chain in M either consists of a single element or of a gift liked by a friend, $g_k \leqslant f_\ell$.

An antichain in M is given by $\{g_1, g_2, \ldots, g_m\}$. We claim that this is a maximal antichain in $(M, \leqslant)$. Indeed, seeking a contradiction, suppose that there is an antichain A in M that contains a gifts and more than $m - a$ friends, so $|A| > m$. The union of the gifts liked by the friends must contain more than $m - a$ gifts, but it cannot contain any of the a gifts contained in A. This contradicts the fact that there are only m gifts. In other words, a maximal antichain contains m elements. By Dilworth's theorem, this means that we can cover M with m chains.

Each gift is either covered by a chain of length 1, which means that the gift is not given to anyone, or it is covered by a chain of length 2, in which case one friend receives the gift. Therefore, after re-indexing any partition of M into chains can be written in the form

$$\{\{g_1 \leqslant f_1\}, \{g_2, \leqslant f_2\}, \ldots, \{g_k \leqslant f_k\}, \{g_{k+1}\}, \ldots, \{g_m\}, \{f_{k+1}\}, \ldots, \{f_n\}\}.$$

In other words, a partition into merely m chains must be of the form

$$\{\{g_1 \preccurlyeq f_1\}, \{g_2, \preccurlyeq f_2\}, \ldots, \{g_k \preccurlyeq f_k\}, \{g_{k+1}\}, \ldots, \{g_m\}\},$$

so it cannot contain any friends that do not receive a gift. □

EXERCISES

6.17. Let $(S, \preccurlyeq)$ be a finite partially ordered set. Suppose that S can be decomposed as a union of n pairwise disjoint chains $C_1, \ldots, C_n$. Show that any antichain in S has at most n elements.

6.18. Let $(S, \preccurlyeq)$ be a finite partially ordered set. Suppose that the largest chain in S has n elements. Show that S can be partitioned into n pairwise disjoint antichains $A_1, \ldots, A_n$.

6.19. Let n be a positive integer. Let $A = (a_1, a_2, \ldots, a_{n^2+1})$ be some permutation of the integers from 1 to $n^2 + 1$. Show that there exist pairwise distinct indices $i_1 < i_2 < \ldots < i_{n+1}$ such that $(a_{i_1}, a_{i_2}, \ldots, a_{i_{n+1}})$ is either increasing or decreasing subsequence of A.

6.20. Let n be a positive integer. Let I denote a set of $n^2 + 1$ open intervals on the set $\mathbf{R}$ of real numbers. Prove that I either contains $n + 1$ intervals that overlap or $n + 1$ intervals that are pairwise disjoint.

6.21. Let $(P, \preccurlyeq)$ be a partial order of height h and width w with $n = |P|$ elements. Show that $n \leqslant hw$.

6.5 Lower and Upper Bounds

Let $(S, \preccurlyeq)$ be a partially ordered set and X a subset of S. An element u in S is called an **upper bound** for X if and only if $x \preccurlyeq u$ holds for all x in X. Notice that the upper bound u does not have to belong to the set X. A subset X of S is called **bounded above** if and only if there exists an upper bound u for X.

The **greatest element** of a subset X of a partially ordered set S is an upper bound that is contained in X itself. The greatest element of X is uniquely determined if it exists.

Example 6.25. Let $S = \{1, 2, 3, 4, 5, 6, 7, 8, 9, 10, 12\}$ be a set of integers partially ordered by divisibility. Two Hasse diagrams for $(S, |)$ are shown as follows.

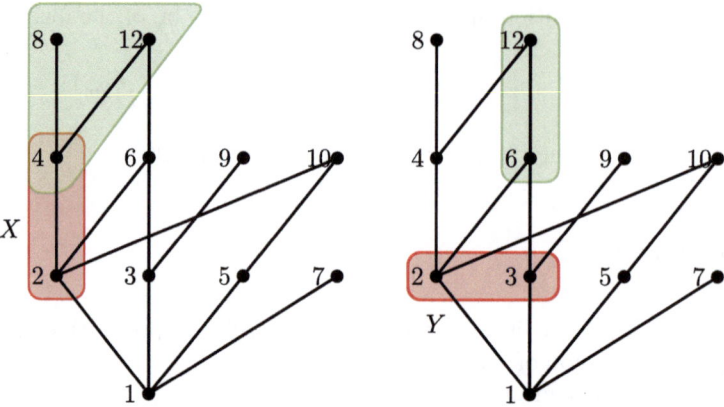

Consider the subset $X = \{2, 4\}$ of S. It is shaded red in the left Hasse diagram. The elements 4, 8, and 12 are upper bounds of X in S. The element 4 is the greatest element of X. On the other hand, consider the subset $Y = \{2, 3\}$ of S in the right Hasse diagram. It has 6 and 12 as upper bounds. However, Y does not have a greatest element, since 2 and 3 do not divide each other.

An element ℓ in S is called a **lower bound** for X if and only if $\ell \leqslant x$ holds for all x in X. A subset X of S is called **bounded below** if and only if there exists a lower bound for X. The **least element** of a subset X of S is a lower bound for X that is contained in X. The least element of a subset X of S does not need to exist, but if it does, then it is uniquely determined.

Example 6.26. We keep the notation of the previous example. We illustrate lower bound in the two Hasse diagrams as follows.

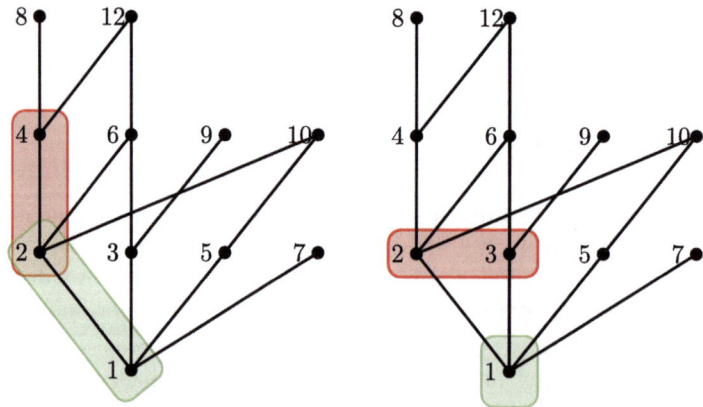

In the Hasse diagram on the left, we consider the set $X = \{2, 4\}$ shown in red. The elements 1 and 2 are lower bounds for $X = \{2, 4\}$ in S, and 2 is a least element of X. In the Hasse diagram on the right, we consider the subset

6.5 Lower and Upper Bounds

$Y = \{2,3\}$ of S. The element 1 is a lower bound for $Y = \{2,3\}$, but Y does not have a least element.

Let X be a subset of a partially ordered set S. We denote by $U(X)$ the set of all upper bounds for X in S. The least element of $U(X)$ is called the **least upper bound** or the **supremum** of X in S. The least upper bound of X in S is uniquely determined if it exists. The least upper bound of X in S is denoted by

$$\sup_S X \quad \text{or} \quad \bigvee_S X.$$

The subscript S is omitted if it is clear from the context. For sets $X = \{x, y\}$ with two elements, we denote $\bigvee X$ also in the infix notation $x \vee y$.

Example 6.27. We keep the notation of the previous two examples. For $X = \{2,4\}$, the set $U(X)$ of upper bounds in S is given by $U(X) = \{4, 8, 12\}$. The least upper bound of X in S is given by the element 4. The least upper bound of $Y = \{2,3\}$ in S is given by 6.

Example 6.28. Let $\mathbf{R}$ be the set of real numbers in its natural order. Let $X = \{x \in \mathbf{Q} \mid x^2 \leqslant 2\}$ be the set of rational numbers whose square is less than 2. Then the least upper bound of X in $\mathbf{R}$ is given by $\sqrt{2}$.

Let X be a subset of a partially ordered set S. We denote by $L(X)$ the set of lower bounds of X in S. The greatest element of $L(X)$ is called the **greatest lower bound** or the **infimum** of X in S. The greatest lower bound of X in S is denoted by

$$\inf_S X \quad \text{or} \quad \bigwedge_S X.$$

The subscript S is omitted if it is clear from the context. If $X = \{x, y\}$, then $\bigwedge X$ is also written as $x \wedge y$.

Example 6.29. Let us keep the notation of Example 6.25. The set $X = \{2,4\}$ has the lower bounds $L(X) = \{1, 2\}$ in S. So the greatest lower bound of X is 2. The greatest lower bound of $Y = \{2,3\}$ in S is 1.

Example 6.30. Let $(S, \preccurlyeq)$ be a partially ordered set. Then the set of upper bounds of the empty set is given by $U(\emptyset) = S$, so $\sup \emptyset$ exists if and only if S has a least element. Similarly, the set of lower bounds of the empty set is given by $L(\emptyset) = S$, and $\inf \emptyset$ exists if and only if S has a greatest element.

Let $(S, \preccurlyeq)$ be a partially ordered set. An element m of S is called **maximal** if and only if for all x in S the relation $m \preccurlyeq x$ implies $m = x$. In other words, a maximal element is not succeeded by any other element. There might exist several different maximal elements or none at all in a partially ordered set.

An element w of S is called **minimal** if and only if for all x in S the relation $x \preccurlyeq w$ implies $x = w$. In other words, a minimal element is not preceded by any other element. Minimal elements, if they exist, are in general not unique.

Example 6.31. Let $S = \{1, 2, 3, 4, 5, 6, 7, 8, 9, 10, 12\}$ be a set of integers partially ordered by divisibility. Two Hasse diagrams for $(S, |)$ are shown below.

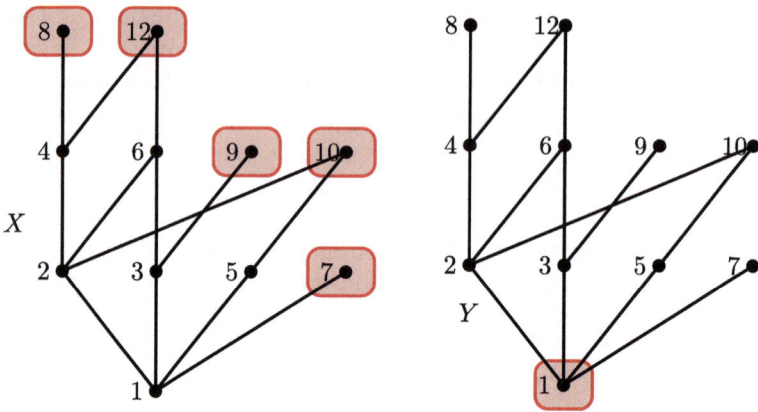

The Hasse diagram on the left shows the maximal elements 7, 8, 9, 10, 12 of the set S. The Hasse diagram on the right shows the minimal element 1 of the set S. In general, a partially ordered set might have more than one minimal element or none at all.

Example 6.32. The following example shown in the Hasse diagram below has two minimal elements a and b, but no least element. The greatest element is ⊤, which is also the unique maximal element.

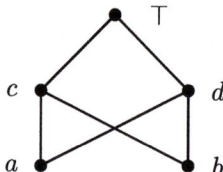

The supremum of c and d is ⊤ $= c \vee d$, but the infimum of c and d does not exist.

EXERCISES

6.22. Prove that a partially ordered set $(P, \leqslant)$ can have at most one least element and at most one greatest element.

6.23. Determine a partially ordered set $(P, \leqslant)$ with nonempty set P of smallest possible cardinality such that $(P, \leqslant)$ does not have a least nor a greatest element. Prove that there is indeed no partial order with fewer elements that has this property.

6.24. Let $(P, \leqslant)$ be a partially ordered set. Show that $(P, \leqslant)$ has a greatest element if and only if every subset of P is bounded above.

6.6 Extensions of Partial Orders

6.25. Show that a finite subset of a partially ordered set does not need to be (a) bounded above nor (b) bounded below.

6.26. Let $(S, \leqslant)$ be a partially ordered set and x and y elements of S such that the supremum $x \vee y$ exists. Show that $x \leqslant x \vee y$ and $y \leqslant x \vee y$.

6.27. Let $(S, \leqslant)$ be a partially ordered set and x and y elements of S. Suppose that the supremum $x \vee y$ exists. Show that $x \vee y = y$ if and only if $x \leqslant y$.

6.28. Let $(S, \leqslant)$ be a partially ordered set, x, y, and z elements of S such that $x \vee z$ and $y \vee z$ exist. Show that if $x \leqslant y$, then $x \vee z \leqslant y \vee z$.

6.29. Show that a finite partially ordered set $(S, \leqslant)$ has at least one minimal element and at least one maximal element.

6.30. Give a (simple) example of a partially ordered set that does not have any minimal nor any maximal elements.

6.31. Determine all minimal and all maximal elements of the set $P = \{x \in \mathbf{Z} \mid x \geqslant 2\}$ ordered by divisibility.

6.32. Let $(S, \leqslant)$ be a partially ordered set. Suppose that X is a subset of S. Let $U(X)$ denote the set of upper bound of X in S. Suppose that b is an upper bound of X that is a minimum in $U(X)$. Does this mean that b is a least upper bound for X? Prove your answer or give a counter example.

6.33. A partially ordered set $(S, \leqslant)$ is called **well-ordered** if and only if every nonempty subset of S has a least element. Show that $(S, \leqslant)$ must be a chain.

6.34. Give an example of a partially ordered set that is a chain but is not well-ordered.

6.6 Extensions of Partial Orders

Given two partial orders P and Q on a set S, we say that Q **extends** P if and only if $P \subseteq Q$ holds. In this section, we are going to show that each partial order can be extended to a total order. These results have numerous applications.

Proposition 6.33. *Let $\leqslant$ be a partial order on a set S. Let a and b be incomparable elements of S. Then there exists a partial order $\sqsubseteq$ extending $\leqslant$ such that $a \sqsubseteq b$.*

Proof. We are going to define the partial order $\sqsubseteq$ on S by including all pairs from the relation $\leqslant$, the pair (a, b), and additional pairs that restore transitivity. Let us define the sets

$$A = \{x \in S \mid x \leqslant a\} \quad \text{and} \quad B = \{x \in S \mid b \leqslant x\}.$$

We define the relation $\ll$ to be the Cartesian product $A \times B$. Hence, the elements in A precede the elements in B under $\ll$; in particular, $a \ll b$. We define $\sqsubseteq$ as the union of the relations $\leqslant$ and $\ll$. In other words, the relation $\sqsubseteq$ consists of the set of pairs

$$\{(x,y) \in S \times S \mid x \leqslant y\} \cup (A \times B).$$

It remains to show that $\sqsubseteq$ is a partial order. The relation $\sqsubseteq$ is reflexive, since it contains the reflexive relation $\leqslant$.

Seeking a contradiction, let us assume that $\sqsubseteq$ is not antisymmetric, that is, there must exist distinct elements x and y in S such that $x \sqsubseteq y$ and $y \sqsubseteq x$. These relations can arise from $x \leqslant y$ or $x \ll y$ and from $y \leqslant x$ or $y \ll x$, so we distinguish four cases. (a) If $x \leqslant y$ and $y \leqslant x$, then $x = y$ by antisymmetry of the relation $\leqslant$, which contradicts that x is distinct from y. (b) If $x \leqslant y$ and $y \ll x$, then $b \leqslant x$, $x \leqslant y$, $y \leqslant a$, so by transitivity $b \leqslant a$, which contradicts the incomparability of a and b. (c) If $x \ll y$ and $y \leqslant x$, then $b \leqslant y$, $y \leqslant x$, $x \leqslant a$, so by transitivity $b \leqslant a$, which contradicts the incomparability of a and b. (d) If $x \ll y$ and $y \ll x$, then $b \leqslant x$ and $x \leqslant a$, so by transitivity $b \leqslant a$, which once again contradicts the incomparability of a and b. Since we arrived at a contradiction in each of the four cases, we can conclude that $\sqsubseteq$ is antisymmetric.

For the transitivity, we need to show that $x \sqsubseteq y$ and $y \sqsubseteq z$ implies $x \sqsubseteq z$. We will distinguish for $x \sqsubseteq y$ the cases $x \leqslant y$ and $x \ll y$ and for $y \sqsubseteq z$ the cases $y \leqslant z$ and $y \ll z$, leading to a total of four cases. (a) If $x \leqslant y$ and $y \leqslant z$, then $x \leqslant z$ by transitivity. (b) If $x \leqslant y$ and $y \ll z$, then $x \leqslant y$, $y \leqslant a$, so by transitivity $x \leqslant a$, which implies that the element x lies in A, and as z is contained in B, we have $x \ll z$, hence, $x \sqsubseteq z$. (c) If $x \ll y$ and $y \leqslant z$, then $b \leqslant y$ and $y \leqslant z$, which implies $b \leqslant z$, so z is an element of B. As x is in A, we have $x \ll z$. (d) If $x \ll y$ and $y \ll z$, then x is in A and z is in B, so $x \ll z$. In all four cases, we can conclude that $x \sqsubseteq z$. □

The next result is known as Zorn's Lemma. It is frequently used to establish the existence of maximal elements in infinite partially ordered sets.

Theorem 6.34 (Zorn's Lemma). *Let $(S, \leqslant)$ be a partially ordered set. If every nonempty chain in S has an upper bound, then the set S has a maximal element.*

Proof. One can use a proof by contradiction to show that Zorn's lemma follows from the ZFC axioms. The shortest proofs use the principle of transfinite induction, but this technique is beyond the scope of this book. You can find a proof for instance in [36]. □

Zorn's lemma is often used as an axiom, since it is equivalent to the axiom of choice. The appeal of Zorn's lemma is that it can be applied quite easily. The proof of the next theorem is a typical example.

Theorem 6.35. *Every partial order can be extended to a total order.*

6.6 Extensions of Partial Orders

Proof. Let P be a partial order on a set S. We can define the family F of partial orders by

$$F = \{Q \mid Q \text{ is a partial order on } S \text{ containing } P \}.$$

The set family F is partially ordered by set inclusion. For every chain C in F, a least upper bound is given by $\bigcup C$. By Zorn's lemma, there must exist a maximal element M in F. It follows from Proposition 6.33 that M must be a total order. □

Example 6.36. We often have a list of tasks that we need to accomplish. For instance, if we get ready in the morning, we need to get dressed before heading out. This means that after we get up, we need to put on a jacket, pants, shoes, shirt, and socks, but preferably not in this order. It is best to put on the pants before wearing shoes. Also, we want to put on socks before wearing the shoes. The partial order of the tasks is shown in the Hasse diagram as follows.

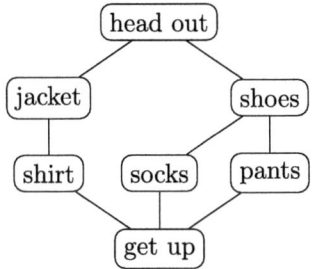

Evidently, it does not matter whether we put on socks first and then pants or the other way around. We can create a total order that is compatible with this partial order by choosing a minimum and then remove it, and repeat this process on the remaining set. This way, we can create for instance the total order

$$\text{get up} \preccurlyeq \text{shirt} \preccurlyeq \text{socks} \preccurlyeq \text{pants} \preccurlyeq \text{jacket} \preccurlyeq \text{shoes} \preccurlyeq \text{head out}.$$

This is not the only choice, since "get up" can be immediately followed by either "shirt," "socks," or "pants" in a total order.

EXERCISES

6.35. Find an extension of the partial order $(P(\{1, 2, 3\}), \subseteq)$ to a total order.

6.36. Let $(P, \preccurlyeq)$ be a finite partial order. (a) Show that there exists a finite set T of total orders extending P such that $\bigcap T = P$. (b) The smallest cardinality of such sets T having the property $\bigcap T = P$ is called the dimension of the partial order. What is the dimension of the partially ordered set $P = \{1, 2, 3, 4\}$ with $\preccurlyeq$ given by $\{(1,1), (2,2), (3,3), (4,4), (1,3), (1,4), (2,3), (2,4)\}$?

6.37. Show that a partially ordered set with n elements can have at most $n!$ linear extensions.

6.38. A function $f\colon S \to S$ from a partially ordered set $(S, \leqslant)$ to itself is called **inflationary** if and only if $x \leqslant f(x)$ holds for all $x \in S$. Suppose that $(S, \leqslant)$ is a partially ordered set such that every nonempty chain has an upper bound. Show that an inflationary function f of S into itself has a fixed point. In other words, show that there exists an element y in S such that $y = f(y)$.

6.7 Monotonic Functions

Let $(A, \leqslant_A)$ and $(B, \leqslant_B)$ be partially ordered sets. A function $f\colon A \to B$ is called **monotonic** if and only if $x \leqslant_A y$ implies $f(x) \leqslant_B f(y)$ for all elements x, y in A. In other words, if x precedes y, then $f(x)$ must precede $f(y)$; however, if x and y are incomparable, then there is no restriction on the ordering of their images $f(x)$ and $f(y)$.

Example 6.37. Let $S = \{1, 2, 3\}$. Suppose that two partial orders on S are given by $\leqslant_A = \{(1,1), (2,2), (3,3), (1,2), (1,3)\}$ and

$$\leqslant_B = \{(1,1), (2,2), (3,3), (1,2), (2,3), (1,3)\}.$$

The identity function $f\colon S \to S$ given by $f(x) = x$ is a monotonic function from $(S, \leqslant_A)$ to $(S, \leqslant_B)$.

Example 6.38. Let X be a set. A monotonic function from the partially ordered set $(P(X), \subseteq)$ to the reversely ordered set $(P(X), \supseteq)$ is given by $f(A) = X \setminus A$. Indeed, if $A \subseteq B$ are subsets of X, then $f(A) = X \setminus A$ contains $f(B) = X \setminus B$.

A function $f\colon A \to B$ is called an **embedding** of the partial order $(A, \leqslant_A)$ into the partial order $(B, \leqslant_B)$ if and only if for all x, y in A, we have $x \leqslant_A y$ if and only if $f(x) \leqslant_B f(y)$.

A function $f\colon A \to B$ is called an **isomorphism** of the partial orders $(A, \leqslant_A)$ and $(B, \leqslant_B)$ if and only if f is a bijective function and an embedding.

The next theorem shows that any partially ordered set $(A, \leqslant)$ can be embedded into the power set $P(A)$ ordered by set inclusion.

Theorem 6.39. *Let $(A, \leqslant)$ be a partially ordered set. Let $f\colon A \to P(A)$ be the function given by*

$$f(y) = \{x \in A \mid x \leqslant y\}.$$

Then f is an embedding of the partially ordered set $(A, \leqslant)$ into $(P(A), \subseteq)$.

Proof. Suppose that y and z are elements of A such that $y \leqslant z$. For all x in A such that $x \leqslant y$, the transitivity of relation $\leqslant$ implies that

$$f(y) = \{x \in A \mid x \leqslant y\} \subseteq f(z) = \{x \in A \mid x \leqslant z\}.$$

6.7 Monotonic Functions

Therefore, the function f is monotonic.

On the other hand, suppose that $f(y) \subseteq f(z)$ holds. By reflexivity of the relation $\leqslant$, the set $f(y)$ contains y. Thus, we can conclude that $y \in f(y) \subseteq f(z)$. Hence, it follows from the definition of the set $f(z)$ that $y \leqslant z$.

Thus, we have shown that $x \leqslant z$ if and only if $f(x) \subseteq f(z)$. □

EXERCISES

6.39. Show that an embedding f of the partial order $(A, \leqslant_A)$ into the partial order $(B, \leqslant_B)$ is an injective function.

6.40. For positive integers a and b, let $f_{a,b}(x) = ax + b$ be a function from the set of positive integers in its natural order into itself. Determine for each pair (a, b) whether $f_{a,b}$ is (a) a monotonic function, (b) an embedding, (c) an isomorphism, or (d) none of the above.

6.41. Let $(P, \leqslant_P)$, $(Q, \leqslant_Q)$, and $(R, \leqslant_R)$ be partial orders. Prove that if $f: P \to Q$ and $g: Q \to R$ are monotone functions, then their composition $g \circ f$ is a monotone function from P to R.

6.42. We can order the Boolean value by declaring $F \leqslant T$. Moreover, we can partially order vectors x and y in $\{F, T\}^n$ by defining $x \leqslant y$ if and only if $x_k \leqslant y_k$ for all k in the range $1 \leqslant k \leqslant n$. A Boolean function $f: \{F, T\}^n \to \{F, T\}$ is **monotone** if and only if $x \leqslant y$ implies $f(x) \leqslant f(y)$. Which of the following Boolean functions are monotone? In each case, explain why or why not the function is monotone.
(a) $x \wedge y$
(b) $x \vee y$
(c) $x \oplus y$
(d) $x \vee \neg y$

6.43. The set $N_1 = \{1, 2, 3, \ldots\}$ of positive integers is partially ordered by the divisibility relation $|$, where $a \mid c$ if and only if there exist an integer b such that $c = ab$. The power set $P(N_1)$ is partially ordered by the subset relation $\subseteq$. Consider the function f that maps a positive integer to its set of prime factors, so, for instance, $f(1) = \emptyset$ and $f(18) = \{2, 3\}$.
(a) Show that $f: (N_1, |) \to (P(N_1), \subseteq)$ is a monotone function
(b) Is f an embedding of partial orders?
(c) Is f an isomorphism of partial orders?

6.44. Prove or disprove the following statements.
(a) A monotone function of partial orders preserves maxima and minima.
(b) An embedding of partial orders preserves maxima and minima.
(c) An isomorphism of partial orders preserves maxima and minima.

6.45. (a) Show that a monotone function maps chains to chains.
(b) Does a monotone function map antichains to antichains? Prove or disprove.

6.8 Lattices

A partially ordered set S in which every pair of elements x and y has a supremum $x \vee y$ and an infimum $x \wedge y$ is called a **lattice**. It follows by induction that every finite set of elements in S has a supremum and an infimum.

A lattice is called **complete** if and only if every subset of the lattice has a supremum and an infimum.

Example 6.40. The set of positive integers ordered by divisibility forms a lattice. Indeed, the supremum of two positive integers x and y is given by the least common multiple $\sup\{x, y\} = \text{lcm}(x, y)$ and their infimum is given by the greatest common divisor $\inf\{x, y\} = \gcd(x, y)$. This lattice is not complete, since the supremum of the subset of all prime numbers does not exist.

Example 6.41. Let X be a set and $S = P(X)$ its power set ordered by inclusion. The supremum of a family F of subsets of X is given by $\sup F = \bigcup F$ and the infimum of this set family is given by $\inf F = \bigcap F$. Thus, $(S, \subseteq)$ is a complete lattice, since the supremum and infimum exist for any family of subsets of X.

Proposition 6.42. *Let $(S, \leqslant)$ be a partially ordered set such that every subset has an infimum. Then S is a complete lattice.*

Proof. Let X be a subset of S and $U(X)$ the set of upper bounds of X in S. The set X belongs to the set of lower bounds of $U(X)$. If we set $b = \inf U(X)$, then $x \leqslant b$ for all x in X, as b is the greatest lower bound of $U(X)$. For any other upper bound d in $U(X)$, we have $b \leqslant d$ by definition of the infimum. Therefore, $b = \sup X$ is the least upper bound of X. Thus, S is a complete lattice, since $\inf X$ and $\sup X$ exist for each subset X of S. □

Lemma 6.43. *Let $X = \{x_1, \ldots, x_n\}$ be a finite nonempty subset of a lattice L. Then the infimum of X exists and is given by*

$$\inf X = (\cdots((x_1 \wedge x_2) \wedge x_3) \wedge \cdots \wedge x_n).$$

Proof. We prove this by induction on the number n of elements in X. If $X = \{x_1\}$ contains a single element, then $\inf X = x_1$. Let us assume that $d = \inf\{x_1, \ldots, x_{n-1}\}$ exists and is of the form $d = (\cdots((x_1 \wedge x_2) \wedge x_3) \wedge \cdots \wedge x_{n-1})$. We form

$$e = (\cdots((x_1 \wedge x_2) \wedge x_3) \wedge \cdots \wedge x_n) = d \wedge x_n.$$

We have $d \leqslant x_1, \ldots, d \leqslant x_{n-1}$, since d is a greatest lower bound, $e \leqslant d$ and $e \leqslant x_n$. By transitivity, we get $e \leqslant x_1, \ldots, e \leqslant x_n$. Let f be any lower bound of $\{x_1, \ldots, x_n\}$. Since f is a lower bound for x_n and $\{x_1, \ldots, x_{n-1}\}$ and d is the greatest lower bound for the latter set, we have $f \leqslant x_n$ and $f \leqslant d$. Therefore,

$$f \leqslant e = d \wedge x_n$$

and we can conclude that $e = \inf\{x_1, \ldots, x_n\}$, as claimed. □

6.8 Lattices

Proposition 6.44. *Every finite lattice is complete.*

Proof. Let L be a finite lattice. By Proposition 6.42, it suffices to show that every subset X of L has an infimum. The infimum of the empty set is given by $\inf \emptyset = \sup L$, see Example 6.30. Furthermore, every nonempty subset X of L has an infimum by Lemma 6.43. Therefore, the lattice L is complete. □

A map F from a partially ordered set into itself is said to have a fixpoint x if and only if $F(x) = x$ holds. Ensuring that a map does have fixed points is often a concern in many applications. The following theorem asserts the existence of fixed points of monotonic maps on complete lattices.

Theorem 6.45 (Knaster-Tarski). *Let $(P, \leqslant)$ be a complete lattice. A monotonic map $f\colon P \to P$ has a nonempty set of fixed points, and a least and greatest fixed point exists.*

Proof. Consider the set $A = \{x \in P \mid f(x) \leqslant x\}$ of elements that are preceded by their image under f and set $\alpha = \inf A$. Then for each x in A, we have $\alpha \leqslant x$, so $f(\alpha) \leqslant f(x) \leqslant x$, and we conclude that $f(\alpha)$ is a lower bound for A. However, then $f(\alpha) \leqslant \alpha = \inf A$, so α is an element in A. Since f is monotonic, we get $f(f(\alpha)) \leqslant f(\alpha)$, whence $f(\alpha)$ is an element of A, which means $\alpha = \inf A \leqslant f(\alpha)$, and we can conclude that $\alpha = f(\alpha)$ is a fixed point.

Similarly, consider the set $W = \{x \in P \mid x \leqslant f(x)\}$ of elements that precede their image under f and set $\omega = \sup W$. Then for each x in A, we have $x \leqslant \omega$, so $x \leqslant f(x) \leqslant f(\omega)$, and $f(\omega)$ is an upper bound for W. Since ω is the least upper bound of W, we have $\omega \leqslant f(\omega)$, and we can conclude that ω is an element of W. Applying the monotonic function f, we get $f(\omega) \leqslant f(f(\omega))$, so $f(\omega)$ is an element of W. Therefore, $f(\omega) \leqslant \omega$ and as $\omega \leqslant f(\omega)$ holds, we can conclude that $\omega = f(\omega)$ is a fixed point.

Since each fixed point p of f is contained in both sets A and W, we have $\alpha \leqslant p \leqslant \omega$. Therefore, α is the least fixed point and ω is the greatest fixed point of f. □

The power set of a set is a complete lattice. As a consequence of the previous theorem, we obtain a relative simple proof[2] of the celebrated Schröder–Bernstein theorem.

Theorem 6.46 (Schröder–Bernstein). *Let A and B be sets. If there exists an injective map $f\colon A \to B$ and an injective map $g\colon B \to A$, then there exists a bijective map $h\colon A \to B$.*

Proof. Suppose that $C \subseteq D$ are two nested subsets of the set A. Then $f(C) \subseteq f(D)$ preserves the inclusion, $B - f(C) \supseteq B - f(D)$ reverses the inclusion, and applying g yields $g(B - f(C)) \supseteq g(B - f(D))$. Complementing once more, the inclusion reverses once again and we get $A - g(B - f(C)) \subseteq A - g(B - f(D))$.

---

[2] Take a look at some of the standard proofs of this theorem to judge for yourself.

Therefore, the map $m\colon P(A) \to P(A)$ given by $m(X) = A - g(B - f(X))$ is monotonic.

Since the power set $P(A)$ is a complete lattice, the function m has a fixed point F by the Knaster–Tarski theorem, so $m(F) = F = A - g(B - f(F))$. We define a map $h\colon A \to B$ by

$$h(x) = \begin{cases} f(x) & \text{if } x \in F, \\ g^{-1}(x) & \text{if } x \in A - F = g(B - f(F)). \end{cases}$$

The function h is surjective. Indeed, consider an element $b \in B$. If $b \in f(F)$, then there exists an $x \in F$ such that $b = f(x) = h(x)$ by the first case of the definition of h. If $b \in B - f(F)$, then $g(b) \in g(B - f(F)) = A - F$, so $b = g^{-1}(g(b)) = h(g(b))$ by the second case of the definition of h.

The function h is injective. Indeed, suppose that x and y are distinct elements of A. If x and y are both elements of F, then $h(x) = f(x) \neq f(y) = h(y)$ by injectivity of f. If x and y are both elements of $A - F$, then $h(x) = g^{-1}(x) \neq g^{-1}(y) = h(y)$ by the injectivity of g^{-1}. If $x \in F$ and $y \in A - F = g(B - f(F))$, then $h(x) \in h(F) = f(F)$ and $h(y) \in h(A - F) = g^{-1}(g(B - f(F))) = B - f(F)$, so $h(x) \neq h(y)$. Hence, for all distinct x and y in A, we have $h(x) \neq h(y)$. □

EXERCISES

6.46. Prove that any totally ordered set $(S, \leqslant)$ is a lattice.

6.47. The set $\mathbf{R}$ of real numbers under its natural order is a lattice, and so are its subintervals. Let x and y denote real numbers such that $x < y$.
(a) Is $\mathbf{R}$ a complete lattice?
(b) Is $(-\infty, x)$ a complete lattice?
(c) Is $(-\infty, x]$ a complete lattice?
(d) Is (x, y) a complete lattice?
(e) Is $[x, y]$ a complete lattice?

6.48. Let $(S, \leqslant)$ be a lattice. Show that for all elements x, y, and z, we have the following laws,
(a) $x \vee y = y \vee x$ (commutative law for suprema),
(b) $x \wedge y = y \wedge x$ (commutative law for infima),
(c) $x \vee x = x$ (idempotent law for suprema),
(d) $x \wedge x = x$ (idempotent law for infima),
(e) $x \vee (y \vee z) = (x \vee y) \vee z$ (associative law for suprema),
(f) $x \wedge (y \wedge z) = (x \wedge y) \wedge z$ (associative law for infima),
(g) $x \vee (x \wedge y) = x$ (absorption law),
(h) $x \wedge (x \vee y) = x$ (absorption law).

6.49. Let $(S, \leqslant)$ be a lattice. Suppose that x, x', y, y' are elements in S satisfying $x \leqslant x'$ and $y \leqslant y'$. Show that $x \vee y \leqslant x' \vee y'$ and $x \wedge y \leqslant x' \wedge y'$ hold.

6.50. Prove that the set $\mathbf{N}_0$ of nonnegative integers is a complete lattice when ordered by the divisibility relation.

6.51. The extended real line $\overline{\mathbf{R}}$ consists of the set of real numbers and two elements $-\infty$ and $+\infty$ such that $-\infty < x < \infty$ for all real numbers x. Show that the extended real line is a complete lattice.

6.52. Use the Schröder–Bernstein theorem to show that the set $\mathbf{R}$ of real numbers has the same cardinality as the open interval $(-1, 1)$.

6.9 Notes

There exist many excellent textbooks on partially ordered sets and lattices that can be used for further reading. Davey and Priestley [17] and Roman [63] give a general overview of partially ordered sets and lattices. Caspard et al. [12] emphasize finite partially ordered sets and some of their applications. Trotter [76] gives a detailed account of dimension theory of partially ordered sets. Many textbooks on algebra or combinatorics contain chapters on partially ordered set and lattices; see, for example, Cohn [14], Jacobson [39], or Jukna [43].

Chapter 7

Floor and Ceiling Functions

Are you really sure that a floor can't also be a ceiling?

— Maurits C. Escher, *On Being a Graphic Artist*

In this chapter, we learn about functions that round real numbers to an integer. The floor function rounds the numbers down, and the ceiling function rounds the numbers up. We will explore a few elementary properties of these functions. Despite their simplicity, the floor and ceiling functions find many applications.

7.1 Rounding Up and Down

The **floor** is a function from the set $\mathbf{R}$ of real numbers to the set $\mathbf{Z}$ of integers that maps a real number x to the greatest integer n not exceeding x. The floor function of a real number x is denoted by $\lfloor x \rfloor$. In other words, the floor function of a real number x is defined to be equal to the integer $n = \lfloor x \rfloor$ if and only if $n \leqslant x < n+1$.

The floor function $\lfloor x \rfloor$ rounds the real number x down to the greatest integer that is smaller or equal to x. If we restrict the floor function to the set of integers, then it is the identity; thus, $\lfloor n \rfloor = n$ for all integers n. The graph of the floor function is shown in Fig. 7.1.

Example 7.1. We have $\lfloor 0.3 \rfloor = 0$, $\lfloor 0.5 \rfloor = 0$, $\lfloor 0.9999 \rfloor = 0$, and $\lfloor 1.01 \rfloor = 1$. For negative numbers, one should pay attention to the fact that rounding down does not mean rounding toward 0. Indeed, $\lfloor -0.3 \rfloor = -1$, $\lfloor -0.7 \rfloor = -1$, and $\lfloor -1.1 \rfloor = -2$.

Similarly, the **ceiling** is a function from the set $\mathbf{R}$ of real numbers to the set $\mathbf{Z}$ of integers that maps a real number x to the smallest integer n greater than or equal to x. The ceiling function of a real number x is denoted by $\lceil x \rceil$. In other words, the ceiling of a real number x is defined to be the integer $n = \lceil x \rceil$ if and only if $n - 1 < x \leqslant n$. If we restrict the ceiling function to the set of

© The Author(s), under exclusive license to Springer Nature Switzerland AG 2025
A. Klappenecker, H. Lee, *Discrete Structures*, Undergraduate Texts in Mathematics, https://doi.org/10.1007/978-3-031-73434-2_7

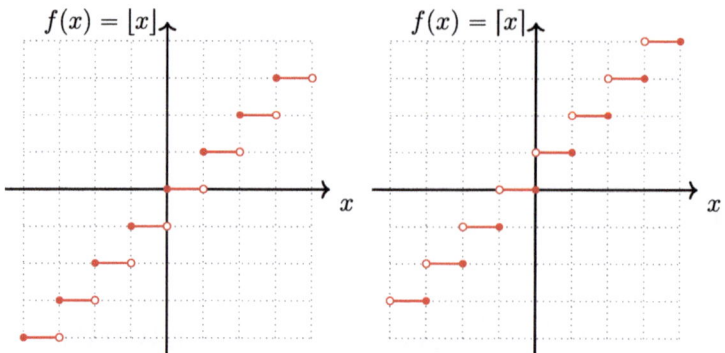

Figure 7.1: The graph on the left depicts the floor function, and the graph on the right depicts the ceiling function

integers, then it is the identity, so $\lceil n \rceil = n$ for all integers n. The graph of the ceiling function is shown in Fig. 7.1.

Example 7.2. We have $\lceil 0.1 \rceil = 1$, $\lceil 0.5 \rceil = 1$, $\lceil 0.999 \rceil = 1$, and $\lceil 1.1 \rceil = 2$. For negative numbers, we have $\lceil -0.3 \rceil = 0$, $\lceil -0.7 \rceil = 0$, and $\lceil -1.1 \rceil = -1$.

The floor and ceiling functions are closely related. The next proposition shows that you can use the floor function to express the ceiling function and vice versa.[1] You should pay close attention to the style of proof that we use here in the argument, since it is typical for proofs of many other properties of these two functions.

Proposition 7.3. *For all real numbers x, we have*
$$\lfloor -x \rfloor = -\lceil x \rceil.$$

Proof. By definition, the floor of the real number $-x$ is equal to the integer $n = \lfloor -x \rfloor$ if and only if $n \leq -x < n+1$. Multiplying the terms in these inequalities by -1 yields $-n \geq x > -(n+1)$, which we can express in terms of the ceiling function as $-n = \lceil x \rceil$. It follows that $\lfloor -x \rfloor = n = -\lceil x \rceil$ holds, as claimed. □

Proposition 7.4. *For any real number x and any integer k, we have*
$$\lfloor x + k \rfloor = \lfloor x \rfloor + k, \quad \text{and} \quad \lceil x + k \rceil = \lceil x \rceil + k.$$

Proof. By definition of the floor function, we have $\lfloor x \rfloor \leq x < \lfloor x \rfloor + 1$. Adding the integer k to each term yields
$$\lfloor x \rfloor + k \leq x + k < \lfloor x \rfloor + k + 1.$$

---

[1] In this case, the coincidence of floor and ceiling is even more real than in the optical illusions of M.C. Escher.

7.1 Rounding Up and Down

Therefore, we can conclude that $\lfloor x + k \rfloor$ is equal to the integer $\lfloor x \rfloor + k$, which proves our first claim.

By the previous proposition and the first claim, we have

$$\lceil x + k \rceil = -\lfloor -x - k \rfloor = -(\lfloor -x \rfloor - k) = -\lfloor -x \rfloor + k = \lceil x \rceil + k,$$

which proves the second claim. □

You should memorize the inequalities characterizing the floor and ceiling functions. If you compare a floor or ceiling with an integer, then the floor or ceiling function might not be needed, as the following proposition shows. It is a simple but instructive exercise to prove this proposition.

Proposition 7.5. *Suppose that x is a real number and n an integer. Then*
(a) $\lfloor x \rfloor < n$ *if and only if* $x < n$,
(b) $n \leqslant \lfloor x \rfloor$ *if and only if* $n \leqslant x$,
(c) $n < \lceil x \rceil$ *if and only if* $n < x$,
(d) $\lceil x \rceil \leqslant n$ *if and only if* $x \leqslant n$.

Proof. See Exercise 7.4. □

When programming, we often have to split an array into two parts that are nearly of equal length. The next proposition shows that taking the first $\lfloor n/2 \rfloor$ elements and the last $\lceil n/2 \rceil$ elements of the array will do the trick.

Proposition 7.6. *For any integer n, we have*

$$n = \left\lfloor \frac{n}{2} \right\rfloor + \left\lceil \frac{n}{2} \right\rceil.$$

Proof. If n is an even integer, then there exists an integer k such that $n = 2k$. It follows that $\lfloor n/2 \rfloor = k = \lceil n/2 \rceil$, so $n = 2k = \lfloor n/2 \rfloor + \lceil n/2 \rceil$.

If n is an odd integer, then there exists an integer k such that $n = 2k + 1$. It follows that $\lfloor n/2 \rfloor = \lfloor k + 1/2 \rfloor = k$ and $\lceil n/2 \rceil = \lceil k + 1/2 \rceil = k + 1$, so

$$n = 2k + 1 = k + (k + 1) = \lfloor n/2 \rfloor + \lceil n/2 \rceil.$$

Since an integer is either even or odd, this proves the claim. □

We conclude this section with a useful result concerning the number of digits in base b representation. Recall that a positive integer a can be represented in base b in the form

$$a = \sum_{k=0}^{n} a_k b^k,$$

where the digits a_k are integers in the range $0 \leqslant a_k < b$, and the leading coefficient $a_n \neq 0$. In base b, the number a can be represented by its digits $a_n a_{n-1} \cdots a_1 a_0$. If the base b is not understood from the context, then it is customary to add the base b as a subscript.

For example, the number $1000 = 10^3$ requires 4 digits in base $b = 10$. The next proposition shows that a representation of 1000 in base 2 requires

$$\lfloor \log_2 1000 \rfloor + 1 = \lfloor 9.9657 \cdots \rfloor + 1 = 10$$

digits. Indeed, the integer 1000 is represented in base 2 by the 10-digit number 1111101000_2. The next proposition allows you to accurately determine the number of digits in base b without calculating a base change.

Proposition 7.7. *The number d of digits of a positive integer n in base b is given by*
$$d = \lfloor \log_b n \rfloor + 1 \quad \text{and} \quad d = \lceil \log_b (n+1) \rceil.$$

Proof. Suppose that a positive integer n has d digits in base b. Then it is of the form
$$n = \sum_{k=0}^{d-1} a_k b^k,$$
where a_k is some integer in the range $0 \leq a_k < b$. It follows that
$$b^{d-1} \leq n < b^d, \tag{7.1}$$
as b^{d-1} is the smallest d-digit integer in base b, and $b^d - 1$ is the largest d-digit integer. Taking the logarithm in base b yields
$$d - 1 \leq \log_b n < d.$$
Therefore, $d - 1 = \lfloor \log_b n \rfloor$ or $d = \lfloor \log_b n \rfloor + 1$, as claimed.

For the version with the ceiling function, we add 1 to the terms in (7.1) and deduce the inequalities
$$b^{d-1} < n + 1 \leq b^d$$
as a consequence. Applying the logarithm to base b yields
$$d - 1 < \log_b(n+1) \leq d.$$
Therefore, $d = \lceil \log_b(n+1) \rceil$, which proves our second claim. □

EXERCISES

7.1. Determine the values of the following floor and ceiling functions.
(a) $\lfloor (2^{32} - 1)/8 \rfloor$
(b) $\lfloor e^{(e^e)} \rfloor$
(c) $\lfloor -1.234 \rfloor$
(d) $\lceil \pi^\pi \rceil$
(e) $\lceil (2^{32} - 1)/33 \rceil$
(f) $\lceil -1.92 \rceil$

7.1 Rounding Up and Down

7.2. Let x be a real number and n an integer. Show that the floor and ceiling function satisfy
(a) $\lfloor x \rfloor = n$ if and only if $x - 1 < n \leq x$.
(b) $\lceil x \rceil = n$ if and only if $x \leq n < x + 1$.

7.3. Characterize the set
$$\{x \in \mathbf{R} \mid \lfloor x \rfloor = \lceil x \rceil\}$$
of arguments where the floor and ceiling functions coincide. Prove your result.

7.4. Prove Proposition 7.5. In other words, show that if x is a real number and n an integer, then
(a) $\lfloor x \rfloor < n$ if and only if $x < n$,
(b) $n \leq \lfloor x \rfloor$ if and only if $n \leq x$,
(c) $n < \lceil x \rceil$ if and only if $n < x$,
(d) $\lceil x \rceil \leq n$ if and only if $x \leq n$.
As a cautionary tale, find **counterexamples** to each of the following statements:
(e) $\lfloor x \rfloor \leq n$ if and only if $x \leq n$,
(f) $n < \lfloor x \rfloor$ if and only if $n < x$,
(g) $n \leq \lceil x \rceil$ if and only if $n \leq x$,
(h) $\lceil x \rceil < n$ if and only if $x < n$.
The moral of the story is that you need to be very careful when you omit floor or ceiling functions in integer comparisons.

7.5. Graph the function $f(x) = \sin\left(\frac{\pi}{2}\left\lceil \frac{x}{3}\right\rceil\right)$ on the interval $[-5, 5]$.

7.6. Show that for all real numbers x, we have
$$2\lfloor x \rfloor \leq \lfloor 2x \rfloor \leq 2\lfloor x \rfloor + 1.$$

7.7. Let n be an integer and m a positive integer. Show that
$$\left\lfloor \frac{n+m-1}{m} \right\rfloor = \left\lceil \frac{n}{m} \right\rceil.$$

7.8. Rep. Quattro Sei plays around with his calculator and observes that
$$\lfloor \sqrt{44} \rfloor = 6, \quad \lfloor \sqrt{4444} \rfloor = 66, \quad \lfloor \sqrt{444444} \rfloor = 666.$$
Show that in general the floor of the square root of the integer consisting of $2n$ repeated digits of 4s is given by the integer consisting of n repeated digits of 6s, so that
$$\lfloor \sqrt{\underbrace{44 \cdots 4}_{\text{4 is repeated } 2n \text{ times}}} \rfloor = \underbrace{66 \cdots 6}_{\text{6 is repeated } n \text{ times}}$$
holds for all positive integers n.

7.9. Show that for any real number x, we have
$$0 \leqslant \lfloor 2x \rfloor - 2\lfloor x \rfloor \leqslant 1.$$

7.10. Let $f \colon \mathbf{N}_1 \to \mathbf{N}_1$ be the function on the set of positive integers given by
$$f(x) = \left\lfloor n + \sqrt{n} + \frac{1}{2} \right\rfloor.$$
Show that if $m \notin \mathrm{ran}(f)$, then m must be a perfect square.

7.11. In January 2016, the largest known prime was
$$p = 2^{74,207,281} - 1,$$
a number with 22,338,618 decimal digits. You join a team building a computer based on ternary arithmetic that is supposed to find even larger primes. How many ternary digits are needed to represent p?

7.12. Show that for any real number x and integer n, we have
$$\lfloor nx \rfloor = \lfloor x \rfloor + \left\lfloor x + \frac{1}{n} \right\rfloor + \left\lfloor x + \frac{2}{n} \right\rfloor + \cdots + \left\lfloor x + \frac{n-1}{n} \right\rfloor.$$

7.13. Evaluate the sum
$$\sum_{k=1}^{n^2-1} \lfloor \sqrt{k} \rfloor.$$
Hint: First determine the range of positive integers k satisfying $a = \lfloor \sqrt{k} \rfloor$.

7.14. (a) Prove by induction that
$$\sum_{k=0}^{n} k 2^k = (n-1)2^{n+1} + 2$$
holds for all nonnegative integers n.
(b) Evaluate the sum
$$\sum_{k=1}^{n} \lfloor \log_2 k \rfloor.$$

7.15. Show that for all nonnegative integers n, we have
$$\sum_{k=0}^{\infty} \left\lfloor \frac{n + 2^k}{2^{k+1}} \right\rfloor = n.$$

(Source: 1968 IMO Problem 6)

7.16. Show that the function $f: \mathbf{R} \to \mathbf{R}$ given by

$$f(x) = \lim_{n \to \infty} \left[(\cos(n!x\pi))^2 \right]$$

satisfies

$$f(x) = \begin{cases} 1 & \text{if } x \in \mathbf{Q}, \\ 0 & \text{if } x \in \mathbf{R}\backslash\mathbf{Q}. \end{cases}$$

(Source: de Koninck, Mercier [18])

7.2 Divisibility and Primes

The next proposition determines the number of positive integer multiples of a positive integer k that do not exceed a bound x.

Proposition 7.8. *Let x be a real number and k a positive integer such that $k \leqslant x$. Then the number m of positive integers that are divisible by k and do not exceed x is given by*

$$m = \left\lfloor \frac{x}{k} \right\rfloor.$$

Proof. By definition of the floor function, we have

$$\left\lfloor \frac{x}{k} \right\rfloor \leqslant \frac{x}{k} < \left\lfloor \frac{x}{k} \right\rfloor + 1.$$

Multiplying the terms of these inequalities by k, we get

$$\left\lfloor \frac{x}{k} \right\rfloor k \leqslant x < \left(\left\lfloor \frac{x}{k} \right\rfloor + 1 \right) k.$$

In other words, multiplying k by $m = \lfloor x/k \rfloor$ does not exceed x, but $(m+1)k$ does. Therefore, the number of positive integers that are divisible by k and do not exceed x are given by m, as claimed. □

For example, the number of integers in the range from 1 to 100 that are multiples of the prime 13 is given by $\lfloor 100/13 \rfloor = 7$. Indeed, the multiples of 13 in this range are explicitly given by

$$13, 26, 39, 52, 65, 78, \text{ and } 91.$$

Let us now turn to a different question. Recall that each positive integer n can be written as a product of primes

$$n = \prod_p p^{v_p(n)}$$

where p ranges over the set of primes, and $k = v_p(n)$ denotes the highest power of p such that p^k divides n, but p^{k+1} does not divide n. We call $v_p(n)$ the **p-adic valuation** of n. If a prime p does not divide n, then $v_p(n) = 0$, hence $p^{v_p(n)} = 1$, so this product really consists of a finite number of factors that are not equal to 1.

Example 7.9. The number $n = 12$ has the factorization $12 = 2^2 \cdot 3$. Therefore, $v_2(12) = 2$, $v_3(12) = 1$, and $v_p(12) = 0$ for all other primes $p > 3$.

For a positive integer n, we now want to focus on the question to determine the highest power of a prime p that divides $n! = 1 \cdot 2 \cdot 3 \cdot \ldots \cdot (n-1) \cdot n$. So we would like to determine

$$v_p(n!) = \sum_{k=1}^{n} v_p(k).$$

It is a good idea to study an example before trying to settle this question.

Example 7.10. Let us determine the 3-adic valuation of $13! = 6,227,020,800$. Fortunately, there is no need to factorize such large numbers, as

$$v_3(13!) = v_3(1) + v_3(2) + v_3(3) + \cdots + v_3(13).$$

Since $v_p(k) = 0$ unless k is a multiple of 3, we have

$$v_3(13!) = v_3(3) + v_3(6) + v_3(9) + v_3(12) = 1 + 1 + 2 + 1 = 5.$$

The 3-adic valuation of $9 = 3^2$ is greater than 1, since it is divisible by 3^2. We do not have any 3-adic valuations greater than 2, since none of the numbers k in the range $1 \leq k \leq 13$ is divisible by $3^3 = 27$.

Proposition 7.11. *Let n be a positive integer and p a prime. Then*

$$v_p(n!) = \sum_{k=1}^{\lfloor \log_p n \rfloor} \left\lfloor \frac{n}{p^k} \right\rfloor.$$

Proof. Since $n!$ is the product of the integers k in the range $1 \leq k \leq n$, we have

$$v_p(n!) = \sum_{k=1}^{n} v_p(k).$$

By Proposition 7.8, the number of integers that contribute a factor of p is given by $\lfloor n/p \rfloor$. Among those numbers, $\lfloor n/p^2 \rfloor$ contribute a second prime factor p. Among those numbers $\lfloor n/p^3 \rfloor$ contribute a third factor, etc.

We have $p^m > n$ if and only if $m > \log n / \log p = \log_p n$ if and only if $m > \lfloor \log_p n \rfloor$. Therefore, the sum

$$\sum_{k=1}^{\lfloor \log_p n \rfloor} \left\lfloor \frac{n}{p^k} \right\rfloor$$

counts the number of times the prime factor p occurs in the product of the numbers from 1 to n. □

We can even use the floor function to construct a function that allows us to characterize primes.

7.2 Divisibility and Primes

Proposition 7.12. *A positive integer n is prime if and only if*

$$\sum_{k=2}^{\lfloor\sqrt{n}\rfloor}\left(\left\lfloor\frac{n}{k}\right\rfloor-\left\lfloor\frac{n-1}{k}\right\rfloor\right)=0. \tag{7.2}$$

Proof. We can express n in the form $n = qk + r$ with $0 \leqslant r < k$. If k does not divide n, then the remainder r satisfies $1 \leqslant r < k$ and we have

$$\left\lfloor\frac{n}{k}\right\rfloor-\left\lfloor\frac{n-1}{k}\right\rfloor=\left\lfloor\frac{qk+r}{k}\right\rfloor-\left\lfloor\frac{qk+r-1}{k}\right\rfloor=q-q=0.$$

If k divides n, then the remainder $r = 0$ and we have

$$\left\lfloor\frac{n}{k}\right\rfloor-\left\lfloor\frac{n-1}{k}\right\rfloor=\left\lfloor\frac{qk}{k}\right\rfloor-\left\lfloor\frac{(q-1)k+k-1}{k}\right\rfloor=q-(q-1)=1.$$

Therefore, the sum (7.2) counts the number of divisors k of n in the range from 2 to $\lfloor\sqrt{n}\rfloor$. The positive integer n is composite if and only if it has a divisor k in the range from $2 \leqslant k \leqslant \lfloor\sqrt{n}\rfloor$. Thus, n is prime if and only if the sum (7.2) is equal to 0. □

EXERCISES

7.17. Show that every real number $x \in \mathbf{R}$ has a *unique* representation in the form
$$x = q + r$$
where q is an integer and r is a real number in the range $0 \leqslant r < 1$. The number $q = \lfloor x \rfloor$ and $r = x - \lfloor x \rfloor$. [You can interpret this as a division by 1 with remainder $r \equiv x \pmod{1}$.]

7.18. Let d be a positive integer. We know that every integer n can be uniquely expressed in the form
$$n = qd + r,$$
where r is an integer in the range $0 \leqslant r < d$. Show that the quotient q is given by $q = \lfloor n/d \rfloor$ and the remainder r by $r = n - \lfloor n/d \rfloor d$.

7.19. Let n and k be nonnegative integers such that $n \geqslant k$, m an integer such that $m > n^k$, and ℓ be the integer given by $\ell = \lfloor (m+1)^n/m^k \rfloor$. Show that the division of ℓ by m yields $\binom{n}{k}$ as a remainder. In other words, if we express ℓ in the form $\ell = qm + r$ with $0 \leqslant r < m$, then $r = \binom{n}{k}$. [Hint: Use the binomial theorem to expand $(m+1)^n$.]

7.20. Determine the number of terminating zeros of 1000!.

7.21. Does there exist a prime p such that $v_p(46!) = 42$?

7.22. Let n be a positive integer and p a prime. Suppose that $n = \sum_{k=0}^{m} n_k p^k$ is the base p expression of n, so the digits n_k are in the range $0 \leqslant n_k < p$ for all indices k, and $m = \lfloor \log_p(n) \rfloor$. Show that

$$v_p(n!) = \frac{n - (n_0 + n_1 + \cdots + n_m)}{p - 1}.$$

7.23. Let n be a positive integer. Show that 2^n does not divide $n!$. [Hint: You can use Exercise 7.22.]

7.24. Let n be an integer such that $n \geqslant 2$. Show that if 2^{n-1} divides $n!$, then n must be a power of 2. [Hint: You can use Exercise 7.22.]

7.25. Let n be an integer such that $n > 1$.
(a) Show that $\binom{2n}{n} \equiv 0 \pmod{2}$.
(b) Show that $\binom{2n}{n} \equiv 0 \pmod{4}$ unless n is a power of 2.
[Hint: You can use Exercise 7.22.]

7.26. Use Proposition 7.12 to show that 113 is a prime number.

7.27. Explain why Proposition 7.12 does not lead to a fast check for primality.

7.28. Let n be a positive integer. Show that

$$\lfloor \sqrt{n} \rfloor - \lfloor \sqrt{n-1} \rfloor = \begin{cases} 1 & \text{if } n \text{ is a perfect square,} \\ 0 & \text{otherwise.} \end{cases}$$

7.3 Functions of Floors and Ceilings

If a function has a floor or ceiling applied to its argument and value, then it requires sometimes delicate arguments to fully analyze the behavior of the resulting function. In other words, if we are given a function $f(x)$, we might be interested in the behavior of

$$\lfloor f(\lfloor x \rfloor) \rfloor, \quad \lfloor f(\lceil x \rceil) \rfloor, \quad \lceil f(\lfloor x \rfloor) \rceil, \quad \text{or} \quad \lceil f(\lceil x \rceil) \rceil.$$

Fortunately, applying the floor (or ceiling) to the argument is sometimes redundant when a floor (or ceiling) is applied to the function value. The next proposition gives a simple example.

Proposition 7.13. *For all nonnegative real numbers, we have*

$$\left\lfloor \sqrt{\lfloor x \rfloor} \right\rfloor = \lfloor \sqrt{x} \rfloor.$$

Proof. Suppose that $n = \left\lfloor \sqrt{\lfloor x \rfloor} \right\rfloor$. This means that

$$n \leqslant \sqrt{\lfloor x \rfloor} < n + 1.$$

7.3 Functions of Floors and Ceilings

Squaring the terms of these inequalities yields
$$n^2 \leq \lfloor x \rfloor < (n+1)^2.$$
Since the bounds are integers, it follows from Proposition 7.5 that
$$n^2 \leq x < (n+1)^2$$
holds. Taking square roots yields
$$n \leq \sqrt{x} < n+1,$$
so $n = \lfloor \sqrt{x} \rfloor$. We can conclude that
$$\lfloor \sqrt{\lfloor x \rfloor} \rfloor = n = \lfloor \sqrt{x} \rfloor,$$
as claimed. □

The next theorem is a vast generalization of the previous proposition. This result is due to McEliece. The domain of the function is an interval $I \subseteq \mathbf{R}$ that is closed under taking floors for the first claim and closed under taking ceilings in the second claim.

Theorem 7.14. *Let f be a continuous, monotonically strictly increasing real-valued function such that $f^{-1}(\mathbf{Z}) \subseteq \mathbf{Z}$. Then*
$$\lfloor f(x) \rfloor = \lfloor f(\lfloor x \rfloor) \rfloor \quad \text{and} \quad \lceil f(x) \rceil = \lceil f(\lceil x \rceil) \rceil.$$

Proof. Let us prove the first claim. If $x = \lfloor x \rfloor$, then the claim evidently holds. If $\lfloor x \rfloor < x$, then $f(\lfloor x \rfloor) < f(x)$, as f is monotonically strictly increasing. Since the floor function is a non-decreasing function, it follows that
$$\lfloor f(\lfloor x \rfloor) \rfloor \leq \lfloor f(x) \rfloor.$$
Seeking a contradiction, let us suppose that $\lfloor f(\lfloor x \rfloor) \rfloor < \lfloor f(x) \rfloor$, so equality does not hold. This implies the inequality $f(\lfloor x \rfloor) < \lfloor f(x) \rfloor$, whence we have $f(\lfloor x \rfloor) < \lfloor f(x) \rfloor \leq f(x)$. Since f is a continuous function, it follows by the intermediate value theorem that there must exist an element y in the range $\lfloor x \rfloor \leq y < x$ such that $\lfloor f(x) \rfloor = f(y)$. As $f(y)$ is an integer, it follows from the hypothesis that y must be an integer. We must have $\lfloor x \rfloor \neq y$, since $\lfloor f(\lfloor x \rfloor) \rfloor$ is supposed to be different from $f(y) = \lfloor f(x) \rfloor$. However, there cannot exist an integer y satisfying $\lfloor x \rfloor < y < x$, which is our desired contradiction. Therefore, we can conclude that we must have $\lfloor f(\lfloor x \rfloor) \rfloor = \lfloor f(x) \rfloor$.

The argument for the second claim is similar, so we omit it here. □

We collect some consequences of this theorem.

Corollary 7.15. *Let n be an integer satisfying $n \geq 2$. Then*
$$\lfloor \sqrt[n]{\lfloor x \rfloor} \rfloor = \lfloor \sqrt[n]{x} \rfloor \quad \text{and} \quad \lceil \sqrt[n]{\lceil x \rceil} \rceil = \lceil \sqrt[n]{x} \rceil.$$

Proof. The function $f(x) = \sqrt[n]{x}$ is continuous and monotonically strictly increasing. If the value $m = \sqrt[n]{x}$ is an integer, then its argument $x = m^n$ is an integer as well. Therefore, the claim follows from the previous theorem. □

Corollary 7.16. *Let b be an integer satisfying $b > 1$. Then*

$$\lfloor \log_b \lfloor x \rfloor \rfloor = \lfloor \log_b x \rfloor \quad \text{and} \quad \lceil \log_b \lceil x \rceil \rceil = \lceil \log_b x \rceil.$$

Proof. The function $f(x) = \log_b x$ is continuous and monotonically strictly increasing. If the value $m = \log_b x$ is an integer, then the argument $x = b^m$ is an integer as well. Therefore, the claim follows from the previous theorem. □

Corollary 7.17. *Suppose that n is a positive integer. Then*

$$\left\lfloor \frac{\lfloor x \rfloor}{n} \right\rfloor = \left\lfloor \frac{x}{n} \right\rfloor \quad \text{and} \quad \left\lceil \frac{\lceil x \rceil}{n} \right\rceil = \left\lceil \frac{x}{n} \right\rceil.$$

Proof. The function $f(x) = x/n$ is continuous and monotonically strictly increasing. If the function value $f(x) = m$ is an integer, then the argument $x = mn$ is an integer as well. Therefore, the claim follows from the previous theorem. □

We can slightly generalize the previous corollary.

Corollary 7.18. *Suppose that n is a positive integer and m an arbitrary integer. Then*

$$\left\lfloor \frac{\lfloor x \rfloor + m}{n} \right\rfloor = \left\lfloor \frac{x + m}{n} \right\rfloor \quad \text{and} \quad \left\lceil \frac{\lceil x \rceil + m}{n} \right\rceil = \left\lceil \frac{x + m}{n} \right\rceil$$

Proof. The function $f(x) = (x+m)/n$ is continuous and monotonically strictly increasing. If the function value $f(x) = k$ is an integer, then the argument $x = kn - m$ is an integer as well. Therefore, the claim follows from the previous theorem. □

We can extend the previous theorem to monotonically strictly decreasing functions. We assume that the domain of the function is an interval $I \subseteq \mathbf{R}$ that is closed under taking floors for the first claim and closed under taking ceilings in the second claim.

Theorem 7.19. *Let f be a continuous, monotonically strictly decreasing real-valued function such that $f^{-1}(\mathbf{Z}) \subseteq \mathbf{Z}$. Then*

$$\lceil f(x) \rceil = \lceil f(\lfloor x \rfloor) \rceil \quad \text{and} \quad \lfloor f(x) \rfloor = \lfloor f(\lceil x \rceil) \rfloor.$$

Proof. Let us prove the first claim. If $x = \lfloor x \rfloor$, then the claim evidently holds. If $\lfloor x \rfloor < x$, then $f(\lfloor x \rfloor) > f(x)$, as f is a monotonically strictly decreasing function. Since $y > y'$ implies $\lceil y \rceil \geq \lceil y' \rceil$, we can conclude that

$$\lceil f(\lfloor x \rfloor) \rceil \geq \lceil f(x) \rceil.$$

7.3 Functions of Floors and Ceilings

Seeking a contradiction, let us suppose that $\lceil f(\lfloor x \rfloor) \rceil > \lceil f(x) \rceil$, so equality does not hold. This implies the inequality $f(\lfloor x \rfloor) > \lceil f(x) \rceil$, whence we have $f(\lfloor x \rfloor) > \lceil f(x) \rceil \geq f(x)$. Since f is a continuous function, it follows by the intermediate value theorem that there must exist an element y in the range $\lfloor x \rfloor \leq y < x$ such that $\lceil f(x) \rceil = f(y)$. As $f(y)$ is an integer, it follows from the hypothesis that y must be an integer. We must have $\lfloor x \rfloor \neq y$, since $\lceil f(\lfloor x \rfloor) \rceil$ is supposed to be different from $f(y) = \lceil f(x) \rceil$. However, there cannot exist an integer y satisfying $\lfloor x \rfloor < y < x$, which is our desired contradiction. Therefore, we can conclude that we must have $\lceil f(\lfloor x \rfloor) \rceil = \lceil f(x) \rceil$.

The argument for the second claim is similar, so we omit it here. □

EXERCISES

7.29. Carl was floored when he saw the following expression in a program:

$$\lfloor \lfloor \lfloor \lfloor x/10 \rfloor /10 \rfloor /10 \rfloor /10 \rfloor.$$

Help him rewrite it using a single floor function.

7.30. Let n be an integer and ξ a real number in the range $0 \leq \xi < 1$. The function $r_+ : \mathbf{R} \to \mathbf{R}$ rounds a real number towards infinity, so

$$r_+(n + \xi) = \begin{cases} n & \text{if } 0 \leq \xi < 1/2, \\ n+1 & \text{if } 1/2 \leq \xi < 1. \end{cases}$$

Show that the rounding function $r_+(x)$ can also be expressed in the following two ways:
(a) $r_+(x) = \lfloor x + \frac{1}{2} \rfloor$ for all $x \in \mathbf{R}$,
(b) $r_+(x) = \lceil \frac{\lfloor 2x \rfloor}{2} \rceil$ for all $x \in \mathbf{R}$.

7.31. The function $r_+(x) = \lfloor x + 1/2 \rfloor$ rounds a real number x to the nearest integer. If two integers are equally near to x, then $r_+(x) = \lfloor x + 1/2 \rfloor$ rounds to the larger of the two integers. Use the floor function to define a function $r_-(x)$ that rounds a real number x to the nearest integer. If two integers are equally near to x, then $r_-(x)$ rounds to the smaller of the two integers.

7.32. Emmy is playing with a calculator. She enters an integer and takes its square root. Then she repeats the process with the integer part of the answer. After the third repetition, the integer part equals 1 for the first time. What is the difference between the largest and the smallest number Emmy could have started with? (Source: Norwegian Math Olympiad 2014–15)

7.33. Show that for all integers m and n, and all real numbers x, we have

$$\left\lfloor \frac{1}{m} \left\lfloor \frac{1}{n} x \right\rfloor \right\rfloor = \left\lfloor \frac{1}{n} \left\lfloor \frac{1}{m} x \right\rfloor \right\rfloor = \left\lfloor \frac{1}{mn} x \right\rfloor.$$

The functions $L_m(x) = \lfloor x/m \rfloor$ are digital lines that are relevant in computer graphics. This exercise claims that the composition of the digital line functions L_m and L_n is commutative, so

$$L_m \circ L_n = L_n \circ L_m = L_{mn}.$$

7.4 Notes

The notations for the floor and ceiling functions were introduced by Iverson and popularized by Knuth. A rich source of information on properties of floor and ceiling functions is Knuth [48] and Graham et al. [30]. The book by Herman et al. [35] contains some interesting problems about floor and ceiling functions.

Chapter 8

Number Theory

> *Mathematics is the queen of the sciences and number theory is the queen of mathematics.*
>
> — Carl Friedrich Gauss

8.1 Divisibility

A large part of number theory is concerned with the multiplicative structure of the integers. We begin by recalling the basics of divisibility. Even though we start from first principles, our approach will be inspired by the more abstract concepts that we learned in the previous chapters.

Given two integers a and b, we write $b \mid a$ to denote that b **divides** a, meaning that there exists an integer c such that $a = bc$. If b divides a, then we call b a **divisor** of a, and a a **multiple** of b. We write $b \nmid a$ if and only if b does not divide a.

For a nonzero integer b, the divisibility relation $b \mid a$ is equivalent to the statement that the rational number a/b is an integer. However, we strongly recommend to use the aforementioned definition, as it generalizes to other domains.

Example 8.1. We have $4 \mid 12$, since $12 = 4 \cdot 3$. Similarly, we have $3 \mid 9$, since $9 = 3 \cdot 3$. However, $6 \nmid 9$, since the equation $9 = 6x$ does not have a solution x in the set of integers.

Example 8.2. For all integers a, we have $1 \mid a$ and $-1 \mid a$, since $1 \cdot a = a$ and $-1 \cdot (-a) = a$. On the other hand, $a \mid 1$ implies that $a = 1$ or $a = -1$, as we will prove shortly.

Example 8.3. The relation $b \mid 0$ holds for all integers b, since $0 = b \cdot 0$. On the other hand, $0 \mid a$ holds if and only if $a = 0$.

Proposition 8.4. *If a and b are integers and $a \neq 0$, then $b \mid a$ implies $1 \leq |b| \leq |a|$.*

Proof. Since $b \mid a$ there must exist an integer c such that $a = bc$. Since $a \neq 0$, we must have $b \neq 0$ and $c \neq 0$. It follows that $|b| \geq 1$ and $|c| \geq 1$. Furthermore, we have $|b| \leq |b||c| = |a|$. Combining these inequalities yields the claim. □

Proposition 8.5. *For all integers a, b, c, we have*

D1. $a \mid a$

D2. $a \mid b$ and $b \mid a$ if and only if $a = \pm b$,

D3. $c \mid b$ and $b \mid a$ implies $c \mid a$,

D4. $c \mid a$ implies $c \mid am$ for all integers m.

D5. $c \mid a$ and $c \mid b$ implies $c \mid (ax + by)$ for all integers x and y.

When restricted to positive integers, the properties **D1**–**D3** ensure that the divisibility relation is a partial order. However, on the set of integers, **D2** is not quite an antisymmetry, since $2 \mid -2$ and $-2 \mid 2$, but $2 \neq -2$. Therefore, the divisibility relation on the set of integers is a preorder but not a partial order.

Proof. Since $a = a \cdot 1$, we have $a \mid a$. Therefore, property **D1** holds.

The proof of property **D2** is more interesting. If $a = \pm b$, then evidently $a \mid b$ and $b \mid a$. Conversely, suppose that $a \mid b$ and $b \mid a$ hold. This immediately implies that if one of a or b is equal to 0, then so is the other. Thus, we may now assume that both a and b are nonzero. By the previous proposition, $a \mid b$ implies $|a| \leq |b|$ and $b \mid a$ implies $|b| \leq |a|$. Therefore, $|a| = |b|$, and we can conclude that $a = \pm b$. This shows that property **D2** holds.

Properties **D3** and **D4** follow directly from the definition of divisibility.

If $c \mid a$ and $c \mid b$, then there exist integers m_a and m_b such that $a = cm_a$ and $b = cm_b$. Therefore, $ax + by = cm_a x + cm_b y = c(m_a x + m_b y)$, so $c \mid ax + by$. Thus, property **D5** holds as well. □

Corollary 8.6. *If an integer a satisfies $a \mid 1$, then $a = 1$ or $a = -1$.*

Proof. For any integer a, we have $1 \mid a$ and by assumption $a \mid 1$, so $a = \pm 1$ by property **D2** of the previous proposition. □

Exercises

8.1. Prove the properties **D3** and **D4** of Proposition 8.5:
(a) For integers a, b, and c, show that if $c \mid b$ and $b \mid a$, then $c \mid a$.
(b) For integers a, c, and m, show that if $c \mid a$, then $c \mid am$.

8.2. Let a and b be integers such that $a \neq 0$. Show that if $a \mid b$, then $(b/a) \mid b$. [Note that the vertical bar $\mid$ denotes the divisibility relation and the slash $/$ denotes the usual division operation.]

8.3. Let a, b, and c be integers. Prove that $a \mid b$ and $a \mid c$ implies that $a^2 \mid bc$.

8.4. Let a and b be odd integers. Prove that $4 \mid (a^2 - b^2)$.

8.5. Let a be an integer. Use only the definition of divisibility to prove that $3 \mid a$ and $4 \mid a$ implies that $12 \mid a$.

8.6. Show that if p is a prime and k is an integer in the range $1 < k < p$, then $p \mid \binom{p}{k}$.

8.7. Prove that for all integers n, we have $120 \mid (n^5 - 5n^3 + 4n)$.

8.8. Let q, m, and n be positive integers.
(a) Show that $(q-1) \mid (q^n - 1)$.
(b) Show that $m \mid n$ implies $(q^m - 1) \mid (q^n - 1)$.

8.9. Give a simple characterization of all integers a such that $10 \mid (a^{10} + 1)$. (Source: de Koninck, Mercier)

8.10. Show that a positive integer is a perfect square if and only if it has an odd number of positive divisors.

8.2 The Greatest Common Divisor

Suppose that a and b are integers, not both zero. The **greatest common divisor** of a and b is an integer d satisfying the following three conditions

G1. d is a common divisor of a and b, so d divides a and d divides b,

G2. if d' is a common divisor of a and b, then d' divides d,

G3. $d > 0$.

The greatest common divisor d is denoted by $\gcd(a, b)$. In texts on number theory, the greatest common divisor is sometimes simply denoted by (a, b).

Conditions **G2** and **G3** ensure that the greatest common divisor is larger than any other common divisor. It is usually formulated as a divisibility condition, since this generalizes to other domains that are not totally ordered.

Example 8.7. Consider the integers $a = 24$ and $b = 108$. The positive divisors of a are given by $1, 2, 3, 4, 6, 8, 12, 24$, and the positive divisors of b are given by $1, 2, 3, 4, 6, 9, 12, 18, 27, 36, 54, 108$. The common positive divisors of a and b are $1, 2, 3, 4, 6$, and 12, so

$$\gcd(24, 108) = 12$$

is the greatest common divisor of 24 and 108.

Listing all divisors of an integer is too tedious. If the prime factorization of a and b is known, then we can easily determine the greatest common divisor, as the following proposition shows.

Proposition 8.8. Let a and b be nonzero integers. Let $p_1, p_2, \ldots, p_n$ denote all prime divisors of a and b. If a is of the form $a = \pm p_1^{a_1} p_2^{a_2} \cdots p_n^{a_n}$ and b is of the form $b = \pm p_1^{b_1} p_2^{b_2} \cdots p_n^{b_n}$, then the greatest common divisor of a and b equals

$$\gcd(a, b) = p_1^{\min\{a_1, b_1\}} p_2^{\min\{a_2, b_2\}} \cdots p_n^{\min\{a_n, b_n\}}.$$

Proof. By definition, $d > 0$, so **G3** is satisfied. Furthermore, d is a common divisor of a and b, so **G1** holds. Finally, if d' is another positive common divisor of a and b, then any prime dividing d' must divide both a and b. Therefore, we can write d' in the form

$$d' = p_1^{c_1} p_2^{c_2} \cdots p_n^{c_n}$$

for some nonnegative integers c_k with $1 \leqslant k \leqslant n$. Since d' is a common divisor of a and b, we must have $0 \leqslant c_k \leqslant \min\{a_k, b_k\}$ for all k in the range $1 \leqslant k \leqslant n$. It follows that d' divides d, so property **G2** is satisfied as well. □

Example 8.9. Suppose that we want to find the greatest common divisor of $a = 24$ and $b = 108$. The prime-power decompositions of a and b are given by $a = 24 = 2^3 \cdot 3$ and $b = 108 = 2^2 3^3$. For each prime dividing a or b, we need to inspect its exponents in the prime power decompositions and choose the smaller exponent for the greatest common divisor. In this case, we have

$$\gcd(a, b) = 2^{\min\{3,2\}} 3^{\min\{1,3\}} = 2^2 \cdot 3 = 12,$$

according to the previous proposition.

One practical drawback of this approach is that finding the factorization of an integer into primes is not an easy task. In fact, no fast algorithms are known that can accomplish this task efficiently on a classical computer. Fortunately, Euclid gave a very simple algorithm to quickly find the greatest common divisor of two integers.

The main idea behind the Euclidean algorithm is the following observation that reduces the calculation of the greatest common divisor of a and b to a simpler problem with smaller numbers.

Proposition 8.10. Let a and b be integers such that $0 < b < a$. If the division of a by b yields a quotient q and remainder r, so $a = qb + r$ with $0 \leqslant r < b$, then

$$\gcd(a, b) = \gcd(b, r).$$

Proof. If d is a common divisor of a and b, then d divides $a - qb = r$. In other words, any common divisor of a and b must be a common divisor of b and r.

Conversely, if d' is a common divisor of b and r, then d' divides $qb + r = a$. Therefore, any common divisor of b and r is a common divisor of a and b.

Since the set of common divisors of a and b and the set of common divisors of b and r are the same, we can conclude that $\gcd(a, b) = \gcd(b, r)$. □

8.2 The Greatest Common Divisor

We can repeatedly use the previous proposition to ease the calculations. The next example illustrates this idea.

Example 8.11. Suppose that we want to calculate the greatest common divisor of 391 and 357. Integer division of 391 by 357 yields

$$391 = 1 \cdot 357 + 34.$$

The previous proposition tells us that $\gcd(391, 357) = \gcd(357, 34)$. Now we simply repeat the process to determine $\gcd(357, 34)$. Integer division of 357 by 34 yields

$$357 = 10 \cdot 34 + 17.$$

Therefore, $\gcd(357, 34) = \gcd(34, 17)$, reducing the size of the numbers substantially. We can determine $\gcd(34, 17)$ by repeating the process once again. Integer division yields

$$34 = 2 \cdot 17 + 0,$$

so the remainder is finally 0. Therefore, we can conclude that

$$\gcd(391, 357) = \gcd(357, 34) = \gcd(34, 17) = \gcd(17, 0).$$

Since $\gcd(17, 0) = 17$, we can conclude that $\gcd(391, 357) = 17$.

Euclid's algorithm simply generalizes this idea. The greatest common divisor of two integers a and b is done by repeatedly applying the previous proposition until a remainder of 0 is reached. In Ruby, we can formulate the Euclidean algorithm as follows:

```
def gcd(a,b)
  return a if b == 0
  gcd(b, a.modulo(b))
end
```

If we start with integers a and b such that $a \geqslant b > 0$, then in each recursive call the value of b gets reduced, eventually reaching 0. Since for any integer m, we have $\gcd(m, 0) = m$, the last step simply needs to return the value contained in the variable a.

Example 8.12. Let us compute the greatest common divisor of 96 and 42. By the Euclidean algorithm, we have

$$\left.\begin{array}{l} 96 = 2 \cdot 42 + 12 \\ 42 = 3 \cdot 12 + 6 \\ 12 = 2 \cdot 6 + 0 \end{array}\right\} \Longrightarrow \left\{\begin{array}{rl} \gcd(96, 42) &= \gcd(42, 12) \\ &= \gcd(12, 6) \\ &= \gcd(6, 0) = 6. \end{array}\right.$$

We can give a nice visual interpretation of these steps of the Euclidean algorithm as follows. Let us draw a rectangle of dimensions $a \times b = 96 \times 42$, corresponding to the arguments of $\gcd(96, 42)$. Our goal is to tile this rectangle with squares, using squares of maximal size and then repeat the process for the remaining area of the rectangle. Since $96 = 2 \cdot 42 + 12$, we can fit two 42×42 squares into this rectangle.

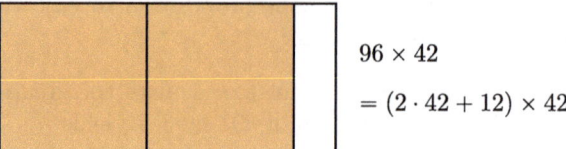

96×42

$= (2 \cdot 42 + 12) \times 42$

There remains a rectangle of dimension 42×12 that is not yet tiled by squares, and the dimension of this rectangle corresponds to the arguments of $\gcd(42, 12)$. Since $42 = 3 \cdot 12 + 6$, we can fit into this rectangle three squares of dimension 12×12.

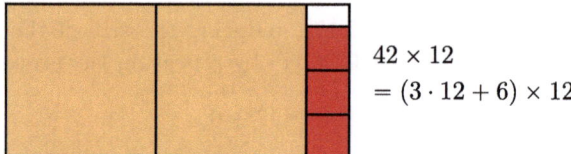

42×12

$= (3 \cdot 12 + 6) \times 12$

There remains an area of dimension 12×6 that is not yet tiled by squares, and this corresponds to the arguments of $\gcd(12, 6)$. As $12 = 2 \cdot 6$, we can use two squares of dimension 6×6 to tile the remaining area.

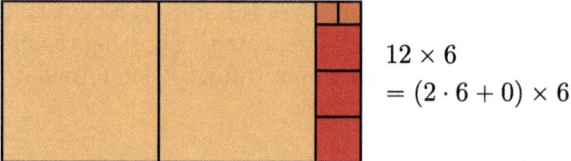

12×6

$= (2 \cdot 6 + 0) \times 6$

The greatest common divisor is the side length of the smallest square that is used in the tiling of the rectangle of dimension $a \times b = 96 \times 42$, so in this case $\gcd(96, 42) = 6$.

You should convince yourself that each calculation of the greatest common divisor of two positive integers has such a visual interpretation of the Euclidean algorithm.

Exercises

8.11. Use Proposition 8.8 to determine the greatest common divisor $\gcd(504, 1386)$.

8.12. Let a be a nonzero integer. Determine (a) $\gcd(a, 1)$ and (b) $\gcd(a, 0)$.

8.13. Use the Euclidean algorithm to compute $\gcd(720, 125)$.

8.14. Use the Euclidean algorithm to compute $\gcd(1751, 1547)$.

8.15. Use the Euclidean algorithm to compute $\gcd(12\,345\,678, 9\,876)$.

8.16. Let a and b be positive integers. Show that if $d = \gcd(a, b)$, then a/d and b/d are integers and $\gcd(a/d, b/d) = 1$.

8.3 Linear Diophantine Equations

8.17. Let a and b be integers that are not both zero. Show that if c is a positive integer, then $\gcd(ca, cb) = c \gcd(a, b)$. [Hint: You may use Exercise 8.16.]

8.18. Let a and b be positive integers. Show that if $\gcd(a, b) = 1$, then $\gcd(a+b, a-b)$ is either 1 or 2. [Hint: You can use facts that were proved in previous exercises.]

8.19. Show that for all nonzero integers a, b, c, we have

$$\gcd(\gcd(a, b), c) = \gcd(a, \gcd(b, c)).$$

8.20. The Ruby implementation of the Euclidean algorithm uses tail recursion. The interpreter can optimize it into an equivalent iterative version, but Ruby may not do this by default. Give an iterative implementation of the Euclidean algorithm.

8.3 Linear Diophantine Equations

A **Diophantine equation** is a polynomial equation in several variables with integral coefficients that are solved for integer solutions (or sometimes nonnegative integer solutions or positive integer solutions). Some Diophantine equations can be difficult to solve. For example, Fermat conjectured that there are no positive integers x, y, and z satisfying the equation

$$x^n + y^n = z^n$$

for any integer value of n greater than 2. A proof remained elusive for 358 years until Andrew Wiles proved Fermat's conjecture.

In this section, we will be concerned with linear Diophantine equations in two variables. Given integers a, b, and c, the goal is to determine integer solutions to the equation

$$ax + by = c.$$

These equations have many applications. Fortunately, solving linear Diophantine equations is not difficult.

We have already encountered Diophantine equations in counting problems.

Example 8.13. Suppose that we want to know in how many different ways we can give change to c cents using nickels and dimes. This amounts to solving the Diophantine equation

$$c = 5n + 10d.$$

In this case, we are looking for nonnegative integer solutions.

We will now investigate when a linear diophantine equation has integer solutions. The next lemma shows when we cannot find solutions.

Lemma 8.14. *Let a, b, and c denote integers. Let $d = \gcd(a, b)$. If d does not divide c, then the linear diophantine equation*

$$ax + by = c$$

does not have any integer solutions $(x, y) \in \mathbf{Z}^2$.

Proof. For any integers x and y, the greatest common divisor of $\gcd(a, b)$ divides $ax + by$. Therefore, if the linear diophantine equation can be solved, then d must divide c. The claim is the contrapositive of this statement. □

It turns out that we can always solve the Diophantine equation $ax + by = c$ when $\gcd(a, b)$ divides the integer c. We first focus on the case when c is equal to $\gcd(a, b)$, and then obtain the general case as a consequence. The proof technique used in this proposition is particularly important, as it takes advantage of the concept of an ideal. You can explore ideals a bit further in the exercises.

Proposition 8.15. *Let a and b be integers that are not both zero. Let $d = \gcd(a, b)$. Then d is the smallest positive linear combination of a and b. This means that there exist integers x_0 and y_0 such that*

$$d = ax_0 + by_0,$$

and d is the smallest positive integer that can be expressed in this form.

Proof. Let us define the set I of all integer linear combination of a and b,

$$I = \{ax + by \mid x, y \in \mathbf{Z}\}.$$

The set I is not empty, since it contains the positive integer $a^2 + b^2$.

Let d denote the smallest positive integer that is contained in I. So d is of the form $d = ax_0 + by_0$ for some integers x_0 and y_0. We claim that $d = \gcd(a, b)$. We prove this claim by establishing the three properties **G1**–**G3** of the greatest common divisor of a and b.

By definition, the integer d is positive, so it satisfies **G3**.

Let d' be a common divisor of a and b. Then there exist integers a_0 and b_0 such that $d'a_0 = a$ and $d'b_0 = b$. Then

$$d = ax_0 + by_0 = d'a_0x_0 + d'b_0y_0 = d'(a_0x_0 + b_0y_0),$$

so d' divides d. Therefore, property **G2** holds.

Using the division algorithm, we can write a and b in the form

$$a = q_a d + r_a, \quad \text{and} \quad b = q_b d + r_b$$

for some integers q_a, r_a and q_b, r_b such that $0 \leqslant r_a < d$ and $0 \leqslant r_b < d$. This means that we can express r_a and r_b as an integer linear combination of a and b. Indeed,

$$r_a = a - q_a d = a - q_a(ax_0 + by_0) = a \cdot (1 - q_a x_0) + b \cdot (-q_a y_0)$$

8.3 Linear Diophantine Equations

and

$$r_b = b - q_b d = b - q_b(ax_0 + by_0) = a \cdot (-q_b x_0) + b \cdot (1 - q_b y_0),$$

so r_a and r_b are nonnegative elements in I that are both smaller than d. Since d was the smallest positive number in I, we can conclude that $r_a = 0$ and $r_b = 0$. In other words, d is a common divisor of a and b, so **G1** holds. □

Corollary 8.16. *Let a, b, and c be integers. If $\gcd(a,b)$ divides c, then*

$$ax + by = c$$

has an integer solution.

Proof. By the previous proposition, there exist integers x_0 and y_0 such that $ax_0 + by_0 = \gcd(a,b)$. Since $\gcd(a,b) \mid c$, there exist an integer m such that $\gcd(a,b)m = c$. So multiplying both sides of the equation $ax_0 + by_0 = \gcd(a,b)$ by m yields

$$a(x_0 m) + b(y_0 m) = c,$$

which proves the claim. □

We will summarize our findings in the next theorem.

Theorem 8.17. *Let a, b, and c be integers, and $d = \gcd(a,b)$. Then the Diophantine equation*

$$ax + by = c$$

has integer solutions if and only if d divides c. If the equation has one solution (x_0, y_0), then all solutions of the equation are given by

$$x = x_0 - \frac{b}{d}z \quad \text{and} \quad y = y_0 + \frac{a}{d}z$$

for $z \in \mathbf{Z}$.

Proof. The first claim follows from the previous results. It remains to show the second claim.

Suppose that (x_1, y_1) is an arbitrary integer solution of the equation $ax + by = c$. Then

$$ax_0 + by_0 = c = ax_1 + by_1.$$

Dividing both sides of the equation by $d = \gcd(a,b)$ yields

$$\frac{a}{d}x_0 + \frac{b}{d}y_0 = \frac{a}{d}x_1 + \frac{b}{d}y_1.$$

Subtracting $(b/d)y_0 + (a/d)x_1$ from both sides yields

$$\frac{a}{d}(x_0 - x_1) = \frac{b}{d}(y_1 - y_0).$$

Since $\gcd(a,b) = d$, we have $\gcd(a/d, b/d) = 1$. As b/d divides $(a/d)(x_0 - x_1)$, we can conclude using Exercise 8.25 that b/d must divide $x_0 - x_1$. So there exists an integer z such that $(b/d)z = x_0 - x_1$. It follows that

$$\frac{a}{d}z\frac{b}{d} = \frac{b}{d}(y_1 - y_0).$$

Dividing both sides by b/d shows that

$$\frac{a}{d}z = y_1 - y_0 \quad \text{or} \quad y_1 = y_0 + \frac{a}{d}z.$$

We can conclude that any integer solution (x_1, y_1) of the Diophantine equation $ax + by = c$ must be of the form

$$x_1 = x_0 - \frac{b}{d}z \quad \text{and} \quad y_1 = y_0 + \frac{a}{d}z,$$

which proves our second claim. $\square$

The previous theorem shows that if we know one solution to a linear Diophantine equation in two variables, then it is easy to find all other solutions. Corollary 8.16 showed that a solution to $ax+by = c$ exists if $\gcd(a,b)$ divides c. The question is how to find such a solution. Fortunately, we can derive a solution directly from the calculations of the Euclidean algorithm. Let us look at an example.

Example 8.18. Let us consider the linear Diophantine equation $96x+42y = 6$. Using the Euclidean algorithm, we can calculate the greatest common divisor $\gcd(96, 42)$ by

$$96 = 2 \cdot 42 + 12 \tag{8.1}$$

$$42 = 3 \cdot 12 + 6 \tag{8.2}$$

$$12 = 2 \cdot 6 + 0 \tag{8.3}$$

It follows that $\gcd(96, 42) = 6$, so the Diophantine equation $96x + 42y = 6$ can be solved. By ignoring the last equation and working our way backward in the Euclidean algorithm, we can actually find the integers x and y. Indeed, it follows from Eq. (8.2) that $6 = 42 - 3 \cdot 12$. Equation (8.1) shows that we can express 12 in the form $12 = 96 - 2 \cdot 42$. Substituting this in the previous equation yields

$$6 = 42 - 3 \cdot 12 = 42 - 3 \cdot (96 - 2 \cdot 42) = 96 \cdot (-3) + 42 \cdot 7.$$

Therefore, $x = -3$ and $y = 7$ is a solution to $96x + 42y = 6$.

8.3 Linear Diophantine Equations

Exercises

8.21. Determine all solutions to the coin change problem given in Example 8.13.

8.22. Follow the proof of Proposition 8.15 and find integers x and y such that $\gcd(504, 1386) = 504x + 1386y$.

8.23. Determine all solutions to the linear Diophantine equation $7x + 9y = 4$ in integers x and y.

8.24. Determine all solutions to the linear Diophantine equation $15x + 21y = 9$ in integers x and y.

8.25. Let a, b, and c be integers such that $\gcd(a, b) = 1$. Show that $a \mid bc$ implies $a \mid c$.

8.26. The Euclidean algorithm applied to integers a and b such that $a \geqslant b \geqslant 0$ yields a chain of divisions with remainders

$$a = b \cdot q_1 + r_1, \quad 0 < r_1 < b,$$
$$b = r_1 \cdot q_2 + r_2, \quad 0 < r_2 < r_1,$$
$$r_1 = r_2 \cdot q_3 + r_3, \quad 0 < r_3 < r_2,$$
$$\vdots$$
$$r_{n-2} = r_{n-1} q_n + r_n, \quad 0 < r_n < r_{n-1},$$
$$r_{n-1} = r_n \cdot q_{n+1} + 0,$$

from which we can deduce that $\gcd(a, b) = r_n$. Show how to obtain integers x and y such that $r_n = \gcd(a, b) = ax + by$ by rewriting these equations.

8.27. A farmer lays out the sum of 1770 crowns in purchasing horses and oxen. He pays 31 crowns for each horse and 21 crowns for each ox, and buys more horses than oxen. How many horses and oxen did the farmer buy? (Source: L. Euler, Elements of Algebra)

An **ideal** I in the integers is a subset I of the set of integers $\mathbf{Z}$ such that (a) I contains 0, (b) if I contains elements a and b, then it contains $a \pm b$, (c) if $m \in \mathbf{Z}$ and $a \in I$, then $ma \in I$.

8.28. Let n be an integer. Show that the set $n\mathbf{Z} = \{nz \mid z \in \mathbf{Z}\}$ of all integer multiples of n is an ideal in $\mathbf{Z}$.

8.29. Let a and b be integers. Show that $a \mid b$ if and only if $a\mathbf{Z} \supseteq b\mathbf{Z}$.

8.30. Let I be an ideal in $\mathbf{Z}$. Show that there exists an integer n such that $I = n\mathbf{Z}$. [An ideal of the form $n\mathbf{Z}$ is called a **principal ideal**. So this exercise shows that all ideals in the integers are principal ideals.]

8.31. Let a and b be integers. Show that the set $I = \{ax + by \mid x, y \in \mathbf{Z}\}$ of integral linear combinations of a and b is an ideal in $\mathbf{Z}$.

8.32. Let a and b be integers such that $ab \neq 0$. Show that
(a) $a\mathbf{Z} + b\mathbf{Z} = \gcd(a,b)\mathbf{Z}$,
(b) $a\mathbf{Z} \cap b\mathbf{Z} = \mathrm{lcm}(a,b)\mathbf{Z}$.

8.4 Linear Congruence Equations

Recall that two integers a and b are called congruent modulo a positive integer m if and only if their difference $a - b$ is divisible by m. We write this relation in the form $a \equiv b \pmod{m}$. A **linear congruence equation** is an equation of the form

$$ax \equiv b \pmod{m},$$

where a and b are given integers, the modulus m is a given positive integer, and our goal is to find an integer x solving this congruence.

Proposition 8.19. *Let a and b be integers, and m a positive integer. Let $d = \gcd(a, m)$. The congruence*

$$ax \equiv b \pmod{m}$$

has a solution if and only if d divides b.

Proof. Suppose that the congruence equation has a solution x_0. Then $ax_0 \equiv b \pmod{m}$ implies that m divides $ax_0 - b$, meaning there must exist an integer k such that $ax_0 - b = km$. Rewriting the latter equation, we get $ax_0 - km = b$. Since d divides the left-hand side of the equation, it follows that d divides b.

Conversely, suppose that $d = \gcd(a, m)$ divides b. This means there exist integers x_0 and y_0 such that $ax_0 + my_0 = b$. Reducing modulo m, we get

$$ax_0 \equiv b \pmod{m},$$

so the congruence equation $ax \equiv b \pmod{m}$ can be solved. $\square$

Example 8.20. The linear congruence equation

$$6x \equiv 9 \pmod{15},$$

has a solution, since $d = \gcd(6, 15) = 3$ divides 9. One solution is given by $x_0 = 4$, as $6 \cdot 4 \equiv 24 \equiv 9 \pmod{15}$.

Example 8.21. The linear congruence equation

$$5x \equiv 1 \pmod{10}$$

does not have a solution, as $d = \gcd(5, 10) = 5$ does not divide 1. Indeed, the least significant digit of an integer multiple of 5 is either 0 or 5. Therefore, we have $5x \equiv 0$ or $5 \pmod{10}$ for all integers x, which means that no integer x satisfies the congruence $5x \equiv 1 \pmod{10}$.

8.4 Linear Congruence Equations

If we are given a solution x_0 to the linear congruence equation $ax \equiv b \pmod{m}$, then any other integer x_1 such that $x_1 \equiv x_0 \pmod{m}$ solves the congruence equation as well. Indeed, we can express x_1 in the form $x_1 = x_0 + km$ for some integer k, so $ax_1 \equiv a(x_0 + km) \equiv ax_0 \pmod{m}$. It follows that
$$ax_1 \equiv ax_0 \equiv b \pmod{m}.$$
In other words, if there exists a solution x_0 to a linear congruence equation, then there exist infinitely many, namely the entire congruence class
$$[x_0] = \{x_0 + km \mid k \in \mathbf{Z}\}.$$
We consider all solutions that belong to the same congruence class modulo m to be equivalent. So this raises the question: How many congruence classes modulo m solve a given linear congruence equation? Before we answer this question, let us examine an example.

Example 8.22. The linear congruence equation
$$6x \equiv 9 \pmod{15}$$
has the solutions $x_0 = 4$, $x_1 = 9$, and $x_2 = 14$. There are no other integer solutions x in the range $0 \leq x \leq 14$. Thus, there are three non-equivalent solutions. Any solution belongs to one of the three congruence classes
$$[4] = \{4 + 15k \mid k \in \mathbf{Z}\}, \quad [9] = \{9 + 15k \mid k \in \mathbf{Z}\}, \quad [14] = \{14 + 15k \mid k \in \mathbf{Z}\}.$$
Notice that all solutions are congruent modulo 5, where $5 = 15/\gcd(6, 15)$.

Proposition 8.23. *Let a and b be integers, and m a positive integer. Let $d = \gcd(a, m)$. If $d \mid b$, then the congruence*
$$ax \equiv b \pmod{m}$$
has d inequivalent solutions. If one solution is given by x_0, then the d inequivalent solutions are given by
$$x_k = x_0 + k\frac{m}{d},$$
where k is an integer in the range $0 \leq k \leq d - 1$.

Proof. The congruence equation $ax \equiv b \pmod{m}$ is equivalent to the linear Diophantine equation
$$ax + my = b.$$
If $d = \gcd(a, m)$ divides b, then there exist integers x_0 and y_0 such that $ax_0 + my_0 = b$. All solutions to the linear Diophantine equation are given by
$$x = x_0 + \frac{m}{d}z, \quad y = y_0 - \frac{a}{d}z,$$
where z is an integer. Reducing modulo m, we obtain the claim. □

Exercises

8.33. Solve the linear congruence equation $6x \equiv 5 \pmod{8}$. If the congruence equation can be solved, then give all inequivalent solutions.

8.34. Solve the linear congruence equation $10x \equiv 12 \pmod{256}$.

8.35. Let a and b be positive integers such that $\gcd(a,b) = 1$. Solve the linear congruence equation $(a+b)x \equiv a^2 + b^2 \pmod{ab}$.

8.36. Let $f(x_1, x_2, \ldots, x_n)$ be a polynomial in n variables with integer coefficients. Let m be a positive integer. Let $a_1, a_2, \ldots, a_n$ and $b_1, b_2, \ldots, b_n$ be integers such that $a_k \equiv b_k \pmod{m}$ holds for all integers k in the range $1 \leqslant k \leqslant n$. Show that

$$f(a_1, a_2, \ldots, a_n) \equiv f(b_1, b_2, \ldots, b_n) \pmod{m}$$

holds.

8.37. (a) Let $f(x_1, x_2, \ldots, x_n)$ be a polynomial in n variables with integer coefficients. Show that the Diophantine equation $f(x_1, x_2, \ldots, x_n) = 0$ does not have a solution if there exists a positive integer m such that the congruence equation $f(x_1, x_2, \ldots, x_n) \equiv 0 \pmod{m}$ cannot be solved. (b) Show that the square of an integer x must satisfy $x^2 \equiv 0 \pmod{3}$ or $x^2 \equiv 1 \pmod{3}$. (c) Use the observations in part (a) and (b) to show that the Diophantine equation $x^2 + y^2 = 3z^2$ does not have a solution.

8.5 The Chinese Remainder Theorem

The Chinese Remainder Theorem concerns the solution of a system of congruence equations. A specific instance of the theorem was already stated in the third century in a book by the Chinese mathematician Sunzi Suanjing. We discuss it in the next example.

Example 8.24. Sunzi Suanjing stated the following problem: *There are certain things whose number is unknown. If we count them by threes, we have two left over; by fives, we have three left over; and by sevens, two are left over. How many things are there?* If we express this in congruence notation, then we look for an integer such that the following congruence equations are satisfied:

$$x \equiv 2 \pmod{3},$$
$$x \equiv 3 \pmod{5},$$
$$x \equiv 2 \pmod{7}.$$

By the first equation, the solution x must be of the form $2, 5, 8, 11, 14, 17, 20, 23$, and so on. The smallest of these solutions that also satisfy the second congruence is 8. Therefore, positive integers that solve the first two congruences equations are given by $8, 23, 38$, and so on, as we need to count by 15 to get

8.5 The Chinese Remainder Theorem

to the same remainders modulo 3 and modulo 5. Now we can check that 23 is the smallest positive integer that satisfies all three congruences. Of course, the solution is not unique. We can add multiples of $105 = \mathrm{lcm}(3,5,7)$ to get other solutions to the congruence equations. So the solution is any integer x such that $x \equiv 23 \pmod{105}$.

The *ad hoc* solution of the previous example is a bit undesirable, since we cannot proceed in the same fashion when the moduli get large. The proof of the Chinese Remainder Theorem tells you how to systematically find solutions to such congruence equations.

Theorem 8.25 (Chinese Remainder Theorem). *Let $m_1, m_2, \ldots, m_n$ be n pairwise coprime integers greater than 1, where n is an integer such that $n \geqslant 2$. Let $r_1, r_2, \ldots, r_n$ be arbitrary integers. Let m denote the product $m = m_1 m_2 \cdots m_n$. Then there exists an integer x such that*

$$x \equiv r_1 \pmod{m_1},$$
$$x \equiv r_2 \pmod{m_2},$$
$$\vdots$$
$$x \equiv r_n \pmod{m_n}.$$

Any two solutions x to this system of congruences are congruent modulo m.

Proof. The integers m/m_i and m_i are coprime, so $\gcd(m/m_i, m_i) = 1$. Therefore, there exist integers a_i and b_i such that

$$\left(\frac{m}{m_i}\right) a_i + m_i b_i = 1.$$

It follows that

$$\left(\frac{m}{m_i}\right) a_i \equiv 1 \pmod{m_i}.$$

If we multiply both sides of the congruence by r_i, we get

$$\left(\frac{m}{m_i}\right) a_i r_i \equiv r_i \pmod{m_i}.$$

We claim that a solution x to the system of congruences is given by

$$x = \sum_{k=1}^{n} \left(\frac{m}{m_k}\right) a_k r_k.$$

Indeed, we have $m/m_k \equiv 0 \pmod{m_i}$ whenever the index k is different from i, since in this case m/m_k is a multiple of m_i. Therefore,

$$x = \sum_{k=1}^{n} \left(\frac{m}{m_k}\right) a_k r_k \equiv \left(\frac{m}{m_i}\right) a_i r_i \equiv r_i \pmod{m_i}$$

for all indices i in the range $1 \leq i \leq n$.

Any two solutions x and x' of the system of congruences satisfy

$$x \equiv x' \pmod{m_i}$$

for all i in the range $1 \leq i \leq n$. Therefore,

$$x \equiv x' \pmod{m},$$

where $m = \text{lcm}(m_1, m_2, \ldots, m_n) = m_1 m_2 \cdots m_n$. □

Example 8.26. Suppose that we are given three moduli $m_1 = 10$, $m_2 = 11$, and $m_3 = 13$. We look for the smallest positive integers satisfying the system of congruences

$$\begin{aligned} x &\equiv 7 \pmod{10}, \\ x &\equiv 8 \pmod{11}, \\ x &\equiv 12 \pmod{13}. \end{aligned}$$

The product m of the moduli is given by $m = 10 \cdot 11 \cdot 13 = 1430$. Since the integers (m/m_k) and m_k are coprime, we can find integers a_k and b_k such that $(m/m_k)a_k + m_k b_k = 1$, namely

$$\left(\frac{m}{m_1}\right) a_1 + m_1 b_1 = 143(-3) + 10 \cdot 43 = 1,$$

$$\left(\frac{m}{m_2}\right) a_2 + m_2 b_2 = 130 \cdot (5) + 11 \cdot (-59) = 1,$$

$$\left(\frac{m}{m_3}\right) a_3 + m_3 b_3 = 110(-2) + 13 \cdot 17 = 1.$$

The residues are given by $r_1 = 7$, $r_2 = 8$, and $r_3 = 12$. We form

$$\begin{aligned} x &= \left(\frac{m}{m_1}\right) a_1 r_1 + \left(\frac{m}{m_2}\right) a_2 r_2 + \left(\frac{m}{m_3}\right) a_3 r_3 \\ &= 143 \cdot (-3) \cdot 7 + 130 \cdot 5 \cdot 8 + 110 \cdot (-2) \cdot 12 \\ &= -443. \end{aligned}$$

The integer $x = -443$ solves the system of congruences, but it is not positive. We know from the previous theorem that all solutions must be congruent modulo $m = 1430$. Therefore,

$$x \equiv -443 + 1430 \equiv 987 \pmod{1430}$$

is the smallest positive integer satisfying $987 \equiv 7 \pmod{10}$, $987 \equiv 8 \pmod{11}$, and $987 \equiv 12 \pmod{13}$.

8.5 The Chinese Remainder Theorem

Exercises

8.38. Determine the smallest positive integer x satisfying the system of congruences

$$x \equiv 3 \pmod{25},$$
$$x \equiv 5 \pmod{26},$$
$$x \equiv 7 \pmod{27}.$$

8.39. Tom has a large collection of quarters. If he arranges them in rows of 10, he will have 3 left over. If he arranges them in rows of 13, then he will have 5 left over. How many quarters does he have, if the total number is less than 130?

8.40. Hilary had a very successful recital. Her friend Mary was very impressed that every single chair was taken. So Mary asked Hilary how many chairs were in the concert hall. Hilary never counted them, but recalled that 2 years ago, they arranged the chairs in rows of twenties, but the last row contained just six chairs. Last year, they arranged it such such each row contained 21 chairs, but then the last row had only five chairs. This year they arranged it such that each row contained 23 chairs, except the last row contained four chairs. The number of chairs was the same each year and according to the fire marshal the concert hall can hold at most 2000 people. Use the Chinese Remainder Theorem to determine how many people attended Hilary's recital.

8.41. Suppose that you are given three pairwise coprime integer moduli m_1, m_2, and m_3, three integer residues r_1, r_2, and r_3, and your goal is to find an integer x satisfying

$$x \equiv r_1 \pmod{m_1},$$
$$x \equiv r_2 \pmod{m_2},$$
$$x \equiv r_3 \pmod{m_3}.$$

Your friend Joe suggests that you should instead solve three much simpler systems of congruences, namely find integers s_1, s_2, and s_3 such that

$$\left.\begin{array}{l} s_1 \equiv r_1 \pmod{m_1} \\ s_1 \equiv 0 \pmod{m_2} \\ s_1 \equiv 0 \pmod{m_3} \end{array}\right\} \quad \left.\begin{array}{l} s_2 \equiv 0 \pmod{m_1} \\ s_2 \equiv r_2 \pmod{m_2} \\ s_2 \equiv 0 \pmod{m_3} \end{array}\right\} \quad \left.\begin{array}{l} s_3 \equiv 0 \pmod{m_1} \\ s_3 \equiv 0 \pmod{m_2} \\ s_3 \equiv r_3 \pmod{m_3} \end{array}\right\}$$

Joe claims that the sum of the three solutions is congruent the solution x of the original system of congruences:

$$x \equiv s_1 + s_2 + s_3 \pmod{m}.$$

(a) Show how to find the solutions s_1, s_2, and s_3 to the simpler system of congruences.

(b) Show that $x \equiv s_1 + s_2 + s_3 \pmod{m}$ is a solution to the original system of congruences.

(c) Use Joe's method to solve the system of congruences

$$x \equiv 1 \pmod{11},$$
$$x \equiv 2 \pmod{13},$$
$$x \equiv 3 \pmod{17}.$$

8.42. Recall that an integer is called **square-free** if and only if it is not divisible by any perfect square other than 1. So $6 = 2 \cdot 3$ is square-free, but $18 = 2 \cdot 3^2$ is not square-free. Show that there exist one million consecutive integers that are *not* square-free.

8.6 The RSA Public Key Cryptosystem

We will discuss in this section the basic principles of the RSA public-key cryptosystem, a system that is used in countless e-commerce applications. The RSA public-key cryptosystem nicely illustrates the number-theoretic principles that we have learned so far.

Suppose that Alice seeks a way that people can send her confidential messages by e-mail. The RSA cryptosystem allows her to publish a key that everyone can use to send her an encrypted message, but that is hard to decipher without a secret that is only known to her.

We need some notation before stating the protocol. Euler's **totient function** $\varphi \colon \mathbf{N} \to \mathbf{N}$ is defined as

$$\varphi(n) = n \prod_{p \mid n} \left(1 - \frac{1}{p}\right),$$

where the product ranges over all primes dividing n. If $n = pq$ is the product of two distinct primes p and q, then $\varphi(n) = (p-1)(q-1)$.

Key Generation:

- Alice selects two distinct large prime numbers p and q, and computes their product $n = pq$.

- She selects an odd integer $e > 0$ such that $\gcd(e, \varphi(n)) = 1$. Thus, there exist integers d and k such that $ed - k\varphi(n) = 1$. Put differently, $\varphi(n) = (p-1)(q-1)$ divides $ed - 1$.

- Alice publishes the pair $P = (e, n)$, her public key. She carefully guards as a secret the factorization of n, the product $\varphi(n) = (p-1)(q-1)$, the integer k, and her secret key $S = (d, n)$.

8.6 The RSA Public Key Cryptosystem

Encryption and Decryption:

- For simplicity, we assume that a message is encoded as an integer M in the range $2 \leqslant M < n$.

- If Bob wants to send a message M to Alice, then he looks up Alice's public key and sends her the number

$$\boxed{C \equiv M^e \pmod{n}.}$$

- Alice uses her secret key to compute

$$\boxed{C^d \equiv M^{ed} \pmod{n}.}$$

It turns out that $M^{ed} \equiv M \pmod{n}$, so she recovers Bob's message.

It might be instructive to look at a simple example that illustrates the RSA protocol.

Example 8.27. Suppose that Alice chooses the primes $p = 991$ and $q = 1087$. Calculating the product $n = pq$ of these primes yields $n = 1,077,217$. The totient function $\varphi(n)$ has the value $\varphi(n) = (p-1)(q-1) = 990 \cdot 1086 = 1,075,140$. As an encoding exponent, she cannot select $e = 3$ or $e = 5$, since they are not coprime to $\varphi(n)$. Let us suppose that she selects $e = 7$ as the encoding exponent. Using the Euclidean algorithm, she determines that

$$\gcd(e, \varphi(n)) = 1 = 7d - \varphi(n)k = 7 \cdot 307,183 - 1,075,140 \cdot 2.$$

Her public RSA key is $(n, e) = (1077217, 7)$. She keeps her decoding exponent $d = 307,183$ and $\varphi(n)$ secret.

If Bob wants to send her the message $M = 127$, then he can calculate

$$173^7 \equiv 955,441 \pmod{1,077,217}.$$

and send Alice the ciphertext $C = 955,441$. Alice can then decode C by raising it do the d-th power modulo n. So she calculates

$$C^{307,183} \equiv 955,441^{307,183} \equiv 127 \pmod{1,077,217}.$$

The choice of e and d ensures that Alice decodes what Bob has sent.

An attacker knows Alice's public key (n, e) and may observe the ciphertext C, but solving the equation $M \equiv C^e \pmod{n}$ for M is believed to be difficult when n is large. Alice can easily solve the equation, since she has extra information such as the decoding exponent d. Since $M \mapsto M^e \pmod{n}$ is easy to compute, but its inverse is difficult to find without extra information, it is called a **trapdoor function**.

Correctness of RSA. We will now prove the claim that Alice always obtains Bob's message. We will need a slightly modified version of Fermat's little theorem to prove this result.

Proposition 8.28 (Fermat). *Let p be a prime. If a is an integer that is not divisible by p, then*
$$a^{p-1} \equiv 1 \pmod{p}.$$

Proof. We showed in Exercise 4.28 that $a^p \equiv a \pmod{p}$ holds for all integers a. The hypothesis implies that $\gcd(a, p) = 1$; hence, there exist integers x and y such that $ax + py = 1$. Therefore, $ax \equiv 1 \pmod{p}$. It follows from $a^p \equiv a \pmod{p}$ that $a^{p-1} \equiv xa^p \equiv xa \equiv 1 \pmod{p}$ holds. □

When Bob intends to send a message M to Alice, he encodes the message M into the ciphertext $C \equiv M^e \pmod{n}$. When Alice receives the ciphertext C, she attempts to decode C by forming $C^d \equiv M^{ed} \pmod{n}$. We will now show that M^{ed} is indeed congruent to M modulo n, so Alice does receive the correct message. The next theorem proves the correctness of this protocol.

Theorem 8.29. *Let $n = pq$ be a product of two distinct primes p and q. Let e, d, and k be positive integers satisfying $ed = 1 + k\varphi(n)$. Then*
$$M^{ed} \equiv M \pmod{n}$$
holds for all integers M.

Proof. It suffices to show that the two congruences
$$M^{ed} \equiv M \pmod{p} \quad \text{and} \quad M^{ed} \equiv M \pmod{q}$$
hold. Indeed, p and q are distinct primes, so $\gcd(p, q) = 1$, and the aforementioned congruences imply $M^{ed} \equiv M \pmod{n}$ by the Chinese Remainder Theorem.

If $M \equiv 0 \pmod{p}$, then certainly $M^{ed} \equiv M \pmod{p}$. If $M \not\equiv 0 \pmod{p}$, then $M^{p-1} \equiv 1 \pmod{p}$ by Proposition 8.28; hence,
$$M^{ed} \equiv M^{1+k\varphi(n)} \equiv M(M^{p-1})^{k(q-1)} \equiv M \, 1^{k(q-1)} \equiv M \pmod{p}.$$

Therefore, $M^{ed} \equiv M \pmod{p}$ holds for all integers M. Replacing p by q in the previous argument shows that $M^{ed} \equiv M \pmod{q}$ for all integers M. □

Exercises

8.43. Let $p = 23$ and $q = 31$ be primes. Can Alice use the public key $(pq, e) = (713, 3)$ as her RSA public key? Explain why or why not.

8.44. Let $p = 23$ and $q = 29$ be primes. Alice decides to use the public key $(pq, e) = (667, 3)$ as her RSA public key. Calculate a valid decoding exponent d. Show all your work.

8.7 Notes

8.45. Alice chooses the primes $p = 79$ and $q = 97$ and forms the product $n = pq$. As an encoding exponent, she chooses $e = 11$. Explicitly determine the following parameters of her RSA cryptosystem.
(a) Compute Alice's public key (n, e).
(b) Determine the value of the totient function $\varphi(n)$.
(c) Compute a decoding exponent d satisfying $ed - k\varphi(n) = 1$ for some integer k using the Euclidean algorithm.
(d) Encode the message $M = 42$ into $C \equiv M^e \pmod{n}$.
(e) Decode the message $C = 24$ using $M \equiv C^d \pmod{n}$.

8.46. Let n be the product of two odd primes p and q. Show that for any choice of encoding exponent e, there are at least nine messages that are not concealed by the RSA public key cryptosystem, meaning that there are nine different messages such that $M^e \equiv M \pmod{n}$. [Hint: Find fixed-points of $x \mapsto x^e \pmod{p}$ and $x \mapsto x^e \pmod{q}$ and use the Chinese Remainder Theorem] (Source: G.R. Blakley and I. Borosh)

8.47. Alice likes to use small encoding exponents such as $e = 3$ for her public RSA key (n, e), so that her friends do not have high computational costs when messaging her. Explain the cryptographic disadvantage of choosing a small encoding exponents. [Hint: Regardless of the exponent, we should exclude 0 and 1 as a message, as $0^e \equiv 0 \pmod{n}$ and $1^e \equiv 1 \pmod{n}$. Larger m are in general not fixpoints of the encoding map $m \mapsto m^e \pmod{n}$, but may be still easily decoded if the encoding exponent is small.]

8.48. Alice tells a friend that she creates the n for her RSA public key (n, e) by looking at a website that gives the 10 largest primes with a given number of bits. She did not realize that one can easily find the prime factors p and q of n using this information. Explain how an attacker can determine a decoding exponent d if the factorization of n into primes p and q is known.

8.49. Alice was not careful and accidentally revealed the value of the totient function $\varphi(n)$ of the integer n in her RSA public key (n, e). Since neither $n = pq$ and $\varphi(n) = (p-1)(q-1)$ directly reveals the primes p and q, she decides not to worry. Her friend Bob tells her that a cryptanalyst can find the primes p and q knowing n and $\varphi(n)$. Explain how to find the primes p and q given the values n and $\varphi(n)$. [Hint: Form the quadratic polynomial $(x-p)(x-q) = x^2 + bx + c$. Show how to express b and c in terms of $\varphi(n)$, n, and constants. Deduce a simple formula for p and q given n and $\varphi(n)$.]

8.50. Let $n = 2491$ and $\varphi(n) = 2392$. Use the method from the previous exercise to find the prime factors p and q of $n = pq$.

8.7 Notes

There are many excellent books on number theory. At an elementary level, there is a very readable and beautifully illustrated introduction to number theory by Weissman [79]. *Elementary number theory* by Dudley [20] is another good option. A bit more advanced but still very accessible is *A Classical Introduction to Modern Number Theory* by Ireland and Rosen [38].

Part II

Summation and Asymptotics

Chapter 9

Sums

> *The art of doing mathematics consists in finding that special case which contains all the germs of generality.*
>
> — David Hilbert

In this chapter, we will learn how to systematically evaluate sums such as

$$1^3 + 2^3 + \cdots + n^3 = \ ?$$

Our main tool will be a discrete analogue of calculus. It offers a difference operator that is a discrete version of the derivative and an antidifference operator that is a discrete version of the indefinite integral. The approach is to take a sum and use the antidifference operator to turn it into a telescoping sum. We begin by illustrating this approach using a simple example.

9.1 A Motivating Example

Suppose that we want to find the value of the sum

$$\sum_{k=1}^{n} \frac{1}{k(k+1)} = \frac{1}{1 \cdot 2} + \frac{1}{2 \cdot 3} + \cdots + \frac{1}{n(n+1)} = \ ?$$

How can we find a closed form solution? We can always try to evaluate the sum for small values of n in the hope to discover a pattern. You might be able to succeed in simple cases such as this one, but your mileage may vary. We use a different approach that transforms the given sum into a telescoping sum. This is everyones favorite kind of sum, since all but two terms cancel.

So let's see how we can transform beastly sums into beautiful telescoping sums. Let us denote the summands of our sum by

$$f(k) = \frac{1}{k(k+1)}.$$

© The Author(s), under exclusive license to Springer Nature Switzerland AG 2025
A. Klappenecker, H. Lee, *Discrete Structures*, Undergraduate Texts in Mathematics, https://doi.org/10.1007/978-3-031-73434-2_9

If we can find a function $F(k)$ such that $f(k)$ can be expressed by its difference

$$f(k) = F(k+1) - F(k),$$

then we can rewrite our sum in the form

$$\sum_{k=1}^{n} f(k) = \sum_{k=1}^{n} (F(k+1) - F(k)).$$

What is the benefit? Well, it turns out that the latter sum is of the telescoping kind. Indeed, using commutativity, we get

$$\sum_{k=1}^{n} (F(k+1) - F(k)) = \sum_{k=1}^{n} F(k+1) - \sum_{k=1}^{n} F(k).$$

Reindexing the first sum of the right-hand side yields

$$\sum_{k=1}^{n} (F(k+1) - F(k)) = \sum_{k=2}^{n+1} F(k) - \sum_{k=1}^{n} F(k).$$

Canceling like terms allows to collapse the sum to merely two remaining terms

$$\sum_{k=1}^{n} (F(k+1) - F(k)) = F(n+1) - F(1).$$

Therefore, we succeeded in transforming the given sum into a nice telescoping sum that is easy to evaluate

$$\sum_{k=1}^{n} \frac{1}{k(k+1)} = \sum_{k=1}^{n} (F(k+1) - F(k)) = F(n+1) - F(1).$$

Of course, it remains to find such a miraculous function $F(k)$ that satisfies

$$\frac{1}{k(k+1)} = F(k+1) - F(k).$$

If we define

$$F(k) = -\frac{1}{k}$$

then we can verify that the difference

$$F(k+1) - F(k) = -\frac{1}{k+1} + \frac{1}{k} = \frac{-k+k+1}{k(k+1)} = \frac{1}{k(k+1)},$$

as claimed. In summary, we can rewrite the sum as follows:

$$\sum_{k=1}^{n} \frac{1}{k(k+1)} = \sum_{k=1}^{n} \left(-\frac{1}{k+1} + \frac{1}{k}\right) = -\frac{1}{n+1} + \frac{1}{1} = \frac{n}{n+1}.$$

9.2 Difference Calculus

Motivation. In the subsequent sections, we show how to generalize this approach. The pattern is similar to what you might know from calculus.[1] The sums correspond to definite integrals. If you want to evaluate a definite integral

$$\int_a^b f(x)\,dx,$$

then you need to determine the indefinite integral

$$F(x) = \int f(x)\,dx$$

of the function f, and then express the definite integral in the form

$$\int_a^b f(x)\,dx = F(b) - F(a).$$

Our approach will be essentially the same. If you want to evaluate a definite sum

$$\sum_{k=a}^b f(k),$$

then you need to find the antidifference $F(x)$ of $f(x)$, which is the analogue of the indefinite integral. Then

$$\sum_{k=a}^b f(k) = F(b+1) - F(a).$$

As expected, this broad-brush analogy has its limits and the devil is in the details. Many of the rules that you know from calculus have their analogon, but are often subtly different. Before we can embark on finding the closed form of sums, we need to learn the analog of the differential operator d/dx. The discrete version of d/dx is the difference operator Δ. We will then show how to find the difference of any polynomial. This will lead us to falling factorials and Stirling numbers. Then we are finally ready to state the Fundamental Theorem of Summation, our main tool in finding the closed form of many sums.

9.2 Difference Calculus

Let f be a function from the set of integers to the set of real numbers. The **difference operator** Δ (or **forward difference operator** to be precise) is defined as

$$\Delta f(x) = f(x+1) - f(x).$$

---
[1] If you have not taken calculus yet, then simply skip to the next section.

The operator Δ serves as a discrete version of the derivative that is studied in calculus, but it has a much simpler definition.

Recall that the differential operator d/dx conceived by Leibniz acts on a function $f\colon \mathbf{R} \to \mathbf{R}$ by

$$\frac{d}{dx}f(x) = \lim_{h \to 0} \frac{f(x+h) - f(x)}{(x+h) - x}.$$

By contrast, the difference calculus that we develop here uses the somewhat crude $h = 1$ approximation to this differential operator. Indeed, we have

$$\frac{f(x+1) - f(x)}{(x+1) - x} = \Delta f(x).$$

Since Δ does not use a limit, it is conceptually much simpler than its cousin d/dx from differential calculus. Geometrically, $\frac{d}{dx}f(x)$ has the interpretation as the slope of the tangent to $f(x)$ at the point x, whereas $\Delta f(x)$ has the interpretation as the slope of a secant line, see Fig. 9.1.

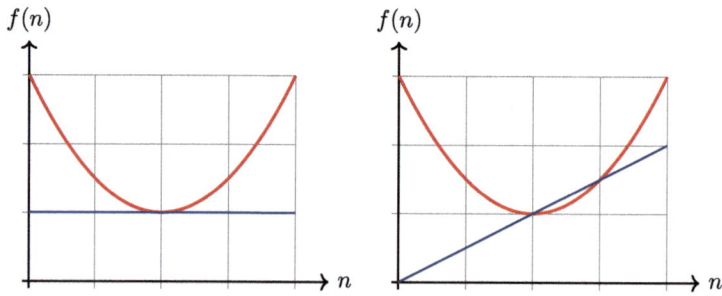

Figure 9.1: The function $f(x) = \frac{1}{2}x^2 - 2x + 3$ has the derivative $\frac{d}{dx}f(x) = x - 2$ and the difference $\Delta f(x) = x - \frac{3}{2}$. The figure on the left shows the tangent line through the point $(2, f(2)) = (2, 1)$ of slope $(\frac{d}{dx}f)(2) = 0$, so this illustrates the geometric interpretation of the derivative. The figure on the right shows a secant line through the points $(2, f(2))$ and $(3, f(3))$ of slope $(\Delta f)(2) = \frac{1}{2}$, so this illustrates the geometric interpretation of the difference operator

Examples. There are many similarities between the differential operator d/dx and the difference operator Δ. For example, the difference and differential operators behave the same on polynomials of degree 1 or less.

Example 9.1. The constant function $f(x) = c$, where c is a real number, has zero difference,
$$\Delta f(x) = f(x+1) - f(x) = c - c = 0.$$

Example 9.2. The function $f(x) = x$ has a constant difference,
$$\Delta x = (x+1) - x = 1.$$

9.2 Difference Calculus

The next few examples stress the difference between the differential operator and the difference operator.

Example 9.3. The function $f(x) = x^2$ has the difference
$$\Delta x^2 = (x+1)^2 - x^2 = 2x + 1.$$

Example 9.4. The function $f(x) = x^3$ has the difference
$$\Delta x^3 = (x+1)^3 - x^3 = 3x^2 + 3x + 1.$$

Example 9.5. Let α be a real number such that $\alpha \neq 1$. Then
$$\Delta \alpha^x = \alpha^{x+1} - \alpha^x = (\alpha - 1)\alpha^x.$$

In particular, $\alpha = 2$ satisfies $\Delta 2^x = 2^x$. Thus, the function 2^x plays the same role in difference calculus as the exponential function e^x in calculus.

Since the difference operator does not use a limiting process, it is defined whenever the function arguments are defined.

Example 9.6. The absolute value $f(x) = |x|$ has the difference
$$\Delta |x| = [x \geq 0] - [x < 0],$$

where $[\cdot]$ denotes the Iverson bracket. In other words, $\Delta |x|$ has the value 1 for all nonnegative integer arguments and the value -1 for all negative integer arguments.

We suggest that you construct more examples to gain familiarity with the difference operator. In particular, you should be familiar with differences of small degree polynomials and differences of exponential functions.

Rules. The difference operator follows rules that are very similar to the rules governing the differential operator from calculus. The first proposition shows that the difference operator Δ is a linear operator.

Proposition 9.7. *Let f and g be functions from the set of integers to the set of real numbers, and let c be a real number. Then*
(a) $\Delta(cf) = c\Delta f$,
(b) $\Delta(f + g) = \Delta f + \Delta g$,

Proof.
(a) The difference operator is homogeneous in the sense that
$$\begin{aligned} \Delta(cf(x)) &= cf(x+1) - cf(x) \\ &= c(f(x+1) - f(x)) = c\Delta f(x) \end{aligned}$$
holds.

(b) The difference operator is additive, since

$$\begin{aligned}
\Delta(f(x)+g(x)) &= (f(x+1)+g(x+1))-(f(x)+g(x)) \\
&= f(x+1)-f(x)+g(x+1)-g(x) \\
&= \Delta f(x)+\Delta g(x)
\end{aligned}$$

holds.

□

Example 9.8. It follows from Examples 9.1–9.4 and the linearity of the difference operator Δ that we can determine the difference $\Delta f(x)$ for any polynomial of degree 3 or less. For instance, if $f(x) = x^3 - 2x + 1$, then

$$\begin{aligned}
\Delta f(x) &= \Delta x^3 - 2\Delta x + \Delta 1 \\
&= (3x^2 + 3x + 1) - 2 \times 1 + 0 = 3x^2 + 3x - 1.
\end{aligned}$$

Perhaps you find the difference of polynomials of degree 2 or 3 a bit confusing. We will see in the next section that there is a simple pattern, even though it might not be apparent now.

The difference operator also satisfies product and quotient rules. However, the form of these rules differs slightly from the versions of differential calculus, as they involve a shift operator. The **shift operator** S is defined by

$$Sf(x) = f(x+1).$$

For instance, $Sx^2 = (x+1)^2 = x^2 + 2x + 1$.

Proposition 9.9. *Let f and g be functions from the set of integers to the set of real numbers. Then the difference operator satisfies the following three product rules:*

$$\Delta(fg) = (\Delta f)g + f\Delta g + (\Delta f)(\Delta g) \qquad (9.1)$$
$$\Delta(fg) = (\Delta f)Sg + f\Delta g \qquad (9.2)$$
$$\Delta(fg) = (\Delta f)g + (Sf)\Delta g \qquad (9.3)$$

These are reminiscent of the product rule for the derivative, but involve either an extra term or a shift operator.

Proof. Let us prove the first product rule. By definition, the left-hand side of Eq. (9.1) yields

$$\Delta(fg) = f(x+1)g(x+1) - f(x)g(x).$$

Expanding the right-hand side of the Eq. (9.1) yields

$$\begin{aligned}
(\Delta f)g + f\Delta g + (\Delta f)(\Delta g) &= (f(x+1)-f(x))g(x) + f(x)(g(x+1)-g(x)) \\
&\quad + (f(x+1)-f(x))(g(x+1)-g(x)) \\
&= f(x+1)g(x+1) - f(x)g(x).
\end{aligned}$$

9.2 Difference Calculus

So both sides are indeed the same, as claimed.

For the second product rule (9.2), we observe that

$$\Delta(fg) = f(x+1)g(x+1) - f(x)g(x)$$
$$= f(x+1)g(x+1) \underbrace{-f(x)g(x+1) + f(x)g(x+1)}_{\text{we inserted an expression equal to 0}} - f(x)g(x)$$

$$= (\Delta f(x))g(x+1) + f(x)\Delta g(x),$$

which proves the product rule given in Eq. (9.2).

We can obtain the third product rule (9.3) from the second product rule (9.2) by exchanging the role of f and g. □

Example 9.10. In Examples 9.2 and 9.5, we established that $\Delta x = 1$ and $\Delta 2^x = 2^x$. By the first product rule of the previous proposition, we have

$$\Delta(x2^x) = (\Delta x)2^x + x(\Delta 2^x) + (\Delta x)(\Delta 2^x)$$
$$= 2^x + x2^x + 2^x$$
$$= 2^{x+1} + x2^x.$$

Of course, we could have used any of the other two product rules to obtain the same result.

Proposition 9.11. *Let f and g be functions from the set of integers to the set of real numbers. Then the difference operator satisfies the quotient rule*

$$\Delta\left(\frac{f}{g}\right) = \frac{g\Delta f - f\Delta g}{gSg}.$$

Proof. By definition

$$\Delta\left(\frac{f}{g}\right) = \frac{f(x+1)}{g(x+1)} - \frac{f(x)}{g(x)} = \frac{f(x+1)g(x) - f(x)g(x+1)}{g(x)g(x+1)}$$
$$= \frac{f(x+1)g(x) - f(x)g(x) + f(x)g(x) - f(x)g(x+1)}{g(x)g(x+1)}.$$

The latter expression implies the claimed quotient rule. □

Example 9.12. Let $f(x) = 2^x$ and $g(x) = x^2$. Thus, we have $\Delta 2^x = 2^x$ and $\Delta x^2 = 2x + 1$. It follows from the quotient rule that

$$\Delta\left(\frac{2^x}{x^2}\right) = \frac{x^2 2^x - 2^x(2x+1)}{x^2(x+1)^2}$$
$$= \frac{2^x(x^2 - 2x - 1)}{x^2(x+1)^2}.$$

Alternatively, we could have used a calculation directly based on the definition,

$$\Delta\left(\frac{2^x}{x^2}\right) = \frac{2^{x+1}}{(x+1)^2} - \frac{2^x}{x^2} = \frac{2^{x+1}x^2 - 2^x(x+1)^2}{x^2(x+1)^2}$$
$$= \frac{2^x(x^2 - 2x - 1)}{x^2(x+1)^2},$$

to arrive at the same conclusion.

Exercises

9.1. Calculate the following differences using linearity of the difference operator Δ and the results derived in the examples of this section.
(a) $\Delta(x^2 + x)$.
(b) $\Delta(2x^2 + 3x + 2)$.
(c) $\Delta(2x^3 + 5x^2)$.
(d) $\Delta(x^3 + x^2 + x + 1)$.

9.2. Let x denote an integer. Recall that the x-th harmonic number H_x is given by $H_x = 1 + \frac{1}{2} + \frac{1}{3} + \cdots + \frac{1}{x}$. Calculate ΔH_x.

9.3. Let c be a nonzero real number. Show that

$$\Delta \log(cx) = \log\left(1 + \frac{1}{x}\right).$$

9.4. Show that

$$\Delta \frac{1}{x} = -\frac{1}{x(x+1)}.$$

9.5. Show that
(a) $\Delta \sin(ax) = 2\cos\left(ax + \frac{a}{2}\right)\sin\left(\frac{a}{2}\right)$.
(b) $\Delta \cos(ax) = -2\sin\left(ax + \frac{a}{2}\right)\sin\left(\frac{a}{2}\right)$.

9.6. Explicitly determine the result of applying the difference operator Δ twice to a function $f(x)$. In other words, determine $\Delta^2 f(x)$. Determine $\Delta^3 f(x)$ as well.

9.7. Use the first product rule to determine $\Delta(x^2 5^x)$.

9.8. Use the second product rule and linearity of Δ to determine $\Delta(H_x x - x)$.

9.9. Use the second product rule and the fact that $\Delta x^2 = 2x + 1$ to calculate $\Delta x^4 = \Delta(x^2 \cdot x^2)$.

9.10. Use the first product rule to determine $\Delta(H_x \cdot H_x)$.

9.11. Use the quotient rule to determine $\Delta(2^x/x)$.

9.3 Falling Factorial Powers

In this section, our goal is to understand the difference operator a bit better. We discover a family of functions that is essentially[2] mapped to itself by the difference operator. The functions in this family are called falling factorial powers. Their main benefit is that the action of the difference operator on polynomials is easy to understand once polynomials are expressed in terms of falling factorial powers. We will also get a useful interpretation of binomial coefficients in terms of falling factorial powers.

Recall that a **polynomial** with real coefficients in the indeterminate x is an expression

$$f(x) = a_k x^k + a_{k-1} x^{k-1} + \cdots + a_1 x + a_0$$

such that the coefficients $a_0, \ldots, a_k$ are real numbers. The **degree** of a nonzero polynomial $f(x)$ is defined as $\deg f(x) = \max\{i \mid a_i \neq 0\}$, and by convention $\deg 0 = -\infty$. The coefficient a_0 is called the **constant term** of the polynomial. The largest index k such that $a_k \neq 0$ is called the **leading coefficient** of $f(x)$ and $a_k x^k$ is called the **leading term**.

If we apply the difference operator Δ to a polynomial $p(x)$ of degree $k > 0$, then

$$\Delta p(x) = p(x+1) - p(x)$$

is a polynomial of degree $k - 1$. Examples 9.3 and 9.4 show that the difference operator Δ applied to a monomial does not lead to a multiple of a monomial. Fortunately, there exists a family of polynomials on which Δ acts in a simple fashion, and we will define these polynomials next.

For a nonnegative integer k, we define $x^{\underline{k}}$ to be the polynomial of degree k given by

$$x^{\underline{k}} = x(x-1) \cdots (x - k + 1),$$

where it is understood that the empty product is 1, meaning that $x^{\underline{0}} = 1$. We call $x^{\underline{k}}$ a **falling factorial power** and pronounce it as "x to the k falling."

Proposition 9.13. *For nonnegative integers k, we have*

$$\Delta x^{\underline{k}} = k x^{\underline{k-1}}.$$

Proof. It follows from the definitions that

$$\Delta x^{\underline{k}} = (x+1)^{\underline{k}} - x^{\underline{k}} = (x+1) x^{\underline{k-1}} - x^{\underline{k-1}}(x - k + 1).$$

Factoring out the term $x^{\underline{k-1}}$ yields

$$\Delta x^{\underline{k}} = ((x+1) - (x - k + 1)) x^{\underline{k-1}} = k x^{\underline{k-1}},$$

which proves the claim. □

---

[2] By "essentially" we actually mean "up to multiplication by a scalar".

The next proposition shows that every polynomial can be expressed as a linear combination of falling factorial powers.

Proposition 9.14. *A polynomial of degree n can be expressed in the form*

$$p(x) = \sum_{k=0}^{n} a_k x^{\underline{k}}$$

for some uniquely determined coefficients $a_0, \ldots, a_n$ in $\mathbf{R}$.

Proof.
(a) *Existence.* We are going to prove the claim by induction on the degree n. If the polynomial $p(x)$ has degree $n \leqslant 0$, then it is a constant and can be written in the form $p(x) = cx^{\underline{0}}$ for some uniquely determined real number c.

Let us assume that the claim holds for all polynomials of degree n or less. We will show that it must hold for polynomials of degree $n+1$. If $p(x)$ is a polynomial of degree $n+1$ with leading coefficient a_{n+1}, then

$$r(x) := p(x) - a_{n+1} x^{\underline{n+1}}$$

is a polynomial of degree n. By induction hypothesis, $r(x)$ can be written in the form $r(x) = \sum_{k=0}^{n} a_k x^{\underline{k}}$ for some coefficients $a_0, \ldots, a_n$. Therefore, we can conclude that $p(x)$ can be written in the form $\sum_{k=0}^{n+1} a_k x^{\underline{k}}$.

(b) *Uniqueness.* Seeking a contradiction, let us suppose that $p(x)$ is a polynomial of smallest possible degree that has two different representations

$$\sum_{k=0}^{n} a_k x^{\underline{k}} = p(x) = \sum_{k=0}^{n} b_k x^{\underline{k}}$$

for some real numbers $a_0, a_1, \ldots, a_n$ and $b_0, b_1, \ldots, b_n$. Then

$$0 = p(x) - p(x) = \sum_{k=0}^{n} (b_k - a_k) x^{\underline{k}}$$

shows that the terms of degree n must cancel, so $b_n = a_n$ must hold. Therefore,

$$\sum_{k=0}^{n-1} a_k x^{\underline{k}} = \sum_{k=0}^{n-1} b_k x^{\underline{k}}$$

Since n was the smallest degree for which representations can differ, we can conclude that $(a_0, a_1, \ldots, a_{n-1}) = (b_0, b_1, \ldots, b_{n-1})$. Thus, the two representation of $p(x)$ are actually the same, contradicting the assumption of non-uniqueness.

□

9.3 Falling Factorial Powers

Remark 9.15. If you are familiar with linear algebra, then you will notice that the previous proposition simply states that the polynomials in the set

$$B = \{x^{\underline{k}} \mid k \text{ is a nonnegative integer }\}$$

forms a basis of the vector space of polynomials in the indeterminate x with real coefficients. The advantage of using the basis B instead of the set

$$\{x^k \mid k \text{ is a nonnegative integer }\}$$

of monomials is that the falling factorial powers are manipulated with ease when using the difference operator Δ. In the next section, we will see how one can explicitly express any monomial x^n in terms of the falling factorial powers. □

We can extend the falling factorial powers to negative exponents by defining

$$x^{\underline{-n}} = \frac{1}{(x+1)(x+2)\cdots(x+n)}$$

for all integers $n > 0$. This definition ensures that

$$\frac{x^{\underline{k+1}}}{x^{\underline{k}}} = x - k$$

holds for all integers k. The next proposition shows that the difference of negative factorial powers is once again a multiple of a negative factorial power.

Proposition 9.16. *For all integers k, we have*

$$\Delta x^{\underline{k}} = k x^{\underline{k-1}}.$$

This extends the claim of Proposition 9.13 to all integers k.

Proof. For $n > 0$, we have

$$\Delta x^{\underline{-n}} = \frac{1}{(x+2)(x+3)\cdots(x+n+1)} - \frac{1}{(x+1)(x+2)\cdots(x+n)}$$
$$= \frac{(x+1) - (x+n+1)}{(x+1)(x+2)\cdots(x+n+1)} = -n x^{\underline{-n-1}}$$

We have $\Delta x^{\underline{0}} = 0 x^{\underline{-1}}$. Thus, the claim holds for all integers k that are not positive. For positive k, the claim follows from Proposition 9.13. □

We can define two important functions with the help of factorial falling powers. The **factorial** $x!$ of a nonnegative integer x is defined as

$$x! := x^{\underline{x}} = x(x-1)\cdots 2 \cdot 1.$$

The **binomial coefficient** is defined as

$$\binom{x}{k} = \frac{x^{\underline{k}}}{k!}.$$

This definition is a bit broader than you might expect, since it even makes sense when x is a real number such as $x = 1/2$. For instance,

$$\binom{\frac{1}{2}}{3} = \frac{\frac{1}{2}\left(\frac{1}{2}-1\right)\left(\frac{1}{2}-2\right)}{3!} = \frac{1}{16}.$$

We will exploit this generality in the definition a bit later in Newton's formula.

The next proposition shows how the difference operator behaves when applied to the top argument of the binomial coefficient.

Proposition 9.17. *We have*

$$\Delta\binom{x}{k} = \binom{x}{k-1},$$

where the difference is formed in the argument x.

Proof. By definition,

$$\Delta\binom{x}{k} = \binom{x+1}{k} - \binom{x}{k} = \frac{(x+1)^{\underline{k}} - x^{\underline{k}}}{k!} = \frac{((x+1)-(x-k+1))x^{\underline{k-1}}}{k!}$$

and the latter expression can be simplified to $\frac{kx^{\underline{k-1}}}{k!} = \binom{x}{k-1}$. □

Corollary 9.18. *The previous proposition implies the recurrence relation*

$$\binom{x+1}{k} = \binom{x}{k} + \binom{x}{k-1}$$

that governs the construction of Pascal's triangle.

Exercises

9.12. Prove that $x^{\underline{k+1}} + kx^{\underline{k}} = xx^{\underline{k}}$.

9.13. Prove that $\dfrac{(x+1)^{\underline{k}}}{k!} = \dfrac{x^{\underline{k}}}{k!} + \dfrac{x^{\underline{k-1}}}{(k-1)!}$.

9.14. The factorial $x!$ of a nonnegative integer x is defined as $x! := x^{\underline{x}}$. (a) Calculate $\Delta x!$, and (b) find a function $F(x)$ such that $\Delta F(x) = x!$.

9.15.
(a) Use the binomial theorem to show that

$$\Delta x^n = \sum_{k=0}^{n-1} \binom{n}{k} x^k.$$

(b) Use part (a) to obtain an explicit expression for Δx^4 by evaluating and simplifying the binomial coefficients.

9.4 Stirling Numbers

(c) Use part (a) to obtain an explicit expression for Δx^5 by evaluating and simplifying the binomial coefficients.

9.16. Let n be a positive integer. Let $p(x) = \sum_{k=0}^{n} a_k x^k$ be a polynomial of degree n with leading coefficient a_n. Show that $\Delta p(x)$ is a polynomial of degree $n-1$ with leading coefficient na_n.

9.17. Let n be a positive integer and $p(x) = \sum_{k=0}^{n} a_k x^k$ a polynomial of degree n with leading coefficient a_n. Let Δ^n denote applying the difference operator n times. Prove by induction that applying the Δ operator n times to $p(x)$ yields the constant polynomial
$$\Delta^n p(x) = n! a_n.$$
This result is sometimes called the **Fundamental Theorem of Difference Calculus**. Hint: Use the previous exercise.

9.18. We can apply the difference operator n times and denote the resulting operator by Δ^n. A direct calculation of $\Delta^n f(x)$ can be tedious. Show that
$$\Delta^n f(x) = \sum_{k=0}^{n} (-1)^{n-k} \binom{n}{k} f(x+k).$$
[Hint: You can express Δ^n in the form $\Delta^n = (S-I)^n$, where S is the shift operator and I is the identity operator.]

9.19. Show that
$$f(x) = \sum_{k=0}^{n} \frac{(\Delta^k f)(a)}{k!} (x-a)^{\underline{k}}$$
holds for all polynomials $f(x)$ of degree at most n. This is Newton's analogue of a Taylor series.

9.20. Determine the differences of the function $f(x) = 4x(x-1)(x-2)(x-3)$.
(a) $\Delta f(x)$.
(b) $\Delta^2 f(x)$.
(c) $\Delta^3 f(x)$.
(d) $\Delta^4 f(x)$.

9.4 Stirling Numbers

In this section, we show how to express a monomial x^n in terms of falling factorial powers. The linear combination is given using the so-called Stirling numbers of the second kind. These numbers are interesting in their own right, but for now their main benefit is that they enable us to change quickly from the usual polynomial representation in terms of monomials to the falling power representation.

In the previous section, we showed that we can express the powers x^n using the falling factorial powers. In fact, it follows from Proposition 9.14 that there exist real numbers $S(n,k)$ such that

$$x^n = \sum_{k=0}^{n} S(n,k) x^{\underline{k}}. \tag{9.4}$$

The numbers $S(n,k)$ are called the **Stirling numbers of the second kind**. The Stirling numbers of the second kind are also denoted as

$$S(n,k) = \left\{ \begin{matrix} n \\ k \end{matrix} \right\}.$$

This notation is a good mnemonic, since these numbers count the number of ways a set with n elements can be partitioned into k nonempty subsets, as we will see later. We will use the more compact notation $S(n,k)$ whenever many formulas need to be manipulated.

The next proposition shows how to quickly calculate the Stirling numbers of the second kind.

Proposition 9.19. *The Stirling numbers of the second kind satisfy* $S(n,n) = 1$ *for all nonnegative integers* n,

$$S(0,k) = S(k,0) = 0 \quad \text{for } k > 0$$

and the recurrence relation

$$S(n,k) = S(n-1, k-1) + k S(n-1, k)$$

for all positive integers n *and all integers* k *in the range* $1 \leq k \leq n-1$.

Proof. The claim has four different parts that we will derive one by one.
(a) A comparison of the coefficients of x^n on both sides of (9.4) shows that $S(n,n) = 1$ holds for all nonnegative n.
(b) A comparison of the constant terms on both sides of (9.4) shows that $S(n,0) = 0$ for $n > 0$, since the constant terms of all polynomials $x^{\underline{k}}$ are 0 when $k > 0$.
(c) Since $x^0 = x^{\underline{0}} = \sum_k S(0,k) x^{\underline{k}}$, we have $S(0,0) = 1$ and $S(0,k) = 0$ for all positive integers.
(d) For the recurrence relation, we are going to express x^n in the form $x \cdot x^{n-1}$. We have

$$x^n = x \cdot x^{n-1} = \sum_{k=0}^{n-1} S(n-1, k) x \, x^{\underline{k}}$$

By definition, $x^{\underline{k+1}} = x^{\underline{k}}(x-k)$, and adding $k x^{\underline{k}}$ to both sides yields $x x^{\underline{k}} = x^{\underline{k+1}} + k x^{\underline{k}}$. It follows that

$$\begin{aligned} x^n &= \sum_{k=0}^{n-1} S(n-1,k) x^{\underline{k+1}} + \sum_{k=0}^{n-1} S(n-1,k) k x^{\underline{k}} \\ &= \sum_{\ell=1}^{n} S(n-1, \ell-1) x^{\underline{\ell}} + \sum_{k=1}^{n-1} S(n-1,k) k x^{\underline{k}} \end{aligned}$$

9.4 Stirling Numbers

Table 9.1: The Stirling numbers of the second kind $S(n,k)$.

$n\backslash k$	0	1	2	3	4	5
0	1	0	0	0	0	0
1	0	1	0	0	0	0
2	0	1	1	0	0	0
3	0	1	3	1	0	0
4	0	1	7	6	1	0
5	0	1	15	25	10	1

Combining the two sums, we get

$$x^n = \sum_{k=1}^{n-1} \Big(S(n-1, k-1) + S(n-1, k)k\Big)x^{\underline{k}} + S(n,n)x^{\underline{n}}.$$

Comparing the coefficients of the latter expression with (9.4) yields the recurrence relation.

□

The recurrence relation from the previous proposition is similar to the recurrence relation of the binomial coefficients that governs the construction of Pascal's triangle. Before computing "Stirling's triangle," let us have a look at a small example.

Example 9.20. By definition of the Stirling numbers, we have

$$x^2 = S(2,2)x^{\underline{2}} + S(2,1)x^{\underline{1}} + S(2,0)x^{\underline{0}}.$$

It remains to explicitly determine the values of these Stirling numbers. By the previous proposition, we have $S(2,2) = 1$ and $S(2,0) = 0$. The recurrence relation allows one to calculate $S(2,1)$ in the form

$$S(2,1) = S(1,0) + 1S(1,1) = 0 + 1 = 1.$$

Therefore, $x^2 = x^{\underline{2}} + x^{\underline{1}}$.

Proposition 9.19 allows one to compute the Stirling numbers for small n and k. We are going to set up a "Stirling's triangle" that contains the coefficients $S(n,k)$ in the n-th row. Given the values of the coefficients $S(n-1,k)$ of row $n-1$, the recurrence $S(n,k) = S(n-1, k-1) + kS(n-1, k)$ allows us to quickly obtain the values of row n. The results are compiled in Table 9.1 for Stirling numbers up to $n = 5$.

In this chapter, the main purpose of Stirling numbers is to express monomials in terms of falling factorial powers. We saw that it is fairly straightforward to calculate the values of Stirling numbers of the second kind. We will discuss combinatorial properties of Stirling numbers in Sect. 11.5.

Exercises

9.21. Calculate the following Stirling numbers of the second kind:
(a) $S(7,3) = \{^7_3\}$,
(b) $S(7,5) = \{^7_5\}$,
(c) $S(8,3) = \{^8_3\}$,
(d) $S(8,5) = \{^8_5\}$.

9.22. Derive from Eq. (9.4) explicit expressions for monomials of small degree in terms of falling factorials by explicitly calculating the Stirling numbers of the second kind for the following monomials:
(a) x^3
(b) x^4
(c) x^5
(d) x^6
(e) x^7

9.23. Prove by induction that $S(n,1) = 1$ holds for all $n \geq 1$.

9.24. Prove by induction that $S(n, n-1) = \sum_{k=1}^{n-1} k$ holds for all positive integers n.

9.25. Verify that the number of different partitions of the set $\{1,2,3,4\}$ into nonempty sets is equal to

$$S(4,1) + S(4,2) + S(4,3) + S(4,4) = \left\{^4_1\right\} + \left\{^4_2\right\} + \left\{^4_3\right\} + \left\{^4_4\right\}.$$

9.26. Prove that the upper bound

$$S(n,k) \leq \frac{1}{2}\binom{n}{k} k^{n-k}$$

holds for all integers n and k satisfying $n \geq 2$ and $1 \leq k \leq n-1$.

9.27. Prove that the lower bound

$$\frac{1}{2}(k^2 + k + 2)k^{n-k-1} - 1 \leq S(n,k)$$

holds for all integers n and k such that $n \geq 2$ and $1 \leq k \leq n$.

9.5 The Fundamental Theorem of Summation

Let f be a function from the set of integers to the set of real numbers, and a and b integers such that $a \leq b$. We want to derive a method to evaluate sums of the form

$$\sum_{k=a}^{b} f(k) = f(a) + f(a+1) + \cdots + f(b).$$

9.5 The Fundamental Theorem of Summation

A function F is called an **antidifference** of f (or the **indefinite sum** of f) if and only if $\Delta F = f$ holds. In other words, an antidifference F of f satisfies

$$\Delta F(x) = F(x+1) - F(x) = f(x).$$

We write $\Delta^{-1} f$ to denote an antidifference of f. Antidifferences play the same role in summation as indefinite integrals in integration.

If we know an antidifference F of the function f, then we can transform a partial sum of f into a nicely summable sum of the telescoping kind. Indeed, we have by definition

$$\sum_{k=a}^{b} f(k) = \sum_{k=a}^{b} \Delta F(k) = \sum_{k=a}^{b} (F(k+1) - F(k)).$$

The sum on the right-hand side contains more terms, but is very easy to evaluate! Indeed, the next theorem shows that all but two terms cancel.

Theorem 9.21 (Fundamental Theorem of Summation). *Let f be a function from the set of integers to the set of real numbers, and let F be an antidifference of f. Then*

$$\sum_{k=a}^{b} f(k) = F(b+1) - F(a).$$

Proof. By definition of an antidifference, we have

$$\sum_{k=a}^{b} f(k) = \sum_{k=a}^{b} (F(k+1) - F(k)) = F(b+1) - F(a),$$

where the intermediate terms $F(a+1), F(a+2), \ldots, F(b)$ all cancel, since they are added once and subtracted once. □

It might be instructive to expand on the aforementioned proof using a small example.

Example 9.22. Suppose that we want to evaluate the sum

$$\sum_{k=1}^{4} f(k) = f(1) + f(2) + f(3) + f(4).$$

If we know an antidifference $F(k)$ of $f(k)$, then by definition

$$\Delta F(k) = F(k+1) - F(k) = f(k).$$

Using the antidifference, we can express the aforementioned sum in the form

$$f(1) + f(2) + f(3) + f(4) =$$
$$(F(2) - F(1)) + (F(3) - F(2)) + (F(4) - F(3)) + (F(5) - F(4)).$$

Reordering the terms on the right-hand side, we obtain

$$f(1) + f(2) + f(3) + f(4)$$
$$= F(5) + F(4) - F(4) + F(3) - F(3) + F(2) - F(2) - F(1)$$

So the entire point of rewriting the sum in terms of antidifferences should now be clear! Indeed, if we remove all the canceling terms in the middle, we get the following simple expression for the sum on the left-hand side

$$f(1) + f(2) + f(3) + f(4) = F(5) - F(1).$$

The aforementioned theorem asserts that the same approach works for sums with many more terms!

So let us see how we can use the theorem to evaluate some actual sums.

Example 9.23. What is the value of the sum

$$1 \cdot 2 \cdot 3 + 2 \cdot 3 \cdot 4 + 3 \cdot 4 \cdot 5 + \cdots + n(n+1)(n+2)?$$

Using falling factorials, we can express this sum in the form

$$\sum_{k=3}^{n+2} k^{\underline{3}} = \sum_{k=3}^{n+2} k(k-1)(k-2).$$

By Proposition 9.13, an antidifference of $f(k) = k^{\underline{3}}$ is given by

$$F(k) = \frac{1}{4} k^{\underline{4}} = \frac{1}{4} k(k-1)(k-2)(k-3).$$

By the Fundamental Theorem of Summation, we have

$$\sum_{k=3}^{n+2} k^{\underline{3}} = F(n+3) - F(3) = \frac{(n+3)(n+2)(n+1)n}{4}.$$

This approach is easy, as long as you can find the antidifference.

How can we find antidifferences? Known difference relations are our main source, at least initially. For instance, we have shown that

$$\Delta x^{\underline{2}} = 2x^{\underline{1}} = 2x.$$

Therefore, it follows that we can choose $\Delta^{-1} x$ to be $\frac{1}{2} x^{\underline{2}} = \frac{1}{2} x(x-1)$. We collect a few such results in Fig. 9.2.

In the remainder of this section, we give more applications of the previous theorem and prove the less straightforward claims of Fig. 9.2.

9.5 The Fundamental Theorem of Summation

Constants. $\Delta^{-1}\alpha = \alpha x$, where $\alpha \in \mathbf{R}$.

Falling Factorial Powers. $\Delta^{-1} x^{\underline{k}} = \dfrac{1}{k+1} x^{\underline{k+1}}$, where $k \in \mathbf{Z}\setminus\{-1\}$, and $\Delta^{-1} x^{\underline{-1}} = H_x$.

Reciprocals. $\Delta^{-1} \dfrac{1}{x} = H_{x-1}$.

Monomials. $\Delta^{-1} x^n = \sum_{k=0}^{n} \dfrac{S(n,k)}{k+1} x^{\underline{k+1}}$, where $n \in \mathbf{N}_1$.

Exponentials. $\Delta^{-1} \alpha^x = \dfrac{\alpha^x}{\alpha - 1}$, where $\alpha \in \mathbf{R}\setminus\{-1\}$, and $\Delta^{-1} \dfrac{1}{2^x} = -\dfrac{1}{2^{x-1}}$.

Alternating Signs. $\Delta^{-1}(-1)^x = \dfrac{(-1)^{x+1}}{2}$.

Binomial Coefficients. $\Delta^{-1} \binom{x}{k} = \binom{x}{k+1}$.

Fibonacci Numbers. $\Delta^{-1} F_x = F_{x+1}$.

Logarithms. $\Delta^{-1} \log x = \log(x-1)!$.

Figure 9.2: The figure shows a selection of antidifferences. We omitted the addition of constants C on the right-hand side

Example 9.24. We begin with the sum of the first n positive integers

$$1 + 2 + 3 + \cdots + n.$$

Most likely you are already aware how to evaluate this sum. However, the simplicity of this example will allow you to focus on the method. We can formulate this problem as the sum

$$\sum_{k=1}^{n} f(k) = \sum_{k=1}^{n} k.$$

As we have noted earlier, an antidifference of $f(x) = x$ is given by $F(x) = x^{\underline{2}}/2 = x(x-1)/2$. By the Fundamental Theorem of Summation, we have

$$\sum_{k=1}^{n} k = F(n+1) - F(1) = \dfrac{(n+1)n}{2} - \dfrac{1 \cdot 0}{2} = \dfrac{(n+1)n}{2}.$$

Incidentally, this is the formula that Karl Friedrich Gauss supposedly derived in elementary school when he was asked by his teacher to sum the first 100 positive integers.

Geometric series are another standard example of sums that you will frequently encounter in practice. We will cover them in the next example.

Example 9.25. Let α be a real number such that $\alpha \neq 1$. For integers a and b satisfying $a \leqslant b$, we want to evaluate the sum

$$\sum_{k=a}^{b} \alpha^k = \alpha^a + \alpha^{a+1} + \alpha^{a+2} + \cdots + \alpha^b.$$

It follows from Example 9.5 that an antidifference of $f(x) = \alpha^x$ is given by $F(x) = \alpha^x/(\alpha - 1)$. By the Fundamental Theorem of Summation, we have

$$\sum_{k=a}^{b} \alpha^k = \frac{\alpha^{b+1} - \alpha^a}{\alpha - 1}.$$

In particular, if $\alpha = 2$ then

$$\sum_{k=0}^{n} 2^k = 2^{n+1} - 1.$$

Such sums of geometric series are for instance frequently used in the analysis of fast Fourier transform algorithms or randomized algorithms.

If we want to handle more elaborate sums, then we need to know more about antidifferences. We know the antidifference of $x^{\underline{k}}$ for all integers k, except when the exponent $k = -1$. We need to introduce harmonic numbers so that we can fill the gap for $k = -1$.

We define the x-th **harmonic number** H_x as

$$H_x = 1 + \frac{1}{2} + \cdots + \frac{1}{x},$$

for all nonnegative integers x. Then

$$\Delta H_x = \frac{1}{x+1} = x^{\underline{-1}}.$$

In other words, H_{x-1} is the discrete analog of the logarithm, which explains why harmonic numbers occur quite frequently in the analysis of algorithms.

Proposition 9.26. *For any integer k, we have*

$$\Delta^{-1} x^{\underline{k}} = \begin{cases} \frac{1}{k+1} x^{\underline{k+1}} + C & \text{if } k \neq -1 \\ H_x + C & \text{if } k = -1 \end{cases}$$

where C is a constant function when restricted to the integers.

Proof. For $k \neq -1$, we have $\Delta(\frac{1}{k+1} x^{\underline{k+1}} + C) = x^{\underline{k}}$ by Proposition 9.16. For $k = -1$, we have $\Delta(H_x + C) = x^{\underline{-1}}$. Applying the antidifference operator yields the result. □

9.5 The Fundamental Theorem of Summation

We can use falling factorial powers to obtain antidifferences of ordinary power functions such as squares x^2 and cubes x^3.

Proposition 9.27. *For any positive integer n, we have*

$$\Delta^{-1} x^n = \sum_{k=0}^{n} \frac{S(n,k)}{k+1} x^{\underline{k+1}} + C$$

where $S(n,k)$ denotes the Stirling number of the second kind and C is a constant function when restricted to the integers.

Proof. We have

$$\Delta \left(\sum_{k=0}^{n} \frac{S(n,k)}{k+1} x^{\underline{k+1}} + C \right) = \sum_{k=0}^{n} \frac{S(n,k)}{k+1} \Delta x^{\underline{k+1}} = \sum_{k=0}^{n} S(n,k) x^{\underline{k}} = x^n,$$

where the last equality follows from the definition of the Stirling numbers. Applying the antidifference operator to both sides of the equation yields the claim. □

Example 9.28. We now solve the motivating problem of this chapter: Find the sum

$$1^3 + 2^3 + \cdots + n^3$$

of the first n cubes. By the previous proposition, the antidifference of x^3 is given by

$$\Delta^{-1} x^3 = \frac{S(3,3)}{4} x^{\underline{4}} + \frac{S(3,2)}{3} x^{\underline{3}} + \frac{S(3,1)}{2} x^{\underline{2}} = \frac{1}{4} x^{\underline{4}} + \frac{3}{3} x^{\underline{3}} + \frac{1}{2} x^{\underline{2}}$$

By the Fundamental Theorem of Summation, we get

$$\sum_{k=1}^{n} k^3 = \frac{1}{4}(n+1)^{\underline{4}} + (n+1)^{\underline{3}} + \frac{1}{2}(n+1)^{\underline{2}} - \left(\frac{1}{4} 1^{\underline{4}} + 1^{\underline{3}} + \frac{1}{2} 1^{\underline{2}} \right).$$

The last term in parentheses is equal to 0. After expanding the remaining terms and a bit of algebra, the right-hand side simplifies to

$$\sum_{k=1}^{n} k^3 = \left(\frac{n(n+1)}{2} \right)^2.$$

A similar approach works sums over functions $f(x)$ that are polynomials of low degree. We will later see a more convenient way of determining sums of high degree powers.

Exercises

9.28. Let f be a function from the set of integers to the set of real numbers. Show that an antidifference $F(x)$ of $f(x)$ is not uniquely determined.

9.29. Let f be a function from the set of integers to the set of real numbers such that $f(k) = 0$ for all negative integers k.
(a) Show that the partial sum

$$F(x) = \sum_{k=0}^{x-1} f(x)$$

is an antidifference of $f(x)$.
(b) Explain why the partial sum $F(x)$ given in part (a) is in general an undesirable version of the antidifference $\Delta^{-1} f(x)$.

9.30. Show that $\Delta^{-1}\left(\dfrac{1}{2^x}\right) = -\dfrac{1}{2^{x-1}} + C$

9.31. Use the Fundamental Theorem of Summation to show that the sum of the first n odd numbers yields n^2,

$$1 + 3 + 5 + \cdots + (2n-1) = n^2.$$

9.32. Use the Fundamental Theorem of Summation to show that

$$1^2 + 2^2 + \cdots + n^2 = \frac{n(n+1)(2n+1)}{6}.$$

9.33. Use the Fundamental Theorem of Summation to show that

$$\frac{1}{1 \cdot 2} + \frac{1}{2 \cdot 3} + \cdots + \frac{1}{n(n+1)} = 1 - \frac{1}{n+1}.$$

9.34. The Fibonacci sequence is defined as $F_1 = 1$, $F_2 = 1$, and $F_n = F_{n-1} + F_{n-2}$ for $n \geqslant 3$. Calculate the sum

$$F_1 + F_2 + \cdots + F_n$$

using the Fundamental Theorem of Summation.

9.35. Determine the sum

$$\sum_{x=0}^{n} \binom{x}{k} = \binom{0}{k} + \binom{1}{k} + \binom{2}{k} + \cdots + \binom{n}{k}$$

in closed form using the Fundamental Theorem of Summation.

9.6 Analysis of Programs

9.36. Madame L'Orange wants to decorate her supermarket by stacking oranges into triangular pyramids. It occurs to her that she should know whether she has enough oranges before attempting to build a pyramid of n layers of oranges. Let P_n denote the number of oranges in a triangular pyramid of n layers of oranges. For a single layer, she figures that a single orange suffices, $P_1 = 1$. For two layers, she reasons that a single orange should be stacked on top of three oranges, so $P_2 = 4$. For three layers, Madame L'Orange pictures a single orange on top of three oranges that are stacked on six oranges, so $P_3 = 10$. Now she recognized the pattern. The number P_n of oranges in n layers is given by

$$P_n = \sum_{k=1}^{n} T_k,$$

where T_k is the triangular number $T_k = k(k+1)/2$.

(a) Find the antidifference of the triangular numbers

$$S_k = \Delta^{-1} T_k.$$

(b) Use the Fundamental Theorem of Summation to express P_n in closed form (that is, express P_n with a formula that does not contain summation signs Σ or ellipses).

9.37. Determine a closed form of the sum

$$1 \cdot 1! + 2 \cdot 2! + 3 \cdot 3! + \cdots + (n-1) \cdot (n-1)!$$

using the Fundamental Theorem of Summation.

9.38.

(a) Show that an antidifference of $\lfloor x/2 \rfloor$ is given by

$$\left\lfloor \frac{x}{2} \right\rfloor \left\lfloor \frac{x-1}{2} \right\rfloor.$$

(b) Find the value of the sum

$$\sum_{k=1}^{n} \left\lfloor \frac{k}{2} \right\rfloor$$

using the Fundamental Theorem of Summation.

*9.6 Analysis of Programs

In this section, we will illustrate how sums occur naturally in the analysis of algorithms.

Example 9.29. In this first example, we consider a simple version of bubblesort, which is given by the following implementation in the programming language Ruby:[3]

```
class Array
  def bubblesort
    n = self.length - 1;
    for i in 1..n
      for k in 0..n - i
        self[k], self[k+1] = self[k+1], self[k] if self[k] > self[k+1]
      end
    end
    self
  end
end
```

The method is called on a given array as follows:

`[4,3,2,1,5,6].bubblesort`

and it returns the sorted array `[1,2,3,4,5,6]`.

The program consists of two nested loops. In the first iteration of the outer loop, the inner loop swaps elements until the largest element in the sorting order is at the correct place. In the second iteration of the outer loop, the elements are swapped by the inner loop until the second largest element is at the correct place, and so on.

Let ℓ be the length of the array, so ℓ is equal to `self.length`. The number B_i of comparisons `self[k]>self[k+1]` in the i-th iteration of the outer loop is given by $B_i = \ell - i$. The total number of comparisons in all iterations is given by

$$\sum_{i=1}^{\ell-1} B_i = \sum_{i=1}^{\ell-1} (\ell - i).$$

Since $\Delta^{-1}(\ell - x) = \ell x^{\underline{1}} - \frac{1}{2}x^{\underline{2}} = \ell x - x(x-1)/2$, we obtain by the Fundamental Theorem of Summation a total of

$$\sum_{i=1}^{\ell-1}(\ell - i) = \ell^2 - \frac{\ell(\ell-1)}{2} - \ell \cdot 1 + \frac{1(1-1)}{2} = \frac{\ell(\ell-1)}{2}$$

comparisons. This does not come as a surprise, since the sum

$$\sum_{i=1}^{\ell-1}(\ell - i) = (\ell - 1) + (\ell - 2) + \cdots + 2 + 1$$

is simply the type of sum that we have studied in Example 9.24 in reverse order.

---

[3] In idiomatic Ruby, one would replace the for loops by iterators over a range, but we opted to use the notation that most programmers will understand, even if they are not familiar with Ruby.

9.6 Analysis of Programs

Example 9.30. We use the well-known quicksort algorithm by C.A.R. Hoare as an example that illustrates how sums emerge in an analysis of the average-case running time. Consider the following simple[4] but complete implementation of quicksort in Ruby:

```
class Array
  def quicksort
    return self if self.length <= 1
    pivot = self.shift
    left, right = self.partition {|e| e < pivot }
    left.quicksort + [pivot] + right.quicksort
  end
end
```

If the input array is of length ≤ 1, then there is nothing to do and the input array is returned. Otherwise, it selects the leftmost element of the array as a pivot element, partitions the array into a left part containing all elements that are smaller than the pivot, and a right part containing all elements that are greater or equal to the pivot. The left and right parts are recursively sorted. The resulting sorted array is composed of the sorted left part, the pivot, and the sorted right part. This extension to the Array class allow us to sort an array in the form

```
[7,5,6,3,4,8,1,2].quicksort
```

and this yields the result

```
[1,2,3,4,5,6,7,8]
```

A simple measure for the expected running time is the expected number C_n of comparisons `e < pivot` with the pivot element. Let us assume that the input is chosen uniformly at random from all possible permutations. The partition operation will use $n-1$ comparisons for an input of length n. If the correct sorting order of the pivot element is the k-th position in the array, then the `left` array will have $k-1$ elements and the `right` array will have $n-k$ elements. Since the input is assumed to be uniformly distributed, the sorting order of the pivot element is uniformly distributed as well. This means that the average number of comparison satisfies the recurrence relation

$$C_n = \begin{cases} n - 1 + \frac{1}{n}\sum_{k=1}^{n}(C_{k-1} + C_{n-k}) & \text{if } n > 0, \\ 0 & \text{if } n = 0. \end{cases}$$

We can simplify this recurrence relation to the form

$$C_n = \begin{cases} n - 1 + \frac{2}{n}\sum_{k=0}^{n-1} C_k & \text{if } n > 0, \\ 0 & \text{if } n = 0. \end{cases}$$

---

[4] In production code, it is recommended to use an implementation of quicksort that uses in-place array manipulations and a better pivot selection. Since our goal is to merely illustrate how summation techniques occur in the analysis of algorithms, this simple version will do.

If we multiply both sides of the recurrence by n, then we get
$$nC_n = n(n-1) + 2\sum_{k=0}^{n-1} C_k$$
and consequently
$$(n-1)C_{n-1} = (n-1)(n-2) + 2\sum_{k=0}^{n-2} C_k.$$
Subtracting these two equations yields the much simpler recurrence
$$nC_n - (n-1)C_{n-1} = 2(n-1) + 2C_{n-1}.$$
After collecting terms, dividing by $n(n+1)$, and substituting various values for n, we get the chain of equations:
$$\frac{1}{n+1}C_n = \frac{2(n-1)}{n(n+1)} + \frac{1}{n}C_{n-1}$$
$$\frac{1}{n}C_{n-1} = \frac{2(n-2)}{(n-1)n} + \frac{1}{n-1}C_{n-2}$$
$$\vdots$$
$$\frac{1}{3}C_2 = \frac{2 \cdot 1}{2 \cdot 3} + \frac{1}{2}C_1$$
Since $C_1 = 0$, we can unroll this chain of equations into a single sum:
$$C_n = 2(n+1)\sum_{k=2}^{n} \frac{k-1}{k(k+1)} = 2(n+1)\sum_{k=1}^{n-1} \frac{k}{(k+1)(k+2)}.$$
We can simplify the summand by rewriting it in the form
$$\frac{k}{(k+1)(k+2)} = \frac{2}{k+2} - \frac{1}{k+1}.$$
This yields
$$C_n = 2(n+1)\sum_{k=1}^{n-1}\left(\frac{2}{k+2} - \frac{1}{k+1}\right) = 2(n+1)\left(\sum_{k=3}^{n+1}\frac{2}{k} - \sum_{k=2}^{n}\frac{1}{k}\right).$$
Since the antidifference of $1/x$ is H_{x-1}, the Fundamental Theorem of Summation yields
$$C_n = 2(n+1)\left((2H_{n+1} - 2H_2) - (H_n - H_1)\right).$$
As $2H_{n+1} = 2H_n + 2/(n+1)$, we can simplify this expression to
$$C_n = 2(n+1)H_n - 4n.$$
The significance of this average number C_n of comparisons with the pivot is that the average running time of quicksort is bounded by a constant times C_n.

9.7 Notes

The techniques described in this chapter have a long history. An early monograph by Boole [10] describing the Calculus of Finite Differences had been published in 1860. Other early works include the comprehensive monographs by Milne-Thomson [61] and Jordán [42], and all are still very valuable today. Many books on discrete mathematics give a brief introduction to the Calculus of Finite Differences, since it has so many applications in the analysis of computer programs. Our exposition has been influenced by Aigner [4], Graham, Knuth, and Patashnik [30], Jordán [42], and Stopple [75]. The recent interest in calculus without limits sparked a new renaissance for the Calculus of Finite Differences under the moniker "quantum calculus," see Kac and Cheung [44].

The material of this chapter finds applications in the analysis of algorithms, see for example Knuth [48–51]. Our analysis of quicksort is adapted from [50], see also [4].

Chapter 10

Asymptotic Analysis

> *We could, of course, use any notation we want; do not laugh at notations; invent them, they are powerful. In fact, mathematics is, to a large extent, invention of better notations.*
>
> — Richard Feynman, *The Feynman Lectures on Physics, Vol. 1*

In this chapter, our goal is to compare a function $f(n)$ with some simple function $g(n)$ so that we can understand its order of growth as n approaches infinity. We discuss asymptotic equality $\sim$, asymptotic tightness Θ, asymptotic upper bounds O and o, and asymptotic lower bounds Ω and ω. These asymptotic notations are widely used in computer science, mathematics, and related disciplines. These notations allow us to quickly determine the order of growth of a function without forcing us to become imprecise. We briefly illustrate the use of asymptotic notations in the analysis of algorithms.

10.1 Asymptotic Equality

Let f and g be functions from the set of positive integers to the set of real numbers. We write $f \sim g$ and say that f is **asymptotically equal** to g if and only if
$$\lim_{n\to\infty} \frac{f(n)}{g(n)} = 1$$
holds. By definition of the limit, this means that for each real number $\epsilon > 0$, there exists a positive integer n_ϵ such that
$$\left| \frac{f(n)}{g(n)} - 1 \right| < \epsilon \qquad (10.1)$$
holds for all $n \geqslant n_\epsilon$. As the notation suggests, the positive integer n_ϵ depends on the choice of ϵ. Informally, this means that for large n, the fraction $f(n)/g(n)$ gets arbitrarily close to 1.

Another way to interpret the inequality (10.1) is that two functions f and g are asymptotically equal if and only if the relative error $(f(n) - g(n))/g(n)$ between these functions vanishes for large n. Essentially, this means that the functions f and g have the same growth for large n.

Let us see why this notation is popular, especially in analysis, combinatorics, and number theory.

Example 10.1. Suppose that we are given the function
$$f(n) = \frac{n \ln n + \sqrt{n} + \ln n + 1517 \ln \ln n}{n}.$$
If we are only interested in the growth of the function f for large n, then it is important to realize that most of the terms of f do not have a significant contribution! Indeed, the three terms
$$\lim_{n \to \infty} \frac{\sqrt{n}}{n} = 0, \quad \lim_{n \to \infty} \frac{\ln n}{n} = 0, \quad \lim_{n \to \infty} \frac{1517 \ln \ln n}{n} = 0$$
all vanish for large n, so only the term $n \ln n / n = \ln n$ has an impact on the growth of the function $f(n)$ for large n. The asymptotic equality
$$f(n) \sim \ln n$$
expresses that $f(n)$ essentially grows like the natural logarithm $\ln n$. Since $\ln n$ is shorter and more familiar than $f(n)$, it is the preferred way to describe the behavior of $f(n)$ for large arguments n.

Another example is the Harmonic number H_n that is defined by a sum that does not have a simple closed form. The asymptotic equality nevertheless allows us to describe the growth of H_n by comparing it to a familiar function.

Example 10.2. The n-th Harmonic number $H_n = 1 + \frac{1}{2} + \cdots + \frac{1}{n}$ is asymptotically equal to the natural logarithm $\ln n$,
$$H_n \sim \ln n.$$
Indeed, since the inequalities $\ln(n+1) \leq H_n \leq 1 + \ln n$ hold, dividing by $\ln n$ and taking the limit yields for the logarithmic terms
$$\lim_{n \to \infty} \frac{\ln(n+1)}{\ln n} = \lim_{n \to \infty} \frac{n}{n+1} = 1 \quad \text{and} \quad \lim_{n \to \infty} \frac{1 + \ln n}{\ln n} = 1,$$
where we used l'Hôpital's rule in the calculation of the first limit. Thus, it follows from the squeeze theorem for limits that
$$\lim_{n \to \infty} \frac{H_n}{\ln n} = 1,$$
which proves that $H_n \sim \ln n$. In other words, the Harmonic numbers grow like the natural logarithm for large n. Relating a function such as H_n to the more familiar logarithmic function $\ln n$ certainly helps one to understand the growth of H_n as n approaches infinity.

10.1 Asymptotic Equality

Perhaps the best known example is given by Stirling's approximation, which relates the function $n!$ to a function that allows one to understand the asymptotic growth much better.

Example 10.3. The Stirling approximation of $n!$ is given by

$$n! \sim \sqrt{2\pi n} \left(\frac{n}{e}\right)^n.$$

The function $f(n) = n!$ grows less rapidly than n^n. Dividing n^n by e^n yields a better approximation of $n!$, but $(n/e)^n$ is always an underestimate for $n!$. By multiplication with the corrective term $\sqrt{2\pi n}$, we obtain the function $g(n) = \sqrt{2\pi n} \left(\frac{n}{e}\right)^n$ that has asymptotically the same growth as the factorial function $f(n) = n!$. The approximation works well even for relatively small numbers n. For instance, if we approximate $f(10) = 10!$ by $g(10) = \sqrt{2\pi \cdot 10}\,(10/e)^{10}$, then the error ϵ of the approximation is

$$\left|\frac{10!}{\sqrt{2\pi 10}\left(\frac{10}{e}\right)^{10}} - 1\right| < \epsilon < 1.009.$$

So the multiplicative error of Stirling's approximation of $10!$ is less than 1%.

One advantage of the asymptotic equality $\sim$ is that the expression can be simplified quite a bit. The next proposition illustrates this in the case of polynomials.

Proposition 10.4. Let $p(x) = \sum_{k=0}^{m} a_k x^k$ be a nonzero polynomial of degree m with real coefficients. Then $p(x)$ is asymptotically equal to its leading term,

$$p(x) \sim a_m x^m.$$

Proof. By assumption, the leading coefficient $a_m \neq 0$. By expanding terms, we get

$$\lim_{x \to \infty} \frac{p(x)}{a_m x^m} = \lim_{x \to \infty} \left(1 + \frac{a_{m-1}}{a_m x} + \cdots + \frac{a_1}{a_m x^{m-1}} + \frac{a_0}{a_m x^m}\right) - 1,$$

which proves the claim. $\square$

The leading term of a polynomial dominates the growth. The advantage of the asymptotic notation $\sim$ is that we can dramatically simplify expressions such as $p(x) = x^3 + 3x^2 + 2x + 1$ when considering large x, since the relation $x^3 + 3x^2 + 2x + 1 \sim x^3$ tells us that the polynomial $p(x)$ grows like x^3. The notation brings clarity and gives us more compact expressions. Naturally, one can derive similar propositions for many other classes of functions.

The definition of $f \sim g$ by a limit means that all the techniques that you have learned in calculus can be applied. You can apply the knowledge that you have gained in calculus to deduce asymptotic equality in situations that are not entirely obvious. As an example, we give a convenient criterion for asymptotic equality of expressions that are of the form $f(n+k) - f(n)$ for some constant k. It is a slight generalization of a lemma given in Koecher [52, Chap. II.1.7]. The use of the mean value theorem in the proof is particularly noteworthy, since you can apply similar methods in other contexts as well.

Proposition 10.5. *Let c be a positive real number. Let f be a continuously differentiable function from the set of positive real numbers to the set of real numbers such that its derivative f' is monotonic, nonzero, and satisfies*

$$\lim_{n \to \infty} f'(n+c)/f'(n) = 1.$$

Then

$$f(n+c) - f(n) \sim cf'(n).$$

Proof. By the mean value theorem of calculus, there exists a real number θ in the range $0 \leqslant \theta \leqslant c$ such that

$$f(n+c) - f(n) = (n+c-n)f'(n+\theta) = cf'(n+\theta).$$

If f' is monotonically increasing (or monotonically decreasing), then

$$cf'(n) \underset{(\geqslant)}{\leqslant} f(n+c) - f(n) \underset{(\geqslant)}{\leqslant} cf'(n+c).$$

Dividing by $cf'(n)$ yields by assumption

$$\lim_{n \to \infty} \frac{cf'(n)}{cf'(n)} = 1 \quad \text{and} \quad \lim_{n \to \infty} \frac{cf'(n+c)}{cf'(n)} = 1.$$

Therefore, by the squeeze theorem for limits, we have

$$\lim_{n \to \infty} \frac{f(n+c) - f(n)}{cf'(n)} = 1,$$

which proves our claim. □

Example 10.6. Let c be a positive constant. Then

$$\sqrt{n+c} - \sqrt{n} \sim \frac{c}{2\sqrt{n}}.$$

Indeed, if we set $f(x) = \sqrt{x}$, then f is a continuously differentiable function on the positive real numbers. Its derivative $f'(x) = 1/(2\sqrt{x})$ is nonzero, monotonically decreasing, and satisfies $\lim_{n \to \infty} f'(n+c)/f'(n) = 1$. The claim follows from Proposition 10.5.

Example 10.7. Let k be a positive real number. Then

$$\ln\left(1 + \frac{1}{n}\right)^k \sim \frac{k}{n}.$$

10.1 Asymptotic Equality

Indeed, choose $f(x) = k\ln(x)$. Then $f'(x) = k/x$. As $f'(x) = k/x$ is monotonically decreasing and nonzero, and

$$\lim_{n\to\infty} \frac{f'(n+1)}{f'(n)} = \lim_{n\to\infty} \frac{n}{n+1} = 1,$$

it follows from Proposition 10.5 that

$$\ln\left(1 + \frac{1}{n}\right)^k = k\ln(n+1) - k\ln(n) \sim \frac{k}{n}.$$

We conclude this section by illustrating how the asymptotic equality can simplify the calculation of limits.

Proposition 10.8. *If $f_1 \sim g_1$ and $f_2 \sim g_2$, then*

$$\lim_{n\to\infty} \frac{f_1(n)}{f_2(n)} = \lim_{n\to\infty} \frac{g_1(n)}{g_2(n)}$$

provided either of these limits exist (the proviso is important!).

Proof. We multiply the limit of f_1/f_2 by 1, here written in the form $1 = (g_1 g_2)/(g_1 g_2)$. This yields

$$\lim_{n\to\infty} \frac{f_1(n)}{f_2(n)} = \lim_{n\to\infty} \frac{f_1(n)}{g_1(n)} \frac{g_1(n)}{g_2(n)} \frac{g_2(n)}{f_2(n)}$$

$$= \left(\lim_{n\to\infty} \frac{f_1(n)}{g_1(n)}\right) \left(\lim_{n\to\infty} \frac{g_1(n)}{g_2(n)}\right) \left(\lim_{n\to\infty} \frac{g_2(n)}{f_2(n)}\right),$$

where we have used in the last equality that the limit of a product is the product of the limits of its factors. Since the first and last limit is 1 by hypothesis, we can conclude that

$$\lim_{n\to\infty} \frac{f_1(n)}{f_2(n)} = \lim_{n\to\infty} \frac{g_1(n)}{g_2(n)},$$

which proves the claim. □

Example 10.9. Let us try to calculate the limit

$$\lim_{n\to\infty} \frac{3n^{15} + 7n^{14} + 8n^2 + \ln\sqrt{n}}{4n^{15} + 2n^7 + 2n\ln\sqrt{n}}.$$

Since the numerator and the denominator are both unbounded, we could apply l'Hôpital's rule 15 times[1] and calculate the limit. If we observe that $3n^{15} + 7n^{14} + 8n^2 + \ln\sqrt{n} \sim 3n^{15}$ and $4n^{15} + 2n^7 + 2n\ln\sqrt{n} \sim 4n^{15}$, then the previous proposition yields the limit

$$\lim_{n\to\infty} \frac{3n^{15} + 7n^{14} + 8n^2 + \ln\sqrt{n}}{4n^{15} + 2n^7 + 2n\ln\sqrt{n}} = \lim_{n\to\infty} \frac{3n^{15}}{4n^{15}} = \frac{3}{4}$$

in a straightforward fashion.

---

[1] Of course, there are also other ways to solve this problem. Focus on the explanation that follows!

We can use similar tricks when the numerator and denominators contain radicals.

Exercises

10.1. Suppose that f, g, and h are functions from the set of positive integers to the set of real numbers that are nonzero for all but a finite number of arguments. Show that
  (i) $f \sim f$,
  (ii) $f \sim g$ if and only if $g \sim f$,
  (iii) $f \sim g$ and $g \sim h$ implies $f \sim h$.
This means that $\sim$ is an equivalence relation on the set of functions that are nonzero for all but a finite number of arguments.

10.2. As an important reminder that $\sim$ is an equivalence relation only on the set of functions that are *nonzero* for all but a finite number of arguments, show that there cannot be a function f such that $f \sim 0$ or $0 \sim f$.

10.3. Ernie and Bert study the functions $f(n) = n^2 + 2n$ and $g(n) = n^2$. Ernie claims that the functions are asymptotically equal, so $f \sim g$. Bert insists that f can never be asymptotically equal to g, since they always differ quite significantly, namely $f(n) - g(n) \geq 2n$ for all $n \geq 1$. Explain why one is right and why the other is wrong.

10.4. Let k and ℓ be positive integers. Show that
$$\frac{1 + n^k}{n^\ell} \sim n^{k-\ell}.$$

10.5. For a positive integer n, let $d(n)$ denote the number of divisors n. For example, $d(6) = 4$, since 6 has the divisors 1, 2, 3, and 6.
(a) Show that
$$d(n) = \sum_{k=1}^{n} \left(\left\lfloor \frac{n}{k} \right\rfloor - \left\lfloor \frac{n-1}{k} \right\rfloor \right).$$
(b) The function $d(n)$ exhibits some very erratic behavior that is difficult to analyze. However, the average of the first n values of the function d is easier to analyze. Show that
$$\frac{1}{n} \sum_{k=1}^{n} d(k) \sim \ln n.$$

10.6. Let a be a real constant satisfying $a < 1$. Show that
$$\exp(n^a) \sim \exp((n+1)^a).$$

10.7. Show that
$$\binom{2n}{n} \sim \frac{4^n}{\sqrt{\pi n}}.$$

10.8. Show that $\sin(1/n)$ is asymptotically equal to $1/n$.

10.9. Show that $\ln(n!)$ is asymptotically equal to $n \ln n$.

10.2 Limit Superior and Limit Inferior

The asymptotic equality is sometimes a bit too strict, especially when trying to work out the asymptotic behavior of sequences that fluctuate between different values. In this case, it is often better to settle for asymptotic bounds. Before embarking on the discussion of asymptotic bounds, we recall the notions of upper and lower limits from calculus. These notions turn out to be useful when trying to establish asymptotic bounds.

Upper and Lower Envelope. Let f be a function from the set of positive integers to the set of real numbers. If f oscillates between different values, then the limit $\lim_{n \to \infty} f(n)$ fails to exist. We can associate with f its upper and lower envelope, which are better behaved with respect to taking limits. The **upper envelope** u of the function f is defined by

$$u_f(n) = \sup\{f(n), f(n+1), f(n+2), \ldots\},$$

and the **lower envelope** ℓ of the function f is defined by

$$\ell_f(n) = \inf\{f(n), f(n+1), f(n+2), \ldots\}.$$

Example 10.10. Let $f(n)$ denote the function given by

$$f(n) = \begin{cases} 2 + 1/n & \text{if } n \text{ is even,} \\ 1 - 1/n & \text{if } n \text{ is odd.} \end{cases}$$

Then the upper envelope $u_f(n)$ of the function $f(n)$ is given by

$$u_f(1) = \sup\{f(1), f(2), f(3), \ldots\} = f(2) = 2 + 1/2,$$
$$u_f(2) = \sup\{f(2), f(3), f(4), \ldots\} = f(2) = 2 + 1/2,$$
$$u_f(3) = \sup\{f(3), f(4), f(5), \ldots\} = f(4) = 2 + 1/4,$$
$$u_f(4) = \sup\{f(4), f(5), f(6), \ldots\} = f(4) = 2 + 1/4, \ldots$$

The general term of the upper envelope $u_f(n)$ of the function f is given by

$$u_f(n) = f(2\lceil n/2 \rceil) = 2 + \frac{1}{2\lceil n/2 \rceil}$$

for all positive integers n.

The lower envelope $\ell_f(n)$ of the function $f(n)$ is given by

$$\ell_f(1) = \inf\{f(1), f(2), f(3), \ldots\} = f(1) = 1 - 1/1,$$
$$\ell_f(2) = \inf\{f(2), f(3), f(4), \ldots\} = f(3) = 1 - 1/3,$$
$$\ell_f(3) = \inf\{f(3), f(4), f(5), \ldots\} = f(3) = 1 - 1/3,$$
$$\ell_f(4) = \inf\{f(4), f(5), f(6), \ldots\} = f(5) = 1 - 1/5, \ldots$$

The general term of the lower envelope $\ell_f(n)$ is given by

$$\ell_f(n) = f(2\lfloor n/2 \rfloor + 1) = 1 - \frac{1}{2\lfloor n/2 \rfloor + 1}$$

for all positive integers n.

Figure 10.1 depicts the function f and its upper and lower envelope.

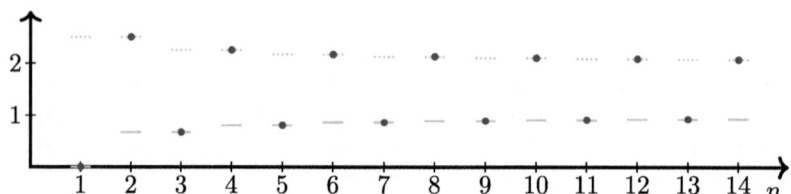

Figure 10.1: The graph shows the function (or sequence) f from Example 10.10 with values depicted by large dots. The values of the upper envelope u_f are shown by small dotted lines, and the values of the lower envelope ℓ_f by small solid lines

A crucial point in the definition of the upper and lower envelope is that they omit more and more initial terms of f from consideration as $n \to \infty$. Therefore, the upper envelope is in general a monotonically decreasing function, and the lower envelope is in general a monotonically increasing function. Consequently, the limit of the upper envelope $u_f(n)$ and the lower envelope $\ell_f(n)$ always exist in the extended real line.

Recall that the extended real line consists of the set of real numbers $\mathbf{R}$ and two symbols $-\infty$ and $+\infty$ such that $-\infty < r < \infty$ for all real numbers r. The extended real line has the convenient property that every subset has a supremum and an infimum, so it is a complete lattice. The extended real line is convenient for denoting the limit of diverging sequences. We will explicitly point out the properties related to the nonfinite values in the subsequent discussion.

Limit Inferior and Limit Superior. The **limit superior** $\limsup_{n \to \infty} f(n)$ is defined as

$$\limsup_{n \to \infty} f(n) = \lim_{n \to \infty} \left(\sup\{f(k) \mid k \geq n\} \right)$$

and the **limit inferior** $\liminf_{n \to \infty} f(n)$ is defined as

$$\liminf_{n \to \infty} f(n) = \lim_{n \to \infty} \left(\inf\{f(k) \mid k \geq n\} \right).$$

In other words, the limit superior is the limit of the upper envelope of f, and the limit inferior is the limit of the lower envelope of f.

Let us have a look at a particularly instructive example.

10.2 Limit Superior and Limit Inferior

Example 10.11. Let us continue the Example 10.10. Recall that the function

$$f(n) = \begin{cases} 2 + 1/n & \text{if } n \text{ is even,} \\ 1 - 1/n & \text{if } n \text{ is odd.} \end{cases}$$

has the upper envelope $u_f(n) = f(2\lceil n/2 \rceil) = 2 + \frac{1}{2\lceil n/2 \rceil}$ and the lower envelope $\ell_f(n) = f(2\lfloor n/2 \rfloor + 1) = 1 - \frac{1}{2\lfloor n/2 \rfloor + 1}$. Then

$$\limsup_{n \to \infty} f(n) = \lim_{n \to \infty} \left(2 + \frac{1}{2\lceil n/2 \rceil}\right) = 2$$

and

$$\liminf_{n \to \infty} f(n) = \lim_{n \to \infty} \left(1 - \frac{1}{2\lfloor n/2 \rfloor + 1}\right) = 1.$$

The example shows that the limit inferior is in general not the smallest value of f, since

$$0 = 1 - 1/1 = \inf\{f(n) \mid n \geqslant 1\} < \liminf_{n \to \infty} f(n) = 1.$$

Likewise, the example also shows that the limit superior is in general not the largest value of f, since

$$2 = \limsup_{n \to \infty} f(n) < \sup\{f(n) \mid n \geqslant 1\} = 2 + 1/2.$$

We will characterize the limit inferior and limit superior as smallest and largest accumulation points of f.

The next proposition stresses the difference between the infimum and limit inferior as well as the difference between the supremum and the limit superior.

Proposition 10.12. *Let f be a function from the set of positive integers to the set of real numbers. Then*

$$\inf\{f(k) \mid k \geqslant 1\} \leqslant \liminf_{n \to \infty} f(n) \leqslant \limsup_{n \to \infty} f(n) \leqslant \sup\{f(k) \mid k \geqslant 1\}.$$

The previous example shows that any of these three inequalities can be strict.

Proof. The first inequality holds, since the lower envelope ℓ_f of f is a monotonically increasing function and $\ell_f(1) = \inf\{f(k) \mid k \geqslant 1\}$. The second inequality follows directly from the definitions. The third inequality holds, since the upper envelope u_f of f is a monotonically decreasing function and $u_f(1) = \sup\{f(k) \mid k \geqslant 1\}$. □

Let us now discuss four different situations when the limit inferior or limit superior do not have finite values. If the function f does not have an upper bound, then the suprema $\sup\{f(n), f(n+1), f(n+2), \ldots\} = \infty$ for all n, and we define

$$\limsup_{n \to \infty} f(n) = +\infty.$$

Example 10.13. Suppose that $f(n) = n$. Then $\limsup_{n\to\infty} f(n) = \infty$.

If the function f approaches $-\infty$ as $n \to \infty$, then we define
$$\limsup_{n\to\infty} f(n) = -\infty.$$

Example 10.14. Suppose that $f(n) = -n$. Then
$$u_f(n) = \sup\{-n, -(n+1), -(n+2), \ldots\} = -n,$$
so $\limsup_{n\to\infty} f(n) = \lim_{n\to\infty} u_f(n) = -\infty$.

Example 10.15. We should stress the limit superior will not be $-\infty$ when merely a subsequence approaches $-\infty$. Indeed, suppose that
$$f(n) = \begin{cases} 0 & \text{if } n \text{ is even,} \\ -n & \text{if } n \text{ is odd.} \end{cases}$$

Then $u_f(n) = \sup\{f(n), f(n+1), f(n+2), \ldots\} = 0$, so the limit superior
$$\limsup_{n\to\infty} f(n) = 0,$$
even though $f(2n+1) \to -\infty$ as $n \to \infty$.

Similarly, if the function f does not have a lower bound, then the infima $\inf\{f(n), f(n+1), f(n+2), \ldots\} = -\infty$ for all n, and we define
$$\liminf_{n\to\infty} f(n) = -\infty.$$

Example 10.16. Let the function $f : \mathbf{N}_1 \to \mathbf{R}$ be given by
$$f(n) = \begin{cases} 0 & \text{if } n \text{ is even,} \\ -n & \text{if } n \text{ is odd.} \end{cases}$$

Then $\liminf_{n\to\infty} f(n) = -\infty$ and $\limsup_{n\to\infty} f(n) = 0$.

If the function f diverges to $+\infty$ as $n \to \infty$, then we define
$$\liminf_{n\to\infty} f(n) = +\infty.$$

Example 10.17. The function $f(n) = n$ has the limit inferior
$$\liminf_{n\to\infty} n = +\infty.$$

Example 10.18. If merely a subsequence converges to $+\infty$, then the limit inferior is in general not $+\infty$. For instance, the function
$$f(n) = \begin{cases} 42 & \text{if } n \text{ is even,} \\ n & \text{if } n \text{ is odd} \end{cases}$$
has the subsequence $f(2n+1) \to \infty$ as $n \to \infty$. However, the limit inferior of $f(n)$ is given by $\liminf_{n\to\infty} f(n) = 42$.

10.2 Limit Superior and Limit Inferior

Accumulation Points. We will now give a characterization of the finite values of the limit superior and limit inferior in terms of accumulation values. Recall that a real number a is an accumulation value of a function $f: \mathbf{N}_1 \to \mathbf{R}$ if and only if for each real number $\epsilon > 0$ there exist an infinite number of function values $f(n)$ such that

$$|f(n) - a| < \epsilon.$$

For example, if the limit $\ell = \lim_{n \to \infty} f(n)$ exists, then ℓ is an accumulation value of f. If a bounded function f does not have a limit as $n \to \infty$, then it still has accumulation points. For instance, the function f from Example 10.10 has two accumulation points, namely $a = 1$ and $a = 2$.

Let f be a function from the set of positive integers to the set of real numbers. The real number u is an **upper accumulation point** of f if and only if the following two conditions are met:

U1 For each real number $\epsilon > 0$ there exist infinitely many positive integers n such that $f(n) > u - \epsilon$,

U2 For each real number $\epsilon > 0$ there exist at most finitely many positive integers such that $f(n) > u + \epsilon$.

Evidently, an upper accumulation point is an accumulation point. If an upper accumulation point of f exists, then it is unique.

The function f is called **bounded above** if and only if there exists a real number u such that $f(n) \leq u$ holds for all positive integers n. If f has an upper accumulation point, then it is bounded above.

Proposition 10.19. *Let f be a function from the set of positive integers to the set of real numbers that is bounded above and does not diverge to $-\infty$. Then the real number u is equal to $\limsup_{n \to \infty} f(n)$ if and only if u is the upper accumulation point of f.*

Proof. ($\Rightarrow$) Suppose that the limit superior of the function f is given by the real number $u = \limsup_{n \to \infty} f(n)$. We are going to show that u must be an upper accumulation point. This means that we need to prove that (i) $f(n)$ satisfies **U1** and that (ii) $f(n)$ satisfies **U2**.
  (i) Seeking a contradiction, suppose that there exists an $\epsilon > 0$ and an n such that $f(k) \leq u - \epsilon$ for all $k \geq n$. Then $u_f(k) \leq u - \epsilon$ for all $k \geq n$, so $\limsup_{n \to \infty} f(n) = \lim_{n \to \infty} u_f(n) \leq u - \epsilon$, contradicting our assumption $u = \limsup_{n \to \infty} f(n)$. Therefore, we can conclude that **U1** holds.
  (ii) Seeking a contradiction, suppose that there exists an $\epsilon > 0$ such that there are infinitely many positive integers n so that $f(n) > u + \epsilon$. This implies that $u_f(n) = \sup\{f(n), f(n+1), f(n+2), \ldots\} > u + \epsilon$ for all $n \geq 1$. Therefore,

$$\limsup_{n \to \infty} f(n) = \lim_{n \to \infty} u_f(n) > u + \epsilon,$$

contradicting our assumption. Consequently, for each real number $\epsilon > 0$ there exist at most finitely many positive integers such that $f(n) > u + \epsilon$, so **U2** holds.

($\Leftarrow$). Suppose that u is an upper accumulation point of f. It follows from **U1** that $u - \epsilon \leq u_f(n)$. Furthermore, it follows from **U2** that there exists a positive integer n such that $u_f(k) \leq u + \epsilon$ for all $k \geq n$. Therefore, we can conclude that for all $\epsilon > 0$, we have

$$\limsup_{n \to \infty} f(n) = \lim_{n \to \infty} u_f(n) \in [u - \epsilon, u + \epsilon],$$

so $\limsup_{n \to \infty} f(n) = u$. $\square$

The limit superior of f can now be characterized as

$$\limsup_{n \to \infty} f(n) = \begin{cases} +\infty & \text{if } f \text{ is not bounded above,} \\ u & \text{if the upper accumulation point } u \text{ of } f \text{ exists,} \\ -\infty & \text{otherwise.} \end{cases}$$

Example 10.20. Let $f(n)$ denote the function given by

$$f(n) = \begin{cases} 2 + 1/n & \text{if } n \text{ is even} \\ 1 - 1/n & \text{if } n \text{ is odd} \end{cases}$$

For any $\epsilon > 0$, all even positive integers n such that $n > 1/\epsilon$ have a function value $f(n) > 2 - \epsilon$. We have $f(n) > 2 + \epsilon$ only when n is an even positive integer satisfying $n < 1/\epsilon$. Therefore, 2 is an upper accumulation point of $f(n)$. This is also illustrated in Fig. 10.2. We can conclude that $\limsup_{n \to \infty} f(n) = 2$.

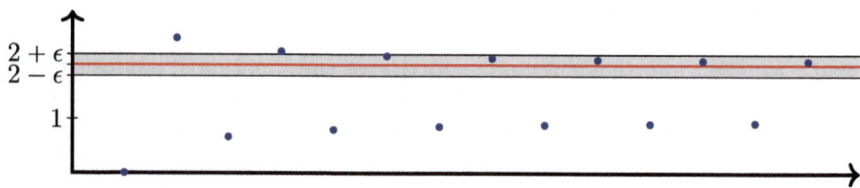

Figure 10.2: The graph shows the function (or sequence) f from Example 10.20. For each $\epsilon > 0$, there are finitely many points above $2 + \epsilon$, but an infinite number above $2 - \epsilon$, so it has an upper accumulation point of value 2. Hence, $\limsup_{n \to \infty} f(n) = 2$

The real number ℓ is called a **lower accumulation point** of f if and only if the following two conditions are met:

L1. For each real number $\epsilon > 0$, there exist infinitely many positive integers n such that $f(n) < \ell + \epsilon$,

10.2 Limit Superior and Limit Inferior

L2. For each real number $\epsilon > 0$, there exist at most finitely many positive integers such that $f(n) < \ell - \epsilon$.

If a lower accumulation point of f exists, then it is unique. Figure 10.3 gives an example of the lower accumulation point for the function f from Example 10.20.

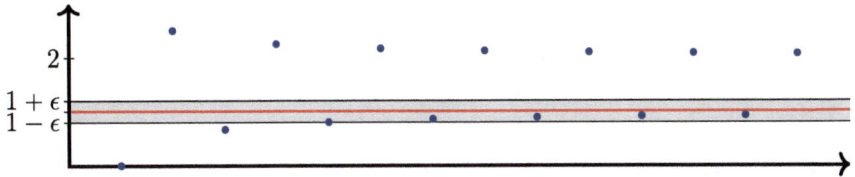

Figure 10.3: The graph shows the function (or sequence) f from Example 10.20. For each $\epsilon > 0$, there are finitely many points below $1 - \epsilon$, but an infinite number below $1 + \epsilon$, so it has a lower accumulation point of value 1. Hence $\liminf_{n\to\infty} f(n) = 1$.

The function f is called **bounded below** if and only if there exists a real number ℓ such that $f(n) \geq \ell$ holds for all positive integers n. If f has a lower accumulation point, then it is bounded below.

Proposition 10.21. *Let f be a function from the set of positive integers to the set of real numbers that is bounded below and does not diverge to ∞. Then the real number ℓ is equal to $\liminf_{n\to\infty} f(n)$ if and only if ℓ is the lower accumulation point of f.*

Proof. Since $\liminf_{n\to\infty} f(n) = -\limsup_{n\to\infty} -f(n)$, the claim is a direct consequence of Proposition 10.19. □

As a consequence of the previous proposition, the limit inferior of f can be characterized as

$$\liminf_{n\to\infty} f(n) = \begin{cases} -\infty & \text{if } f \text{ is not bounded below,} \\ \ell & \text{if the lower accumulation point } \ell \text{ of } f \text{ exists,} \\ +\infty & \text{otherwise.} \end{cases}$$

Exercises

10.10. Find all accumulation points of the following functions, and determine their upper and lower accumulation points.
(a) $f(n) = (-1)^n$,
(b) $f(n) = 4 + (-1)^n n/(n+10)$
(c) $f(n) = ((-1)^n + (-1)^{\lfloor n/2 \rfloor})(1 + 1/n)$.

10.11. Construct a function $f(n)$ that has all positive integers as accumulation points.

10.12. Determine $\limsup_{n\to\infty} f(n)$ and $\liminf_{n\to\infty} f(n)$ when $f(n)$ is given as

(a) $f(n) = (-1)^n$,
(b) $f(n) = \frac{1}{n}$,
(c) $f(n) = (1 + (-1)^n)n$,
(d) $f(n) = (-1)^n(2n+1)/(n+1)$.

10.13. Show that if $\limsup_{n\to\infty} f(n) = L$ and $\liminf_{n\to\infty} f(n) = L$, then the limit $\lim_{n\to\infty} f(n)$ exists and is equal to L.

10.14. Show that every bounded function $f: \mathbf{N}_1 \to \mathbf{R}$ has an accumulation point. This result is due to Bolzano and Weierstrass.

10.3 Asymptotically Tight Bounds

The asymptotic equality is often a bit too strict. Sometimes it is desirable to relax the constraints and consider the growth up to a constant factor and without the need for the existence of a limit. In this section, we discuss so-called asymptotically tight bounds that provide both upper and lower asymptotic bounds for a function, yet often allow one to use very simple functions to describe the growth.

Let f and g denote functions from the set of positive integers to the set of real numbers. We say that f and g have the **same order of growth** and write $f \asymp g$ if and only if there exist positive real constants c and C and a positive integer n_0 such that
$$c|g(n)| \leq |f(n)| \leq C|g(n)|$$
holds for all $n \geq n_0$. The notation $f \asymp g$ goes back to Hardy [34] and is popular in mathematics. Computer scientists like to express this in the form $f \in \Theta(g)$, where
$$\Theta(g) = \{f : \mathbf{N}_1 \to \mathbf{R} \mid f \asymp g\}$$
is the set of functions that have the same order of growth as g, see Knuth [47]. If $f \in \Theta(g)$ or $f \asymp g$, then we also say that g is an **asymptotically tight bound** for f.

> In the literature, you will frequently find that $f \in \Theta(g)$ is written in the form $f = \Theta(g)$. Evidently, a function f cannot be "equal" to a set $\Theta(g)$. In the early days, the notation $\Theta(g)$ simply meant a given but unspecified representative of the set $\Theta(g)$. Even though $\Theta(g)$ is now understood as a set, the notation $f = \Theta(g)$ is commonly used instead of the more proper notation $f \in \Theta(g)$. You will need to get used to this idiosyncratic notation.

Example 10.22. As a first example, let us consider the function
$$f(n) = \frac{1}{9}n^2 + \frac{12}{10n}.$$
We compare this function with two scalar multiples of n^2, namely $\frac{1}{10}n^2$ and $\frac{1}{8}n^2$. Then
$$\frac{1}{10}n^2 \leq f(n) \leq \frac{1}{8}n^2$$

10.3 Asymptotically Tight Bounds

holds for all $n \geq 5$. Indeed, the lower bound is clear. For the upper bound, it suffices to note that $n \geq 5$ implies $n^3 \geq 8 \cdot 9 \cdot 12/10$, and dividing both sides by $8 \cdot 9 \cdot n$ yields $\frac{1}{8}n^2 - \frac{1}{9}n^2 = \frac{n^2}{8 \cdot 9} \geq \frac{12}{10n}$, so adding $\frac{1}{9}n^2$ to both sides yields the upper bound $\frac{1}{8}n^2 \geq f(n)$. The graphs of these functions are shown in Fig. 10.4.

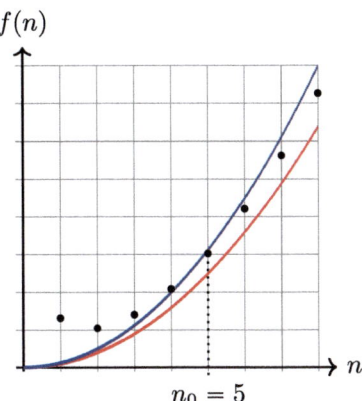

Figure 10.4: The function $f(n) = \frac{1}{9}n^2 + \frac{12}{10n}$ is shown with dots. The lower bound is given by $\frac{1}{10}n^2$ and the upper bound by $\frac{1}{8}n^2$, both shown with solid graphs even though only the values at integer arguments are relevant. The inequality $\frac{1}{10}n^2 \leq f(n) \leq \frac{1}{8}n^2$ holds for all $n \geq 5$. Therefore, $f(n) \in \Theta(n^2)$. It is okay that the first few function values $f(1), f(2), f(3)$, and $f(4)$ are not within the bounds, since the inequality is required to hold only for large arguments

When we want to prove that $f \in \Theta(g)$ holds, we need to find three constants n_0, c, and C such that

$$c|g(n)| \leq |f(n)| \leq C|g(n)|$$

holds for all $n \geq n_0$. It can be a bit cumbersome to verify this definition. The next proposition gives a convenient sufficient condition for establishing $f \in \Theta(g)$, but unfortunately it is not always applicable.

Proposition 10.23. *Let f and g be functions from the set of positive integers to the set of real numbers. If g is a positive function and the limit*

$$d = \lim_{n \to \infty} \frac{|f(n)|}{|g(n)|}$$

exists and is a nonzero real number d, then $f \in \Theta(g)$.

Proof. It follows from the definition of the limit that for each $\epsilon > 0$ there exists a positive integer n_ϵ such that

$$d - \epsilon \leq \frac{|f(n)|}{|g(n)|} \leq d + \epsilon$$

for all $n \geqslant n_\epsilon$. In other words, for the constants $c = d - \epsilon$ and $C = d + \epsilon$ there exists an n_ϵ such that $c|g(n)| \leqslant |f(n)| \leqslant C|g(n)|$ holds for all $n \geqslant n_\epsilon$, which proves $f \in \Theta(g)$. □

If f and g are asymptotically equal, then the limit of their quotient is 1. We record this simple observation in the next corollary.

Corollary 10.24. *If two functions f and g are asymptotically equal, $f \sim g$, then they have the same order of growth, so $f \asymp g$.*

Example 10.25. Recall that the expected number of comparisons in randomized Quicksort is given by $2(n+1)H_n - 4n$. Since $H_n \sim \ln n$, it follows that

$$\lim_{n \to \infty} \frac{2(n+1)H_n - 4n}{n \ln n} = \lim_{n \to \infty} \frac{2(n+1)\ln n - 4n}{n \ln n}$$
$$= \lim_{n \to \infty} \frac{2n \ln n + 2 \ln n - 4n}{n \ln n} = 2.$$

It follows from Proposition 10.23 that the expected number of comparisons of Quicksort is given by $\Theta(n \ln n)$. Even though the expression $2(n+1)H_n - 4n$ is more exact, it is probably better to use the more straightforward bound $\Theta(n \ln n)$ when communicating to fellow programmers.

The next example shows that even when the limit of the quotient of two functions does not exist, they might still have the same order of growth.

Example 10.26. Let $f(n) = (2 + (-1)^n)n^2$ and $g(n) = n^2$. Then the limit

$$\lim_{n \to \infty} \frac{|f(n)|}{|g(n)|}$$

does not exist, as the quotient fluctuates between 3 and 1, but

$$|g(n)| \leqslant |f(n)| \leqslant 3|g(n)|$$

holds for all $n \geqslant 1$; hence, $f \in \Theta(g)$.

We call a function f from the set of positive integers to the set of real numbers **eventually nonzero** if and only if there exists an integer n_0 such that $f(n)$ is nonzero for all $n \geqslant n_0$.

The next proposition gives a necessary and sufficient condition for asymptotically tight bounds.

Proposition 10.27. *Let f be a function from the set of positive integers to the set of real numbers, and g an eventually nonzero function from the set of integers to the set of real numbers. Then $f \in \Theta(g)$ if and only if*

$$\liminf_{n \to \infty} \frac{|f(n)|}{|g(n)|} > 0 \quad \text{and} \quad \limsup_{n \to \infty} \frac{|f(n)|}{|g(n)|} < \infty.$$

10.3 Asymptotically Tight Bounds

Proof. If $f \in \Theta(g)$, then there exist positive constants c and C and a positive integer n_0 such that $c \leq |f(n)|/|g(n)| \leq C$ holds for all $n \geq n_0$. This implies that

$$\liminf_{n \to \infty} \frac{|f(n)|}{|g(n)|} \geq c > 0 \quad \text{and} \quad \limsup_{n \to \infty} \frac{|f(n)|}{|g(n)|} \leq C < \infty$$

hold.

Conversely, suppose that f and g are functions satisfying

$$c := \liminf_{n \to \infty} |f(n)|/|g(n)| > 0 \quad \text{and} \quad C := \limsup_{n \to \infty} |f(n)|/|g(n)| < \infty.$$

By definition of the limit superior and inferior, for any ϵ in the range $0 < \epsilon < c$ there exists a positive integer n_0 such that

$$0 < c - \epsilon \leq \frac{|f(n)|}{|g(n)|} \leq (C + \epsilon)$$

holds for all $n \geq n_0$. Multiplying these inequalities by $|g(n)|$ shows that $f \in \Theta(g)$ holds. □

Example 10.28. On input of a positive integer n, a program executes a block of code $T(n)$ times, where

$$T(n) = \begin{cases} 2^n & \text{when } n \text{ is even,} \\ 2^{n+10+(\log_2 n)/n} & \text{when } n \text{ is odd.} \end{cases}$$

We claim that $T(n) \in \Theta(2^n)$. Indeed, let us calculate the limit inferior and limit superior of $T(n)/2^n$. We have

$$(T(1)/2^1, T(2)/2^2, T(3)/2^3, \ldots) = $$
$$(2^{10+\log_2(1)}, 1, 2^{10+\log_2(3)/3}, 1, 2^{10+\log_2(5)/5}, 1, 2^{10+\log_2(7)/7}, \ldots).$$

In other words, the quotient $T(n)/2^n = 1$ when n is even and $T(n)/2^n = 2^{10+(\log_2 n)/n}$ when n is odd. Therefore, we get

$$\liminf_{n \to \infty} \frac{T(n)}{2^n} = 1 \quad \text{and} \quad \limsup_{n \to \infty} \frac{T(n)}{2^n} = 2^{10}.$$

Since the limit inferior and the limit superior are both positive real numbers, it follows from Proposition 10.27 that $T(n) \in \Theta(2^n)$.

A characteristic of an asymptotically tight bound $f \in \Theta(g)$ is that $f(n)$ cannot fluctuate too much in comparison to $g(n)$. In general, we want to keep the functions g simple. For instance, $g(n) = n^k$ for some positive integer k, $g(n) = b^n$ for some positive integer b, or similar elementary functions. A given function f might not be as nicely behaved as the simple functions that we use for comparison. Then we might be forced to use separate simple functions as upper and lower asymptotic bounds. We describe the appropriate asymptotic bounds in the next sections.

Exercises

10.15. Suppose that f, g, and h are functions from the set of positive integers to the set of real numbers. Show that
   (i) $f \asymp f$,
   (ii) $f \asymp g$ if and only if $g \asymp f$,
   (iii) $f \asymp g$ and $g \asymp h$ implies $f \asymp h$.
In other words, show that $\asymp$ is an equivalence relation.

10.16. Show that a polynomial $p(x)$ of degree m satisfies $p(x) = \Theta(x^m)$.

10.17. Let $f(n) = 2n^3 + 3n + 2$ and $g(n) = 5n^4 + 3n^3 + 2n + 1$. Express $\Theta(fg)$ in the simplest possible terms.

10.18. Let b and d be positive real numbers that are not equal to 1.
   (a) Show that $\Theta(\log_b n) = \Theta(\log_d n)$, so one can write $\Theta(\log n)$ using a baseless logarithm without causing confusion.
   (b) Prove or disprove: Does $\Theta(n^{\log_b n}) = \Theta(n^{\log_d n})$ hold in general?

10.19. Let a be a real number, $a > 1$. Show that
$$\sum_{k=0}^{n} a^k = \Theta(a^n).$$

10.20. Show that for all positive integers k, we have
$$1^k + 2^k + \cdots + n^k = \Theta(n^{k+1}).$$

10.4 Asymptotic Upper Bounds

The asymptotic equality and asymptotic tight bounds classify functions of similar growth. In this and the next section, we are concerned with relating functions of different growths.

Let f and g be functions from the set of positive integers to the set of real numbers. We say that g is an **asymptotic upper bound** for f and write $f \in O(g)$ if and only if there exists a positive real constant C and a positive integer n_0 such that
$$|f(n)| \leq C|g(n)|$$
holds for all $n \geq n_0$.

We denote by $O(g)$ the set of all functions $f: \mathbf{N}_1 \to \mathbf{R}$ such that g is an asymptotic upper bound for f.

The relation $f \in O(g)$ is often written as $f = O(g)$ in the literature. Even worse, subset relations such as $O(n) \subseteq O(n^2)$ are often denoted as $O(n) = O(n^2)$. These idiosyncratic notations are here to stay, so you need to get used to them. If you read statements in the literature such as $\ln n = O(n) = O(n^2)$, simply translate the equalities into the usual "element of" and "subset relations", such as $\ln n \in O(n) \subseteq O(n^2)$.

10.4 Asymptotic Upper Bounds

The asymptotic upper bound relation $f \in O(g)$ allows us to compare the growth of functions for large arguments n. Figure 10.5 shows the graphs of three functions. While it is possible to gain some insights into the growth of functions, it is easy to be misled by such graphs, as they show the values for the functions for some sets of small values of n. Yet the asymptotic upper bound is concerned with the behavior of the functions at large values of their arguments. For this reason, it is preferable to give a proof for the bounds.

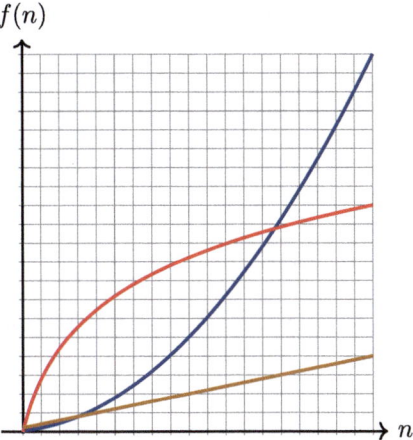

Figure 10.5: The graphs of the functions $\frac{1}{5}n$, $4\ln n$, and $\frac{1}{20}n^2$ on the interval $[1, 20]$. The quick growth of the function $\frac{1}{20}n^2$ suggests that $\frac{1}{5}n \in O(\frac{1}{20}n^2)$ and $4\ln n \in O(\frac{1}{20}n^2)$. However, graphs are not a good tool to judge such relationships. For example, $4\ln n$ exceeds $\frac{1}{5}n$ throughout this graph, but it is not difficult to show that $4\ln n \in O(\frac{1}{5}n)$, whereas $\frac{1}{5}n \notin O(4\ln n)$

Example 10.29. We have $n \in O(n^2)$. Indeed, for all integers $n \geqslant 1$, multiplying both sides by n yields $n \leqslant n^2$. Therefore, we have

$$|n| \leqslant 1|n^2|.$$

for all $n \geqslant 1$. The same argument shows that $|n| \leqslant 1|-n^2|$, so $n \in O(-n^2)$. For positive functions, there is no need to use absolute values.

Example 10.30. Suppose that k and ℓ are integers such that $1 \leqslant k < \ell$. For all integers $n \geqslant 1$, raising both sides to the $(\ell - k)$-th power yields $n^{\ell-k} \geqslant 1$. If we then multiply both sides by n^k, we get $n^\ell \geqslant n^k$. Therefore, we can conclude that $n^k \in O(n^\ell)$.

The next proposition gives a convenient necessary and sufficient condition to establish an asymptotic upper bound $f \in O(g)$.

Proposition 10.31. *Let f be a function from the set of positive integers to the set of real numbers, and g an eventually nonzero function from the set of positive integers to the set of real numbers. Then $f(n) \in O(g(n))$ if and only if*

$$\limsup_{n \to \infty} \frac{|f(n)|}{|g(n)|} < \infty.$$

Proof. The proof is straightforward, but instructive if you are not yet familiar with the notion of the limit superior, see Exercise 10.23. □

Corollary 10.32. *Let f and g be functions from the set of positive integers to the set of real numbers. If the limit*

$$\lim_{n \to \infty} \frac{|f(n)|}{|g(n)|}$$

exists and is finite, then $f \in O(g)$.

Example 10.33. Let us consider the functions $\ln n$ and n. We claim that $\ln n \in O(n)$. Since both $\ln n \to \infty$ and $n \to \infty$ as $n \to \infty$, we can use l'Hôpital's rule to deduce that

$$\lim_{n \to \infty} \frac{\ln n}{n} = \lim_{n \to \infty} \frac{1}{n} = 0.$$

Therefore, it follows from Corollary 10.32 that $\ln n \in O(n)$, as claimed.

Figure 10.6 gives some rules that are convenient when trying to establish asymptotic upper bounds. For example, it follows from the constant rule that $2n \in O(2n) = O(n)$. You can gain familiarity with asymptotic upper bounds by proving these rules.

Let f and g be functions from the set of positive integers to the set of real numbers. We say that g is a **strict asymptotic upper bound** for f and write $f \in o(g)$ if and only if for every real number $\epsilon > 0$ there exists a positive integer n_ϵ such that

$$|f(n)| \leq \epsilon |g(n)|$$

holds for all $n \geq n_\epsilon$. By definition, $f \in o(g)$ implies that $f \in O(g)$.

If $f \in o(g)$, then f is asymptotically much smaller than g. This property is nicely explained by the next proposition.

Proposition 10.34. *Let f and g be functions from the set of positive integers to the set of real numbers such that g is eventually nonzero. Then $f \in o(g)$ if and only if*

$$\lim_{n \to \infty} \frac{f(n)}{g(n)} = 0 \qquad (10.11)$$

holds.

10.4 Asymptotic Upper Bounds

Constants. If c is a nonzero constant, then

$$cO(f(n)) = O(f(n)), \qquad (10.2)$$
$$O(cf(n)) = O(f(n)). \qquad (10.3)$$

Idempotency. The Big Oh operator is idempotent, meaning that

$$O(O(f(n))) = O(f(n)). \qquad (10.4)$$

Multiplications. The multiplication of Big Oh expressions follows the rules

$$O(f(n))O(g(n)) = O(f(n)g(n)), \qquad (10.5)$$
$$O(f(n)g(n)) = f(n)O(g(n)). \qquad (10.6)$$

Absorption. We can simplify Big Oh expressions using the rule

$$O(f(n)) + O(g(n)) = O(g(n)) \text{ provided that } f(n) \in O(g(n)). \quad (10.7)$$

Powers. For all positive integers k, we have

$$(f(n) + g(n))^k \in O\big((f(n))^k\big) + O\big((g(n))^k\big). \qquad (10.8)$$

Linear Combinations. If $f(n) \in O(h(n))$ and $g(n) \in O(h(n))$, then

$$af(n) + bg(n) \in O(h(n)) \quad \text{for all } a, b \in \mathbf{R}. \qquad (10.9)$$

Swap. The next rule allows you to swap Big Oh terms.

If $f(n) \in g(n) + O(h(n))$ then $g(n) \in f(n) + O(h(n))$. $\qquad (10.10)$

Figure 10.6: Some simple rules for the manipulation of upper bounds

Proof. Suppose that (10.11) holds. By definition of the limit, this means that for any $\epsilon > 0$ there exists a positive integer n_ϵ such that

$$\left|\frac{f(n)}{g(n)}\right| < \epsilon$$

holds for all $n \geqslant n_\epsilon$. This is equivalent to the condition that for each $\epsilon > 0$ there exists an n_ϵ such that

$$|f(n)| \leqslant \epsilon |g(n)|$$

holds for all $n \geqslant n_\epsilon$. In other words, (10.11) is equivalent to $f \in o(g)$. $\qquad \square$

Corollary 10.35. *Let f and g be functions from the set of positive integers to the set of real numbers. Suppose that $f \in o(g)$. Then*

$$g + f \in O(g).$$

Example 10.36. We claim that $n \ln n \in o(n^2)$, so $n \ln n \in O(n^2)$. Indeed, it follows from l'Hôpital's rule that

$$\lim_{n\to\infty} \frac{n \ln n}{n^2} = \lim_{n\to\infty} \frac{\ln n}{n} = \lim_{n\to\infty} \frac{1}{n} = 0.$$

Therefore, we can deduce from Proposition 10.34 that $n \ln n \in o(n^2)$.

Exercises

10.21. Let $f(n) = 4 \ln n$, $g(n) = \frac{1}{5}n$, and $h(n) = \frac{1}{20}n^2$. Show that
(a) $f \in o(g)$,
(b) $g \in o(h)$.
(c) Conclude that $f \in O(g)$ and $g \in O(h)$.

10.22. Let f and g be positive functions. Show that $f \in o(g)$ implies that $g \notin O(f)$.

10.23. Prove Proposition 10.31: Let f be a function from the set of positive integers to the set of real numbers, and g an eventually nonzero function from the set of positive integers to the set of real numbers. Then $f(n) \in O(g(n))$ if and only if

$$\limsup_{n\to\infty} \frac{|f(n)|}{|g(n)|} < \infty.$$

10.24. Prove the idempotency rule given by Eq. (10.4).

10.25. Prove the multiplication rules (10.5) and (10.6).

10.26. Prove the absorption rule (10.7) and the power rule (10.8).

10.27. Prove the linear combination rule given in (10.9).

10.28. Prove the rule (10.10).

10.29. Find at least five examples of an erroneous usage of the Big Oh notation in recent papers or books of Computer Science.

10.30. Suppose that $p(n) = a_0 + a_1 n + \cdots + a_m n^m$ is a polynomial of degree m with complex coefficients. Show that $p(n) = O(n^k)$ for all $k \geq m$, and $p(n) \neq O(n^\ell)$ for $0 \leq \ell < m$.

10.31. Show that for fixed k, we have

$$\binom{n}{k} \in \frac{n^k}{k!} + O(n^{k-1}) \quad \text{and} \quad \binom{n+k}{k} \in \frac{n^k}{k!} + O(n^{k-1}),$$

where $\binom{n}{k} = n(n-1)(n-2)\cdots(n-k+1)/k!$ is the binomial coefficient.

10.5 Asymptotic Lower Bounds

10.32. Show that $n/(n+1) \in 1 + O(1/n)$.

10.33. Let f, g, and h be functions from the set of positive integers to the set of real numbers. We write $f \leq g$ if and only if $f \in O(g)$. Show that
(i) $f \leq f$
(ii) $f \leq g$ and $g \leq h$ implies $f \leq h$
(iii) there exist functions f and g such that neither $f \leq g$ nor $g \leq f$ holds.
In the language of set theory, properties (i) and (ii) mean that $\leq$ is a preorder. The third property shows that the asymptotic growth of functions is not always comparable.

10.34. Prove or disprove: $e^n \in O(2^n)$.

10.35. Prove or disprove: $n^{\ln n} \in O(e^{(\ln n)^2})$.

10.36. Show that $f \in O(g)$ and $f \notin o(g)$ does not imply that $f \in \Theta(g)$.

10.37. Show that $f(x) \in g(x) + O(1)$ implies that $\exp(f(x)) \asymp \exp(g(x))$.

10.5 Asymptotic Lower Bounds

In this section, we define asymptotic lower bounds, which complement the notions that we have defined in the previous section.

Let f and g be functions from the set of positive integers to the set of real numbers. We say that g is an **asymptotic lower bound** to f and write $f \in \Omega(g)$ if and only if there exists a positive constant c and a positive integer n_0 such that
$$c|g(n)| \leq |f(n)|$$
holds for all $n \geq n_0$. This formalizes the notion that $f(n)$ grows at least as fast as a constant multiple of $g(n)$ for large n.

This asymptotic lower bound is related to the asymptotic upper bound in the following way.

Proposition 10.37. *Let f and g be functions from the set of positive integers to the set of real numbers. We have $f \in \Omega(g)$ if and only if $g \in O(f)$.*

Proof. We have $f \in \Omega(g)$ if and only if there exists a positive constant c and a positive integer n_0 such that $c|g(n)| \leq |f(n)|$ holds for all $n \geq n_0$. Dividing both sides by c shows that there exist a positive constant $C = 1/c$ and a positive integer n_0 such that $|g(n)| \leq \frac{1}{c}|f(n)| = C|f(n)|$ holds for all $n \geq n_0$. However, this is nothing but the definition of $g \in O(f)$. □

Let f and g be functions from the set of positive integers to the set of real numbers. We say that g is a **strict asymptotic lower bound** to f and write $f \in \omega(g)$ if and only if for all positive constants c there exists a positive integer n_0 such that
$$c|g(n)| \leq |f(n)|$$
holds for all $n \geq n_0$.

Example 10.38. The function n^2 is in $\omega(n)$, since for a given positive constant c, the inequality $cn = c|n| \leq |n^2| = n^2$ holds for all positive integer $n \geq c$. On the other hand, n is not in $\omega(n)$, since there does not exist any positive integer n for which $2n = 2|n| \leq |n| = n$ holds.

Proposition 10.39. *Let f and g be functions from the set of positive integers to the set of real numbers, and assume that g is eventually nonzero. Then we have $f \in \omega(g)$ if and only if*
$$\lim_{n \to \infty} \frac{|f(n)|}{|g(n)|} = \infty.$$

Proof. We have $f \in \omega(g)$ if and only if for all positive constants c there exists a positive integer n_0 such that $c \leq |f(n)|/|g(n)|$ holds for all $n \geq n_0$, so $|f(n)|/|g(n)|$ grows without bound. By definition of the limit, this is equivalent to
$$\lim_{n \to \infty} \frac{|f(n)|}{|g(n)|} = \infty,$$
which proves the claim. □

10.6 Analysis of Algorithms

In this section, we will give a brief exposition of the analysis of algorithms. The purpose of the analysis is to get some insights into the running time or the space requirements of the computation. Since too many details of the compiler and computer architecture affect the precise time or space requirements, one has to settle for the more modest goal of obtaining reasonable bounds on these quantities. We will focus on the analysis of the running time of an algorithm.

Since we are settling for bounds on the running time rather than a precise estimate, we can use a simplified model of computation such as the random access machine. A random access machine is a very simple single-processor machine that executes instructions one after another without pipelining, speculative execution, branch prediction, or parallelism. It even lacks a memory hierarchy. We could now precisely define the instructions of the random access machine and their costs, as in Aho, Hopcroft, and Ullman [1]. However, we follow the approach taken in Cormen, Leiserson, Rivest, and Stein [16] and assume that elementary operations take a constant—yet unspecified—amount of time. Then there is no need to translate the pseudocode into random access machine instructions, which simplifies matters.

The running time of an algorithm is measured as a function of the size of the input. What is the size of the input? Unfortunately, the answer to this question depends on the application. We assume that we are given a function s that associates to an input x its **size** $s(x)$. For instance, if the input to a sorting algorithm is an array x with n elements, then $s(x) = n$ would be a natural measure for the size of the input. If the input x is an integer, then $s(x) = \lfloor \log_{10} x \rfloor + 1$ may be a good measure of input size for algorithms that deal

10.6 Analysis of Algorithms

with long integer arithmetic. If the integers remain bounded, then $s(x) = 1$ would be a reasonable choice.

Let I denote the set of all possible inputs to an algorithm A. If we denote by $t_A(x)$ the running time of the algorithm A on an input x, then the **worst-case running time** of the algorithm A is defined as

$$\sup \{t_A(x) \mid x \in I, s(x) = n\}.$$

This measures the worst-case performance of the algorithm for all inputs of length n. We will use asymptotic notations to simplify the resulting worst-case running time.

So how do we determine the worst-case running time of an algorithm in practice? The answer is that we simply analyze the algorithm line-by-line and essentially count operations. The main complications are conditional statements, loops, and function or procedure calls. We now put forth the axioms that govern the run-time analysis. The list is incomplete, but you will see the pattern, and it should be easy to extend this list by further language constructs.

A1. Elementary operations have constant running time.

Examples of elementary operations include assignments, arithmetic with machine size words, Boolean operations, comparisons, and array access, among others. The axiom says that they all have $\Theta(1)$ running time.

The running time of a compound statement is bounded by the sum of the running times of the individual component statements.

A2. If the statements S_1 and S_2 respectively have worst-case running time $T(S_1)$ and $T(S_2)$ on an input of size n, then the compound statement $S_1; S_2$ has worst-case running time

$$T(S_1; S_2) \leq T(S_1) + T(S_2)$$

The next axiom gives a coarse-grained view of the running-time of a conditional statement. We should add that sometimes it makes sense to distinguish cases depending on the value of the condition C.

A3. The conditional statement

$$\text{if } C \text{ then } S_1 \text{ else } S_2$$

has worst-case running time $T(C) + \max\{T(S_1), T(S_2)\}$ when the conditional statement C and the statements S_1 and S_2 have worst case running time $T(C)$, $T(S_1)$ and $T(S_2)$ on an input of size n, respectively.

The next axiom gives the estimate for the running-time of a for loop, which iterates through all integers k in the range $a \leq k \leq b$, where a and b are integers. The notation $(a..b)$

A4. The running-time of a for-loop statement S given by

$$\textbf{for } k \textbf{ in } (a..b) \textbf{ do}$$
$$S_1$$
$$\textbf{end}$$

has worst-case running time $T(S) \leqslant c + \sum_{k=a}^{b}(T(S_1, k) + c)$, where $T(S_1, k)$ is the worst-case running time of statement S_1 during iteration k and c is the cost of evaluating the loop condition.

In the worst case, such a for loop iterates $b-a+1$ times and checks the loop condition $b-a+2$ times (as the loop verifies that k is out of bounds before it stops the iteration). One special case is worth noting. If $T(S_1) = T(S_1, k))$ is nonzero and independent of k, then this simplifies to

$$T(S) \leqslant (b-a+1)\Theta(T(S_1)) + (b-a+2)c.$$

There exist numerous variations of for loops. We will leave it to the reader to modify the axiom for these variations.

The running-time of a do-while loop (sometimes also called a repeat-until loop) is given in the next axiom. This style of loop executes the statement S_1 at least once before checking the condition C. It will repeat while the condition C is met.

A5. The running-time of a do-while loop statement S given by

$$\textbf{begin}$$
$$S_1$$
$$\textbf{end while } C$$

has worst-case running time $T(S) \leqslant \sum_{k=1}^{m(n)}(T(C)+T(S_1, k))$, where $m(n)$ is the maximal number of loop iterations on an input of size n and $T(C)$ and $T(S_1, k)$ are the worst-case running times of the conditional statement C and the statement S_1 during iteration k, respectively.

We illustrate how to obtain an estimate for the worst-case running-time with a small example.

Example 10.40. Suppose that you want to write a program to evaluate a polynomial
$$p(x) = a_n x^n + a_{n-1} x^{n-1} + \cdots + a_1 x + a_0,$$
so for a given input x, you would like to know the value $y = p(x)$. You can save multiplications by using the Horner scheme
$$p(x) = (\cdots((a_n x + a_{n-1})x + a_{n-2})x + \cdots a_1)x + a_0.$$

The following algorithm is based on the Horner scheme:

10.6 Analysis of Algorithms

horner(a, n, x)	cost	times
res = a[n]	c_1	1
for k = n−1 down to 0 do	c_2	n+1
res = res * x + a[k]	c_3	n
end	c_4	n
return res	c_5	1
end	c_6	1

The worst-case running time of the algorithm is given by

$$T(n) = c_1 + c_2(n+1) + c_3 n + c_4 n + c_5 + c_6 = \Theta(n).$$

This compares favorably with the naive $\Theta(n^2)$ algorithm that evaluates the polynomial $p(x)$ by computing each power independently by repeated multiplication.

Exercises

10.38. Analyze the running time of the following algorithm using a step count analysis (as in the Horner scheme).

```
// search a key in an array a[1..n] of length n
search(a, n, key)                 cost  times
   for k in (1..n) do
      if a[k]=key then
         return k
   end
   return false
```

Determine the worst-case complexity of this algorithm.

10.39. Analyze the running time of the following algorithm using a step count analysis (as in the Horner scheme).

```
// determine the number of digits of an integer num
binary_digits(num)                cost  times
   int cnt = 1
   while (num>1) do
      cnt = cnt + 1
      num = floor(num/2.0)
   end
   return cnt
```

(a) Give an axiom for the running time of a while loop.
(b) Determine the asymptotic running-time of this algorithm.

10.40. Determine how many times the block `Block(a,b)` is executed in the following code fragment:

```
n = 100
for a in (0..n-1) do
  for b in (0..n) do
    Block(a,b)
  end
end
```

You can assume that `Block(a,b)` does not exit the for loops, nor does it manipulate the loop variables a and b.

10.41. Determine how many times the block `Block(a,b)` is executed in the following code fragment:

```
n = 100
for a in (0..n) do
  for b in (0..a) do
    Block(a,b)
  end
end
```

You can assume that `Block(a,b)` does not exit the encapsulating for loops and does not manipulate the loop variables a and b.

10.7 Notes

Asymptotic notations were introduced by Bachmann [7] and further popularized by Landau [55]. Graham, Knuth, and Patashnik [30] give a more in-depth treatment of asymptotic notations. Spencer [71] dedicates an entire book to asymptotic notations and gives numerous surprising applications. Books on the analysis of algorithms contain numerous examples of asymptotic estimates of the running time; see, for example, Aho, Hopcroft, and Ullman [1] or Cormen, Leiserson, Rivest, and Stein [16].

Part III

Combinatorics

Chapter 11

Counting

> As I was going to St. Ives,
> I met a man with seven wives,
> Each wife had seven sacks,
> Each sack had seven cats,
> Each cat had seven kits:
> Kits, cats, sacks, and wives,
> How many were there going to St. Ives?
>
> — Anonymous, *In: The Oxford Dictionary of Nursery Rhymes*

Enumerative combinatorics is concerned with counting the number of combinatorial objects such as permutations, combinations, multisets, partitions of sets, partitions of numbers, and the like. The counting techniques are simple but are often applied in a skillful way to great effect. The very large number of applications of enumerative combinatorics in computer science led to a resurgence of the field.

11.1 Fundamental Counting Principles

In combinatorics, one does not count the number of objects by listing them, but rather by using an abstract method to determine their number. The basic principles are simple, but choosing the correct approach can be a challenge at times.

We begin with a simple counting principle that is inspired by set theory. Recall that if we are given two sets A and B, then the cardinality of their union satisfies
$$|A \cup B| = |A| + |B| - |A \cap B|.$$

If A and B are disjoint sets, then the term $|A \cap B| = 0$ and can be omitted. The next principle is a straightforward generalization of this simple case.

> **Summation Principle** If a finite set S is the union of m pairwise disjoint sets $S_1, S_2, \ldots, S_m$, then $|S| = \sum_{k=1}^{m} |S_k|$.

We can rephrase the summation principle as follows. If there are m mutually exclusive tasks and the k-th task can be done in n_k different ways, then the number of ways to do one of these tasks is given by

$$n_1 + n_2 + \cdots + n_m.$$

The key is that the tasks are mutually exclusive, so if you choose to do the first task, then there are n_1 ways to do it, but none of the other tasks can be done.

The next two examples illustrate the summation principle.

Example 11.1. Emily wants to buy some nice loaf of bread. She can go to a nice downtown bakery where they offer three different kinds of bread that she likes. Alternatively, she can go to a nearby bakery that has two different kinds of artisanal breads that she fancies. By the summation principle, she can choose from $3 + 2 = 5$ loafs of bread.

Example 11.2. How many strings of six decimal digits have exactly five 7s? Apparently, the strings are of the form

$$x77777, \quad 7x7777, \quad 77x777, \quad 777x77, \quad 7777x7, \quad 77777x,$$

where x can represent any of the nine digits in the set $\{0, 1, \ldots, 6\} \cup \{8, 9\}$.

We can now count the possibilities as follows. Let S_k be the set of strings of length six that have a 7 in each position except at position k. Since there are 9 choices of digits other than 7, we have $|S_k| = 9$. Therefore, the number of strings with six decimal digits that have exactly five 7s is given by summation principle as

$$\sum_{k=1}^{6} |S_k| = \sum_{k=1}^{6} 9 = 54.$$

The advantage of the set theoretic approach is that it easily scales to large problem sizes.

> **Multiplication Principle I** If a finite set S is the Cartesian product $S = S_1 \times S_2 \times \cdots \times S_m$ of m sets, then $|S| = \prod_{k=1}^{m} |S_k|$.

We can use this principle as follows. Suppose that a task can be broken up into m consecutive independent subtasks. If the k-th subtask can be done in n_k ways, then the entire task can be done in

11.1 Fundamental Counting Principles

$$n_1 \times n_2 \times \cdots \times n_m$$

ways.

The next two examples illustrate how to use the multiplication principle.

Example 11.3. Sam has decided on a smartphone. He can choose from the four colors $S_1 = \{\text{silver, gold, red, blue}\}$, three different memory sizes $S_2 = \{16\,\text{GB}, 64\,\text{GB}, 128\,\text{GB}\}$, and sim cards from four different network providers $S_3 = \{P_1, P_2, P_3, P_4\}$. From how many different smartphone configurations can he choose? By the multiplication principle, he can choose from $4 \times 3 \times 4 = 48$ different configurations.

Example 11.4. Tera wants to travel from Houston to Vancouver, and she thought that it might be nice to have a layover in New York. She found three different flights from Houston to New York that are affordable, and four different flights from New York to Vancouver on the next day. By the multiplication principle, she can choose from a total of $3 \times 4 = 12$ different travel itineraries.

Sometimes we need to consider dependencies among the choices. The selection for the first task might influence the choices for the second task. For instance, if we select Amy to be the winner of a science fair, then the runner up should be one of the other students. If we select Bert to be the winner, then the runner up should be someone other than Bert. However, if the number of choices remains the same for each selection, then the multiplication principle still holds in this more general situation.

> **Multiplication Principle II** Let S denote a set consisting of sequences $(s_1, s_2, \ldots, s_m)$ that are constructed as follows. There are n_1 different ways to choose s_1. For each k in the range $1 \leq k < m$, after having chosen $s_1, s_2, \ldots, s_k$, there are precisely n_{k+1} choices for the sequence element s_{k+1}. Then the number $|S|$ of such sequences is given by
>
> $$|S| = n_1 \times n_2 \times \cdots \times n_m.$$

Example 11.5. Suppose that you want to order a pizza. The pizza parlor offers an attractive deal when you choose two different toppings from

$$\{\text{pepperoni, olives, mushrooms}\},$$

where the amount for the first topping is about twice the amount of the second topping. How many different pizzas are covered by this deal? We can visualize the choices using a decision tree, see Fig. 11.1. For the first (and dominant) topping, there are three different choices. If you pick, say, *olives* as the first topping, then there are two choices for the second topping, namely *pepperoni* or *mushrooms*. Since there are three choices for the first topping, and two choice for the second topping (regardless of your choice for the first topping), there

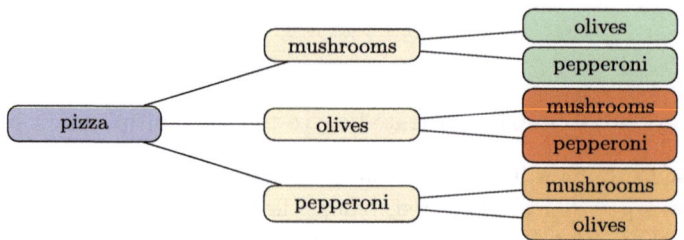

Figure 11.1: The decision tree for the ingredient selection. There are three choices for the first dominant topping. For the second nondominant topping, there remain two choices for each selection of the first topping

are $3 \times 2 = 6$ choices for the two-toppings pizza. We emphasize that the second topping selection choices are not from a fixed set (as in the Multiplication Principle I), but rather from different sets.

Example 11.6. Suppose that a student organization has nine active members. In how many ways can they elect a president, vice-president, secretary, and treasurer as officers, assuming that everyone can hold only one officer position? There are nine ways to choose someone as president. From the remaining eight students, one can choose the vice-president. After president and vice-president have been chosen, there are seven ways to select the secretary. Finally, the treasurer can be chosen from the remaining six students. Therefore, there are a total of
$$9 \times 8 \times 7 \times 6 = 3024$$
different ways of filling the four officer positions from the nine active members of the student organization.

In many counting problems, the solution may require a combination of the multiplication and summation principles, as the next two examples show.

Example 11.7. A restaurant offers on Mother's Day on their prix fixe menu either a three-course meal or a four-course meal. In the three-course meal, one can choose from three different appetizers, four main courses, and three desserts. In the four-course meal, there are three choices for the appetizer course, five choices for the second course, three choices for the third course, and three choices for the dessert course. By the multiplication principle, there are $3 \times 4 \times 3 = 36$ possible three-course meals, and $3 \times 5 \times 3 \times 3 = 135$ possible four-course meals. By the summation principle, the prix fixe menu offers a total of $36 + 135 = 171$ different choices.

Example 11.8. In Sesame street, Season 1, Episode 45, the "As I was going to St. Ives" nursery rhyme quoted at the beginning of this chapter was made into a song. A purple Anything Muppet boy sings the song with the nursery rhyme and challenges a green Anything Muppet girl to figure out how many really go to St. Ives. He then reveals that the answer is 1, as the others go the

11.1 Fundamental Counting Principles

other way. The amusing part about the skit is that the green girl turns the tables and asks him how many go the other way, which he cannot answer. She correctly calculates using the multiplication principle that there is 1 man, 7 wives, $7^2 = 49$ sacks, $7^3 = 343$ cats, and $7^4 = 2401$ kittens. By the summation principle, there are $1 + 7 + 49 + 343 + 2401 = 2801$ going the other way.[1]

Subtraction Principle Let S be a finite set with a subset T. Then $|S \setminus T| = |S| - |T|$.

In the subtraction principle, one counts the number $|T|$ of elements in the subset T of unwanted elements and subtracts this from the total $|S|$. The subtraction principle is also known as **complementary counting**. This method can be beneficial when it is tedious to count the elements in the set $S \setminus T$ directly.

The next two examples illustrate the flavor of this method.

Example 11.9. Suppose that you want to determine the number of positive integers in the set $S = \{1, 2, \ldots, 100\}$ that are not divisible by 6. A direct count does not seem to be so obvious. However, the set

$$T = \{x \in S \mid x \text{ is a multiple of } 6\}$$

of unwanted elements has cardinality $\lfloor 100/6 \rfloor$. Since $|S| = 100$, we can conclude that the number of elements in S that are not divisible by 6 is given by

$$100 - \lfloor 100/6 \rfloor = 100 - 16 = 84.$$

The subtraction method led to a straightforward solution.

Example 11.10. How many integers $x \in \{1, 2, \ldots, 1000\}$ have at least two different digits? We will use the subtraction principle. Therefore, we need to determine how many integers in the set $S = \{1, 2, \ldots, 1000\}$ have all digits equal to the same value? The set T of all such integers is given by

$$T = \{1, 2, \ldots, 9, 11, 22, \ldots, 99, 111, 222, \ldots, 999\},$$

so there are $|T| = 27$. Therefore, $|S \setminus T| = |S| - |T| = 1000 - 27 = 973$ integers have at least two different digits.

Exercises

11.1. How many bit strings of length 12 are there?

11.2. Anna wants to go on a vacation. She packs merely five different shirts, three different pairs of pants, and two different pairs of shoes. In how many different ways can she combine her outfit?

---

[1] She was also counting the sacks to conform to the nursery rhyme. Do you recall a better way to evaluate this geometric series?

11.3. Suppose that you have an alphabet with m different symbols. How many strings of length n can you form that have symbols from this alphabet in each position?

11.4. Suppose that a state has license plates of the following form: three letters followed by four digits. How many different license plates are there if no further restrictions are imposed?

11.5. The selection of a password of a computer account has the following restrictions: The password must be 6, 7, or 8 characters long. A character can be lower case letter or a decimal digit. The first character must be a lowercase letter. Determine the total number of possible passwords with the given restrictions.

11.6. Deduce the subtraction principle from the summation principle.

11.7. The North American Numbering Plan NANP regulates the form of phone numbers in North America. A 10-digit phone number is of the form NXX-NXX-XXXX, where N denotes any digit in the range 2–9, and X denotes any digit in the range 0–9. The first three digits cannot be of the form N11, as these are reserved for special services. Determine how many 10-digit NANP phone numbers there are according to the restrictions given here (there are a few more, which we ignore for the sake of this exercise).

11.8. Andrea wants to create a game that requires many game pieces. She paints a wooden cube red on all sides and plans to cut it on a bandsaw into k^3 smaller congruent cubes. The small cubes will serve as the game pieces. What is the smallest k such that she gets at least 100 small cubes that have one or more red sides?

11.9. Bob suffers from aibohphobia, the irrational fear of palindromes. Calculate for him the number of words of length n over an alphabet with k letters that are not palindromes.

11.10. How many different relations exist on a set with n elements.

11.11. Count the number of reflexive relations on a set with n elements.

11.12. Count the number of irreflexive relations on a set with n elements.

11.13. Count the number of asymmetric relations on a set with n elements.

11.14. Count the number of antisymmetric relations on a set with n elements.

11.15. Count the number of symmetric relations on a set with n elements.

11.16. Suppose that we are given m pairwise disjoint finite sets $S_1, S_2, \ldots, S_m$ of cardinality $c_k = |S_k|$ for all k in the range $1 \leqslant k \leqslant m$. Show that the number of subsets of $S_1 \cup S_2 \cup \cdots \cup S_m$ containing at most one element from each set S_k with $1 \leqslant k \leqslant m$ is given by

$$(c_1 + 1)(c_2 + 1) \cdots (c_m + 1).$$

11.17. Alice, Joe, and Bert are naming integers a, j, and b in the range $0 \leq a, j, b < 1000$. Joe finds it amusing to pretend to be an "average Joe," so he manipulates Alice and Bert to name integers a and b such that he can choose the average
$$j = \frac{a+b}{2}$$
of their values a and b as his integer. How many triples are in the set
$$\{(a, \frac{a+b}{2}, b) \in \mathbf{Z}^3 \mid 0 \leq a, b < 1000\}?$$

11.2 Permutations and Combinations

In this section, we introduce partial permutations and combinations. The counting coefficients for the number of partial permutations and combinations are helpful when solving counting problems. We identify the binomial coefficients as the counting coefficients of combinations. The combinatorial interpretation of binomial coefficients has many interesting consequences that are explored in the subsequent sections.

Let S be a finite set with n elements. Recall that a bijective function on S is also called a permutation. If $S = \{1, 2, 3\}$, then there are six different permutations, namely

$$\begin{pmatrix} 1 & 2 & 3 \\ 1 & 2 & 3 \end{pmatrix}, \begin{pmatrix} 1 & 2 & 3 \\ 2 & 1 & 3 \end{pmatrix}, \begin{pmatrix} 1 & 2 & 3 \\ 3 & 2 & 1 \end{pmatrix},$$

$$\begin{pmatrix} 1 & 2 & 3 \\ 1 & 3 & 2 \end{pmatrix}, \begin{pmatrix} 1 & 2 & 3 \\ 2 & 3 & 1 \end{pmatrix}, \begin{pmatrix} 1 & 2 & 3 \\ 3 & 1 & 2 \end{pmatrix}.$$

The set of all permutations on a set with n elements is denoted by S_n. We have $|S_n| = n!$, as we will prove shortly.

We can generalize the concept of a permutation as follows. An injective function from a set with k elements to a set S with n elements is called a **k-permutation**. In other words, a k-permutation is a partial permutation. If $S = \{a, b, c\}$, then all 2-permutations are given by

$$\begin{pmatrix} 1 & 2 \\ a & b \end{pmatrix}, \begin{pmatrix} 1 & 2 \\ a & c \end{pmatrix}, \begin{pmatrix} 1 & 2 \\ b & a \end{pmatrix}, \begin{pmatrix} 1 & 2 \\ b & c \end{pmatrix}, \begin{pmatrix} 1 & 2 \\ c & a \end{pmatrix}, \begin{pmatrix} 1 & 2 \\ c & b \end{pmatrix}.$$

It is common to identify the injective functions with ordered sequences that simply list the images. In this interpretation, the six 2-permutations of the set $S = \{a, b, c\}$ are given by

$$(a, b), \quad (a, c), \quad (b, a), \quad (b, c), \quad (c, a), \quad (c, b).$$

Example 11.11. Suppose that eight runners qualified for the final 100 meter sprint race. If we denote the set of runners by $S = \{A, B, C, D, E, F, G, H\}$, then the 3-permutations describe the possible placings for gold, silver, and bronze medals, assuming that there are no ties. The 3-permutation (B, F, G) can be interpreted as B winning the gold medal, F winning the silver medal, and G winning the bronze medal. There are eight choices for the potential gold medal winner, seven choices for the silver medal winner from the remaining runners, and six choices for the bronze medal winner. Therefore, it follows from the multiplication principle that $8 \times 7 \times 6 = 336$ is the total number of possible race outcomes in terms of olympic medals.

We denote by $P(n, k)$ the number of k-permutations. The next proposition shows that the answer can be easily formulated in terms of the falling factorial

$$n^{\underline{k}} = n(n-1) \cdots (n-k+1),$$

where $n^{\underline{0}} = 1$.

Proposition 11.12. *The number of k-permutations of a set with n elements is given by*

$$P(n, k) = n^{\underline{k}} = \frac{n!}{(n-k)!}.$$

Proof. We prove the claim by induction on k. For $k = 0$, there exists just a single k-permutation, namely the empty sequence. So the claim $P(n, 0) = 1 = n^{\underline{0}}$ is correct.

Suppose that the claim holds for $P(n, k-1)$. A $(k-1)$-permutation of a set with n elements can be extended to a k-permutation by selecting one of the $n - (k-1)$ elements of S that do not yet occur in the sequence. By the multiplication principle,

$$P(n, k) = P(n, k-1)(n - k + 1) = n^{\underline{k-1}}(n - k + 1) = n^{\underline{k}},$$

as claimed. $\square$

In a k-permutation, the order of the selected k elements matters. If we want to ignore the order, then we arrive at the notion of a k-combination. Specifically, a k-**combination** of a finite set S with n elements is a subset of S of cardinality k. For instance, the set $S = \{a, b, c\}$ has the following three 2-combinations

$$\{a, b\}, \quad \{a, c\}, \quad \{b, c\}.$$

We denote by $C(n, k)$ the number of k-combinations of a set with n elements. Since any k-permutation can be obtained by ordering a k-combination, and there exist $k!$ different ways to order a set of k elements, we have

$$P(n, k) = k!C(n, k).$$

The next proposition reveals that the number of k-combination of a set with n elements is given by a binomial coefficient.

11.2 Permutations and Combinations

Proposition 11.13. *The number of k-combination of a set with n elements is given by*
$$C(n,k) = \binom{n}{k}.$$

Proof. Since $k!C(n,k) = P(n,k)$ and $P(n,k) = n^{\underline{k}}$, we can deduce that
$$C(n,k) = \frac{n^{\underline{k}}}{k!} = \frac{n!}{k!(n-k)!} = \binom{n}{k},$$
as claimed. □

Exercises

11.18. A tennis team consists of 12 players. The team lines up for a group photo, where the players all stand next to each other in a single line. In how many different ways can the photographer arrange the players for the photo?

11.19. Tom has a book shelf containing 20 novels and 7 math books. In how many ways can he arrange the books on the shelf, assuming he wants to keep all the math books next to each other?

11.20. John is moving and has packed all his possessions into six large boxes.
(a) John can load three boxes in his car. In how many ways can John select three of the six boxes?
(b) John makes two trips, transporting three boxes in each trip. How many ways are there to transport the six boxes?
(c) John moved into his new apartment. He labels the boxes from 1 to 6 and forms three stacks of boxes. The first stack has three boxes, the second has two boxes, and the third has one box. In how many ways can he stack the six boxes in this way?
In your answers, you should use counting coefficients $P(n,k)$ or $C(n,k)$ whenever appropriate. Then evaluate your expressions.

11.21. There are 12 guests at a party. Bob observed that each pair of guests clinked glasses precisely once. How many times did the glasses clink at the party?

11.22. A student organization consists of 20 students of mathematics and 30 students of engineering. They need to select two students of mathematics and three students of engineering to help at a middle school science fair. In how many different ways can they select the students for this task?

11.23. Let S be a set of n distinct points on a circle. How many different nondegenerate triangles can be formed from these points?

11.24. Zeta wants to write down all seven digit nonnegative integers that can be formed with digits from the set $\{0, 1, 2, \ldots, 9\}$, but without using a digit twice. She follows the usual conventions for writing down integers. How many different integers would she need to write down?

11.25. Let S be a set of n distinct points such that $n-m$ points lie on a circle and m points lie on a line that does not meet any of the points on the circle. How many different nondegenerate triangles can be formed from these points?

11.3 Combinatorial Proofs

In the previous section, we learned that the number of subsets with k elements of a set with n elements are counted by the binomial coefficient

$$\binom{n}{k}.$$

This combinatorial interpretation suggests to use a combinatorial argument for proofs. A combinatorial argument establishes an equality $A = B$ of two formulas either by the bijection principle or by double counting. In this section, we will explain both proof principles.

The idea underlying the bijection principle is simple. The bijection principle finds a set S such that $A = |S|$ and a set T such that $B = |T|$. If a bijection can be established between the sets S and T, then $A = B$ is proved.

> **Bijection Principle** If S and T are finite sets and there exists a bijection from S onto T, then $|S| = |T|$.

At first glance, the bijection principle might appear to be almost frivolous. However, finding an appropriate bijection to solve a counting problem can lead to insightful proofs. The next example gives a first taste.

Example 11.14. A computer science teacher wants to treat her students, since they did very well in a competition. There are 10 students, and she promises a slice of pie for everyone from a nearby bakery. The students can choose from *apple pie, cherry pie, key lime pie,* and *pecan pie*. The teacher records a apple pies, c cherry pies, k key lime pies, and p pecan pies in the short form (a, c, k, p). So the possible orders of the pies can be described by the set

$$P = \{(a, c, k, p) \mid 0 \leqslant a, c, k, p \leqslant 10, a + c + k + p = 10\}.$$

While they were waiting for their pies to arrive, the students are wondering how many different orders are possible. However, it was not immediately apparent to them how to determine $|P|$.

11.3 Combinatorial Proofs

The students are baffled when their teacher mentions that the pie orders correspond to the set B of bit strings of length 13 that contain precisely three 1s. Indeed, she explains that it is possible to map (a, c, k, p) to the bit string

$$\underbrace{0\cdots 0}_{a \text{ times}} 1 \underbrace{0\cdots 0}_{c \text{ times}} 1 \underbrace{0\cdots 0}_{k \text{ times}} 1 \underbrace{0\cdots 0}_{p \text{ times}}$$

that consists of a initial zeros, followed by a 1, then c zeros, followed by a 1, k zeros, followed by a 1, and finally, p zeros. As every string in B is 13 bits long and contains three 1s, there must be 10 zeros. Evidently, this is a bijection from P onto B. While it might not have been obvious how to count P, it is clear that B contains $\binom{13}{3}$ bit strings, as these are the number of ways to choose the positions of the three bits equal to 1. By the bijection principle, there are

$$|P| = |B| = \binom{13}{3} = 286$$

different ways to order 10 pies from a selection of four different kinds of pies.

In the algebraic definition of binomial coefficients, it is not too difficult to arrive at the equality

$$\binom{n}{k} = \binom{n}{n-k}.$$

In the combinatorial interpretation of binomial coefficients, this equality claims that the number of ways to select subsets with k elements from a set with n elements is equal to the number of ways to select subsets with $n - k$ elements from a set with n elements. The proof of the next proposition shows how to use the bijection principle to establish this fact.

Proposition 11.15. *Let n be a nonnegative integer and k an integer in the range $0 \leqslant k \leqslant n$. Then*

$$\binom{n}{k} = \binom{n}{n-k}.$$

Proof. Let S be a set with n elements. Let $F_k = \{K | K \subseteq S \text{ and } |K| = k\}$ denote the set of subsets of S that have cardinality k and F_{n-k} the set of subsets of S that have cardinality $n - k$. Since the complement A^c of a set A with k elements has $n - k$ elements, and $(A^c)^c = A$, taking the complement maps F_k bijectively onto F_{n-k}. Therefore, $|F_k| = |F_{n-k}|$. The claim follows from the fact that $\binom{n}{k} = |F_k|$ and $\binom{n}{n-k} = |F_{n-k}|$. □

The next counting principle is even more straightforward. Given a set, if one way to count its elements yields A, and another way to count its elements yields B, then we must have $A = B$.

Double Counting If two formulas count the elements of the same set in two different ways, then they must be equal.

The next example illustrates the double counting technique.

Example 11.16. Suppose that we want to prove the equation

$$\sum_{k=1}^{n} k = \frac{n(n+1)}{2}.$$

This time we will not use a proof by induction, but rather a double counting argument. We will count the points in the set

$$S = \{(x, y) \in \mathbf{Z}^2 \mid 0 \leqslant x, y \leqslant n\}$$

in two different ways. Evidently, the set S contains $(n+1)^2$ points, namely each point (x, y) with integer coordinates in the range $0 \leqslant x, y \leqslant n$. However, we can count the set also by diagonals, see Fig. 11.2. The number of points on the diagonals below the main diagonal are given by

$$\sum_{k=1}^{n} k.$$

We obtain the same formula for the number of points on diagonals above the main diagonal of the quadratic grid. In addition, there are $n+1$ points on the main diagonal. Summing over all diagonals, we get

$$2 \left(\sum_{k=1}^{n} k \right) + (n+1) = (n+1)^2.$$

Subtracting $(n+1)$ from both sides and dividing by 2 yields

$$\sum_{k=1}^{n} k = \frac{(n+1)^2 - (n+1)}{2} = \frac{n(n+1)}{2},$$

as claimed.

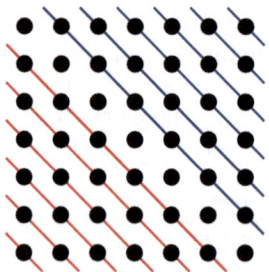

Figure 11.2: The figure illustrates the set S containing $(n+1)^2$ points with integer coordinates. The case $n = 6$ is shown that contains 7×7 points. Counting the points on the diagonals yields $2(1 + 2 + \cdots + 6) + 7 = 7^2$, which implies $1 + 2 + \cdots + 6 = (7^2 - 7)/2 = 6 \cdot 7/2$

11.3 Combinatorial Proofs

In the next proposition, we illustrate the double counting principle by counting the elements of the power set of a set with n elements in two different ways.

Proposition 11.17. *The sum of the binomial coefficients*
$$\sum_{k=0}^{n} \binom{n}{k} = 2^n.$$

Proof. Let S be a set with n elements. The power set $P(S)$ contains subsets of S of cardinality $0, 1, 2, \ldots, n$. The left-hand side
$$\sum_{k=0}^{n} \binom{n}{k}$$
counts *the number of elements in $P(S)$ according to their cardinality k*.

Each of the n elements of S can be either included in or excluded from a subset, so there are two choices per element. There are 2^n choices overall by the product rule. Therefore, we have a total of 2^n subsets of S, so $|P(S)| = 2^n$. □

The previous result would have been cumbersome to prove using just the definition of the binomial coefficients. In the combinatorial proof, we simply counted the number of elements of the power set of S in two different ways.

We will now give another proof of the binomial theorem. We proved it by induction, but the following proof provides additional insights.

Theorem 11.18 (Binomial Theorem). *For a nonnegative integer n, we have*
$$(x+y)^n = \sum_{k=0}^{n} \binom{n}{k} x^k y^{n-k}.$$

Proof. If we expand the left-hand side, we get a product of n terms
$$(x+y)^n = \underbrace{(x+y)}_{1} \underbrace{(x+y)}_{2} \cdots \underbrace{(x+y)}_{n}.$$

From each of these n terms, we can select either x or y. Suppose that we select x from k of these n terms (and y from the remaining $n - k$ terms), then we obtain the term $x^k y^{n-k}$. We can do this in $\binom{n}{k}$ ways. The right-hand side collects these terms for all possible k, which proves our theorem. □

Let us illustrate the idea of the proof for the case $n = 3$. Expanding
$$(x+y)^3 = (x+y)(x+y)(x+y)$$
yields
$$xxx + xxy + xyx + xyy + yxx + yxy + yyx + yyy.$$
Since the latter expression is equal to
$$x^3 + x^2y + x^2y + xy^2 + x^2y + xy^2 + xy^2 + y^3,$$

collecting terms with like x-exponents yields

$$x^3 + 3x^2y + 3xy^2 + y^3 = \binom{3}{3}x^3 + \binom{3}{2}x^2y + \binom{3}{1}xy^2 + \binom{3}{0}y^3.$$

The following important special case is obtained by setting $y = 1$ in the binomial theorem.[2]

Corollary 11.19. *For all nonnegative integers n, we have*

$$(x+1)^n = \sum_{k=0}^{n} \binom{n}{k} x^k.$$

Binomial coefficients are used in a myriad number of expressions. Let us collect some of the simplest consequences of the binomial theorem. If we substitute $x = 1$ and $y = 1$ in the binomial theorem, then we obtain yet another proof of the following fact

$$2^n = (1+1)^n = \sum_{k=0}^{n} \binom{n}{k}.$$

Substituting $x = -1$ and $y = 1$ in the binomial theorem reveals that binomial coefficients with alternating signs sum to 0, since

$$0 = (-1+1)^n = \sum_{k=0}^{n} (-1)^k \binom{n}{k}.$$

Substituting $x = 2$ and $y = 1$ in the binomial theorem yields

$$3^n = (2+1)^n = \sum_{k=0}^{n} \binom{n}{k} 2^k.$$

Proposition 11.20 (Vandermonde's Identity). *For nonnegative integers m, n, and k, we have*

$$\binom{m+n}{k} = \sum_{i=0}^{k} \binom{m}{i}\binom{n}{k-i}.$$

Proof. Let S denote a set with m elements and T a set with n elements such that $S \cap T = \emptyset$. We can choose a subset of cardinality k from the union $S \cup T$ in $\binom{n+m}{k}$ ways, since $|S \cup T| = m + n$.

On the other hand, we can choose a subset of S that has cardinality i with $0 \leq i \leq k$ in $\binom{m}{i}$ ways. By choosing $k - i$ elements from T, we can complement this to a subset of cardinality k of $S \cup T$. Since we can choose the $k - i$ elements from T in $\binom{n}{k-i}$ ways, the product principle shows that there

---

[2] This is the generating function for the binomial coefficients $\binom{n}{k}$ for fixed n that we will discuss later.

11.3 Combinatorial Proofs

are $\binom{m}{i}\binom{n}{k-i}$ subsets of cardinality k of $S \cup T$ that have i elements from S and $k - i$ elements from T. Consequently, the sum

$$\sum_{i=0}^{k} \binom{m}{i}\binom{n}{k-i}$$

counts all subsets of size k from $S \cup T$. □

Proposition 11.21 (Pascal's Identity). *The recurrence*

$$\binom{n}{k} = \binom{n-1}{k} + \binom{n-1}{k-1}$$

holds for all k in the range $1 \leq k \leq n$.

Proof. We will prove the claim by double counting. Let S denote the set $S = \{1, 2, \ldots, n\}$ of n elements. This set has $\binom{n}{k}$ subsets with k elements.

Let x denote a fixed element of S. The subsets of S with k elements either contain the element x or do not contain this element. If we include x in the subset, then there are $\binom{n-1}{k-1}$ ways to choose the remaining elements. If we do not include the element x, then there are $\binom{n-1}{k}$ ways to choose k elements from the remaining $n - 1$ elements of S.

Comparing these two ways of counting the k-subsets of S, we can conclude that

$$\binom{n}{k} = \binom{n-1}{k-1} + \binom{n-1}{k},$$

as claimed. □

Pascal's identity gives a recursive way to compute binomial coefficients. Indeed, $\binom{n}{0} = 1$ and $\binom{n}{n} = 1$ hold for all nonnegative integers n. Therefore, Pascal's identity can be used to calculate all other coefficients $\binom{n}{k}$. This recursion underlies Pascal's triangle, see Fig. 11.3.

$$
\begin{array}{ccccc}
\binom{0}{0} & & & & \\
\binom{1}{0} & \binom{1}{1} & & & \\
\binom{2}{0} & \binom{2}{1} & \binom{2}{2} & & \\
\binom{3}{0} & \binom{3}{1} & \binom{3}{2} & \binom{3}{3} & \\
\binom{4}{0} & \binom{4}{1} & \binom{4}{2} & \binom{4}{3} & \binom{4}{4} \\
\end{array}
\qquad
\begin{array}{ccccc}
1 & & & & \\
1 & 1 & & & \\
1 & 2 & 1 & & \\
1 & 3 & 3 & 1 & \\
1 & 4 & 6 & 4 & 1 \\
\end{array}
$$

Figure 11.3: Pascal's triangle. An entry $\binom{n}{k}$ can be obtained by adding the entry $\binom{n-1}{k}$ above it and the entry $\binom{n-1}{k-1}$ above it to the left. The left part of the figure shows the binomial coefficients, and the right part of the figure shows their evaluation. For example, $\binom{4}{2}$ is obtained by adding $\binom{3}{1} = 3$ and $\binom{3}{2} = 3$

Exercises

11.26. A very simple use of the bijection principle is the accurate counting of the number of elements in a finite arithmetic sequence. For instance, if we are given a set $S = \{a, a+d, a+2d, \ldots, a+(n-1)d\}$, then adding $d-a$ to each element and dividing by d yields a bijection onto the set $\{1, 2, 3, \ldots, n\}$, so $|S| = n$. For instance, if $T = \{1, 4, 7, \ldots, 100\}$ is a set whose elements form an arithmetic sequence with difference 3, then adding 2 to each element and dividing by 3 maps T bijectively onto the set $\{1, 2, 3, \ldots, 34\}$, so $|T| = 34$. Use the bijection principle to determine the number of elements for each of the following sets.
(a) $A = \{27, 28, 29, \ldots, 333\}$,
(b) $B = \{4, 6, 8, \ldots, 2802\}$,
(c) $C = \{55, 58, 61, \ldots, 421\}$,
(d) $D = \{128, 121, 114, \ldots, -61\}$.
The approach outlined earlier can help you to avoid off-by-one errors when quickly counting the number of elements of each set.

11.27. Prove the absorption property of binomial coefficients
$$\binom{n}{k} k = n \binom{n-1}{k-1}.$$

(a) using the definition of the binomial coefficients
(b) using a combinatorial proof (count the number of ways to choose a team of fixed size and their team captain).

11.28. Prove the subset of a subset identity
$$\binom{n}{m}\binom{m}{k} = \binom{n}{k}\binom{n-k}{m-k}.$$

(a) using the definition of the binomial coefficients,
(b) using a combinatorial proof.

11.29. Prove that

(a) $E(n) = \displaystyle\sum_{\substack{k=0 \\ k \text{ even}}}^{n} \binom{n}{k} = 2^{n-1}$,

(b) $O(n) = \displaystyle\sum_{\substack{k=0 \\ k \text{ odd}}}^{n} \binom{n}{k} = 2^{n-1}$.

holds for all $n \geqslant 1$.

11.30. Show that
$$\sum_{k=1}^{n} k \binom{n}{k} = n 2^{n-1}$$
holds. [Hint: Use the absorption identity.]

11.3 Combinatorial Proofs

11.31. Let n be a positive odd integer. Show that
$$\sum_{k=0}^{(n-1)/2} \binom{n}{k} = 2^{n-1}.$$

11.32. Deduce Pascal's identity
$$\binom{n}{k} = \binom{n-1}{k} + \binom{n-1}{k-1}$$
from Vandermonde's identity.

11.33. Show that
$$\binom{2n}{n} = \sum_{k=0}^{n} \binom{n}{k}^2.$$

11.34. Show that the n-th row of Pascal's triangle satisfies
$$\binom{n}{0} < \binom{n}{1} < \cdots < \binom{n}{\lfloor n/2 \rfloor} = \binom{n}{\lceil n/2 \rceil} > \cdots > \binom{n}{n-1} > \binom{n}{n}.$$

If n is even, then the two coefficients in the middle coincide. If n is odd, then the two coefficients in the middle are different but happen to have the same value. A sequence of coefficients that first increases and then decreases is called a **unimodal sequence**.

11.35. Show that
$$\binom{n}{k} \leq \frac{n^n}{(n-k)^{n-k} k^k}$$
holds for all integers k in the range $0 \leq k \leq n$.

11.36. Deduce from Stirling's approximation $n! \sim \sqrt{2\pi n}\left(\frac{n}{e}\right)^n$ the following asymptotic estimate of the central binomial coefficient
$$\binom{n}{n/2} \sim \frac{2^{n+1}}{\sqrt{2\pi n}}$$
for even n.

11.37. Let n be a nonnegative integer and m an integer in the range $0 \leq m \leq n$. Show that
$$\sum_{k=0}^{m} \binom{n}{k} \leq \left(\frac{en}{m}\right)^m.$$

[Hint: Recall that for all real numbers x, we have $\exp(x) = \sum_{k=0}^{\infty} \frac{x^k}{k!}$. Therefore, $e^m = \exp(m) \geq \sum_{k=0}^{m} \frac{m^k}{k!}$.]

11.4 Selections with Repetitions

In Sect. 11.2, we discussed the enumeration of k *distinct* elements from a set S with n elements (using k-permutations for ordered selections and k-combinations for unordered selections). In this section, we will discuss the ordered and unordered selection of k elements from S that do not need to be distinct.

Ordered Selection with Repetition. Let $K = \{1, 2, \ldots, k\}$ and S a set with n elements. An **ordered k-selection with repetition** over a set S with n elements is a function $f \colon K \to S$.

The definition is easily understood by comparing it with k-permutations. Recall that a k-permutation over S is an injective function from K to S. Since we now want to allow repetitions, we give up injectivity and consider a function from K to S.

If we fix the natural order $1 < 2 < \cdots < k$ on K, then we can identify a function $f \colon K \to S$ with the sequence

$$s = (f(1), f(2), \ldots, f(k)).$$

Since the sequence contains all function values of f on K, we can recover f from the sequence. We will use a function f and its corresponding sequence s interchangeably when studying ordered k-selections with repetitions.

Example 11.22. The functions from $K = \{1, 2\}$ to $S = \{a, b, c\}$ yield the sequences

$$(a, a),\ (a, b),\ (a, c),\ (b, a),\ (b, b),\ (b, c),\ (c, a),\ (c, b),\ (c, c).$$

A sequence (x_1, x_2) corresponds to the function $f \colon K \to S$ with $f(1) = x_1$ and $f(2) = x_2$, so functions and ordered sequences are in one-to-one correspondence.

Let $R(n, k)$ denote the number of ordered k-selections with repetition over a set with n elements. In other words, we want to count the number of functions from K to S.

Proposition 11.23. *Let K be a finite set with k elements and S a finite set with n elements. Then the number of functions from K to S is given by n^k. Thus, the number of ordered k-selections of elements from S with repetitions allowed is given by $R(n, k) = n^k$.*

Proof. For an argument $x \in K$, there are n possible images. By the product rule, we get a total of n^k different functions. □

Unordered Selection with Repetition. We will now turn to the more challenging problem of counting the number of unordered selections with k elements from a set of n elements with repetitions allowed.

11.4 Selections with Repetitions

Example 11.24. Suppose that you want to order $k = 3$ scoops of ice cream from a set $S = \{c, s, v\}$ of three different flavors, where we use c to denote chocolate, s to denote strawberry, and v to denote vanilla flavor. Then there are 10 different choices

$$\{c,c,c\}, \quad \{c,c,s\}, \quad \{c,c,v\}, \quad \{c,s,s\}, \quad \{c,s,v\},$$
$$\{c,v,v\}, \quad \{s,s,s\}, \quad \{s,s,v\}, \quad \{s,v,v\}, \quad \{v,v,v\}.$$

Here $\{c, c, v\}$ means that you get two scoops of chocolate ice cream and one scoop of vanilla. Since this is an unordered selection, $\{c, c, v\}$ is the same as $\{c, v, c\}$. Note that $\{c, c, v\}$ is not the same as $\{c, v\}$, since in the latter case you would only receive two scoops of ice cream.

In the previous example, we used the concept of a multiset. A **multiset** is a generalization of the concept of a set, where the order of the elements does not matter (as in the case of sets), but each element can occur multiple times. For instance, $M = \{a, a, b, c, c, c\}$ is a multiset that contains the element a twice, the element b once, and the element c three times. The multiset M coincides with $\{a, b, c, a, c, c\}$ but differs from $\{a, b, c\}$. The number of elements of a multiset is the sum of the multiplicities of its elements, so M contains $6 = 2 + 1 + 3$ elements.

An **unordered k-selection with repetitions** over a set S with n elements is a multiset with k elements from S. The number $M(n,k)$ of unordered k-selections with repetitions is determined in the next proposition.

Proposition 11.25. *The number $M(n,k)$ of multisets with k elements taken from a set with n elements is given by*

$$M(n,k) = \binom{n+k-1}{k}.$$

Proof. Without loss of generality, we may assume that $S = \{1, 2, \ldots, n\}$. Let A denote the set of all multisets with k elements from S, and let B denote the set of all subsets with k elements of the set $\{1, 2, \ldots, n + k - 1\}$.

A multiset $T = \{m_1, m_2, \ldots, m_k\} \in A$ can be arranged in nondecreasing order so that $m_1 \leqslant m_2 \leqslant \cdots \leqslant m_k$. Define a map $\sigma: A \to B$ by

$$\sigma(T) = \{m_1, m_2 + 1, \ldots, m_k + k - 1\}.$$

Adding $j - 1$ to m_j ensures that $m_1 < m_2 + 1 < \cdots < m_k + k - 1$, so $\sigma(T)$ is a set with k distinct elements.

The map σ is injective and surjective. Indeed, we note that its inverse function $\sigma^{-1}: B \to A$ is given by

$$\sigma^{-1}(\{n_1, n_2, \ldots, n_k\}) = \{n_1, n_2 - 1, n_3 - 2, \ldots, n_k - (k-1)\},$$

where $n_1 < n_2 < \ldots < n_k$. Therefore, $|A| = |B|$.

Since the cardinality of the set B of k element subsets of $\{1, 2, \ldots, n+k-1\}$ is given by
$$|B| = \binom{n+k-1}{k},$$
the claim follows. $\square$

Corollary 11.26. *The number of integer solutions to the equation*
$$x_1 + x_2 + \cdots + x_n = k \qquad (11.1)$$
such that $x_m \geq 0$ holds for all m in the range $1 \leq m \leq n$ is given by
$$M(n,k) = \binom{n+k-1}{k}.$$

Proof. Let $S = \{1, 2, \ldots, n\}$. Each solution to the equation $x_1 + x_2 + \cdots + x_n = k$ in the nonnegative integers specifies a multiset M with k elements that contains the element $m \in S$ exactly x_m times for $1 \leq m \leq n$.

Conversely, every multiset M with k elements from S yields a solution to the equation by setting x_m to the multiplicity of the element $m \in S$ in the multiset M.

By the bijection principle, the number of nonnegative integer solutions to the Eq. (11.1) coincides with the number $M(n,k)$ of multisets with k elements from S, and the claim follows from the previous theorem. $\square$

Permutations of Multisets. Let M be a multiset that contains element a_1 with multiplicity m_1, element a_2 with multiplicity m_2, ..., and element a_n with multiplicity m_n. Thus, counting all repetitions, the multiset contains $m = m_1 + m_2 + \cdots + m_n$ elements.

We can rearrange the elements of M in $m!$ ways, but not all permutations can be distinguished. Indeed, we can permute the elements a_k among themselves in $m_k!$ different ways, but the rearrangements are not distinguishable. The number of multiset permutations of M that can be distinguished is given by the **multinomial coefficient**
$$\binom{m}{m_1, m_2, \ldots, m_n} = \frac{m!}{m_1! m_2! \cdots m_n!}.$$

Let us have a look at an example.

Example 11.27. Suppose that we have 3 blue tokens, 2 green tokens, and 4 red tokens. We want to line up the tokens in a single row and count in how many ways we can rearrange these three different types of tokens. Abstractly, we can model the tokens as elements of the multiset $M = \{b, b, b, g, g, r, r, r, r\}$. Although there are $9! = 362880$ different ways to arrange these nine tokens in a row, we can merely distinguish
$$\binom{9}{3, 2, 4} = \frac{9!}{3! 2! 4!} = \frac{362880}{6 \times 2 \times 24} = 1260$$

11.4 Selections with Repetitions

of them. Indeed, for every color pattern in a row, there are 3!2!4! = 288 different rearrangements that look exactly the same, as there are 3! ways to reorder the blue tokens, 2! ways to reorder the green tokens, and 4! ways to reorder the red tokens.

The next example paves the way to the multinomial theorem that generalizes the binomial theorem.

Example 11.28. Consider the variables x_1, x_2, and x_3. Expanding the expression $(x_1 + x_2 + x_3)^4$ in product form yields

$$(x_1 + x_2 + x_3)(x_1 + x_2 + x_3)(x_1 + x_2 + x_3)(x_1 + x_2 + x_3).$$

If we want to fully expand this expression, then we obtain a sum of terms of the form $x_1^{n_1} x_2^{n_2} x_3^{n_3}$ such that $n_1 + n_2 + n_3 = 4$. There are $\binom{4}{n_1, n_2, n_3}$ different ways to obtain the term $x_1^{n_1} x_2^{n_2} x_3^{n_3}$ from the product of four terms. For instance, there are

$$\binom{4}{3, 1, 0} = \frac{4!}{3!1!0!} = 4$$

different ways to form the term $x_1^3 x_2$. Calculating the multinomial coefficient for each term, we obtain the expansion

$$(x_1+x_2+x_3)^4 = x_1^4 + 4x_1^3 x_2 + 4x_1^3 x_3 + 6x_1^2 x_2^2 + 6x_1^2 x_3^2 + 12x_1^2 x_2 x_3 + 4x_1 x_2^3$$
$$+ 4x_1 x_3^3 + 12 x_1 x_2 x_3^2 + 12 x_1 x_2^2 x_3 + x_2^4 + x_3^4 + 4 x_2 x_3^3 + 6 x_2^2 x_3^2 + 4 x_2^3 x_3.$$

By expanding the more general term $(x_1 + x_2 + \cdots + x_m)^n$, we can obtain the multinomial theorem.

Proposition 11.29 (Multinomial Theorem). *For positive integers n and m, and variables $x_1, x_2, \ldots, x_m$, we have*

$$(x_1 + x_2 + \cdots + x_m)^n = \sum_{n_1+n_2+\cdots+n_m=n} \binom{n}{n_1, n_2, \ldots, n_m} x_1^{n_1} x_2^{n_2} \cdots x_m^{n_m}.$$

Proof. Expand the term $(x_1 + x_2 + \cdots + x_m)^n$. The number of ways to obtain the term $x_1^{n_1} x_2^{n_2} \cdots x_m^{n_m}$ is given by $\binom{n}{n_1, n_2, \ldots, n_m}$. □

Exercises

11.38. Determine in how many different ways one can permute the following words:
(a) lillypilly
(b) spoonfeed
(c) subbookkeeper
(d) honorificabilitudinitatibus
(e) supercalifragilisticexpialidocious

11.39. Determine the coefficient of $x^2 y^3 z^4$ of $(3x - 2y + 4z)^9$.

11.40. Prove that for all positive integers n and all nonnegative integers $n_1, \ldots, n_m$ such that $n_1 + \cdots + n_m = n$, one can express the multinomial coefficient $\binom{n}{n_1, n_2, \ldots, n_m}$ as a product of binomial coefficients as follows:

$$\binom{n}{n_1, n_2, \ldots, n_m} = \binom{n}{n_1}\binom{n-n_1}{n_2}\binom{n-n_1-n_2}{n_3} \cdots \binom{n - \sum_{k=1}^{m-1} n_k}{n_m}.$$

11.41. Prove that for all positive integers n and all nonnegative integers $n_1, \ldots, n_m$ such that $n_1 + \cdots + n_m = n$, the following generalization of Pascal's identity

$$\binom{n}{n_1, n_2, \ldots, n_m} = \sum_{k=1}^{m} \binom{n-1}{n_1, \ldots, n_{k-1}, n_k - 1, n_{k+1}, \ldots, n_m}$$

holds. [Hint: Use the multinomial theorem in two different ways.]

11.42. We can use the multinomial theorem to give another proof of Fermat's Little Theorem. Recall that Fermat's Little Theorem asserts that $n^p \equiv n \pmod{p}$ holds for all integers n and all primes p.

(a) Let p be a prime and $p_1, p_2, \ldots, p_n$ nonnegative integers such that $p_1 + p_2 + \cdots + p_n = p$. When is the multinomial coefficient

$$\binom{p}{p_1, p_2, \ldots, p_n},$$

divisible by p? What is the value of this multinomial coefficient modulo p, when the multinomial coefficient is not divisible by p?

(b) Let p be a prime and n an integer. Show that

$$n^p \equiv (1 + 1 + \cdots + 1)^p \equiv (1 + 1 + \cdots + 1) \equiv n \pmod{p},$$

where each sum of 1s consists of n terms.

11.5 Set Partitions

The number of partitions of a set S with n elements into k nonempty parts is denoted by

$$\left\{ {n \atop k} \right\}.$$

Furthermore, we define

$$\left\{ {0 \atop 0} \right\} = 1 \quad \text{and} \quad \left\{ {k \atop 0} \right\} = \left\{ {0 \atop k} \right\} = 0$$

for all integers $k > 0$.

We illustrate the definition with a small example.

11.5 Set Partitions

Example 11.30. We have $\left\{{4 \atop 2}\right\} = 7$, since all the partitions of $S = \{1,2,3,4\}$ into two nonempty parts are given by

$$\{\{1\}, \{2,3,4\}\}, \quad \{\{2\}, \{1,3,4\}\}, \quad \{\{3\}, \{1,2,4\}\}, \quad \{\{4\}, \{1,2,3\}\},$$
$$\{\{1,2\}, \{3,4\}\}, \quad \{\{1,3\}, \{2,4\}\}, \quad \{\{1,4\}, \{2,3\}\}.$$

One can similarly show that $\left\{{4 \atop 1}\right\} = 1$, $\left\{{4 \atop 3}\right\} = 6$, and $\left\{{4 \atop 4}\right\} = 1$.

Example 11.31. We have $\left\{{n \atop n}\right\} = 1$ for all positive integers n, since there is only one partition of $S = \{1,2,\ldots,n\}$ with n nonempty parts, namely the partition $\{\{1\},\{2\},\ldots,\{n\}\}$ of S into singleton sets.

Proposition 11.32. *For all positive integers n and all integers k in the range $1 \leqslant k \leqslant n-1$, we have*

$$\left\{{n \atop k}\right\} = \left\{{n-1 \atop k-1}\right\} + k \left\{{n-1 \atop k}\right\}.$$

Proof. Let S be a set with n elements. We fix an element $x \in S$. The number of partitions in which $\{x\}$ is a singleton block in a partition of S with k nonempty parts is given by $\left\{{n-1 \atop k-1}\right\}$.

The number of partitions in which x is part of block with more than one element is given by $k\left\{{n-1 \atop k}\right\}$, since we can take any partition of $S\setminus\{x\}$ with k nonempty parts and add x to any of the k blocks of the partition.

Therefore, we can conclude that the number of partitions $\left\{{n \atop k}\right\}$ of a set with n elements into k nonempty parts can be also expressed as $\left\{{n-1 \atop k-1}\right\} + k\left\{{n-1 \atop k}\right\}$. $\square$

Remark 11.33. The claim of the previous proposition holds for $k = n$ as well, since $k\left\{{n-1 \atop n}\right\} = 0$. For $k > n$, the left-hand side and the right-hand side are both equal to 0. Therefore,

$$\left\{{n \atop k}\right\} = \left\{{n-1 \atop k-1}\right\} + k \left\{{n-1 \atop k}\right\}$$

holds for all positive integers k.

The number of set partitions $\left\{{n \atop k}\right\}$ of a set with n elements into k nonempty parts satisfies the same recurrence that we have encountered when studying the Stirling numbers $S(n,k)$ of the second kind. Recall that the Stirling numbers $S(n,k)$ of the second kind were defined as the connection coefficients when changing from the basis of falling factorials $x^{\underline{k}}$ to the monomial powers x^n,

$$x^n = \sum_{k=0}^{n} S(n,k) x^{\underline{k}}.$$

Let us now verify that the Stirling numbers $S(n,k)$ of the second kind coincide with the number of set partitions of a set with n elements into k nonempty parts.

Proposition 11.34. *For all nonnegative integers n and k, we have*
$$S(n,k) = \begin{Bmatrix} n \\ k \end{Bmatrix}.$$

Proof. We will prove this by induction on n. Let $P(n)$ denote that the equality
$$S(n,k) = \begin{Bmatrix} n \\ k \end{Bmatrix}$$
of the connecting coefficients $S(n,k)$ and the number of partitions of a set with n elements into k nonempty parts holds for all nonnegative integers k.
Induction Basis For $n = 0$, we have $S(0,0) = 1 = \begin{Bmatrix} 0 \\ 0 \end{Bmatrix}$ and $S(0,k) = 0 = \begin{Bmatrix} 0 \\ k \end{Bmatrix}$ for all integers $k > 0$. Thus, the claim $P(n)$ holds for $n = 0$.
Inductive Step For all positive integers n, we will show that $P(n-1)$ implies $P(n)$. Assuming that $P(n-1)$ holds means that the equality $S(n-1,k) = \begin{Bmatrix} n-1 \\ k \end{Bmatrix}$ holds for all nonnegative integers k. It follows that

$$S(n,k) = S(n-1, k-1) + k(n-1, k) = \begin{Bmatrix} n \\ k-1 \end{Bmatrix} + k\begin{Bmatrix} n-1 \\ k \end{Bmatrix} = \begin{Bmatrix} n \\ k \end{Bmatrix}$$

holds for all positive integers k. Since $S(n,0) = 0 = \begin{Bmatrix} n \\ 0 \end{Bmatrix}$ holds for $n > 0$, we can conclude that $S(n,k) = \begin{Bmatrix} n \\ k \end{Bmatrix}$ holds for all nonnegative integers k, so $P(n)$ holds.

We can conclude by induction that for all nonnegative integers n, we have $S(n,k) = \begin{Bmatrix} n \\ k \end{Bmatrix}$ for all nonnegative integers k. $\square$

Exercises

11.43. Give a simple formula for $\begin{Bmatrix} n \\ 2 \end{Bmatrix}$ and prove your result by a combinatorial argument (in other words, by counting suitable sets).

11.44. Determine the number of equivalence classes on a set with n elements.

11.45. Show that
$$\begin{Bmatrix} n \\ k \end{Bmatrix} = \sum_{i=k-1}^{n-1} \binom{n-1}{i} \begin{Bmatrix} i \\ k-1 \end{Bmatrix}.$$

[Hint: Partition the set $\{1, 2, \ldots, n\}$ into k parts by first forming the block containing the element n and then partitioning the rest into $k-1$ blocks.]

11.6 The Inclusion-Exclusion Principle

Given a finite set S and properties $P_1, \ldots, P_m$ that its elements may or may not have, it is often an interesting question to determine how many elements of S do not have any of the m properties. If we denote by S_k the subset of S comprising all elements that satisfy property P_k, that is, $S_k = \{x \in S \mid x \text{ has property } P_k\}$,

11.6 The Inclusion-Exclusion Principle

then the question can be reformulated as the task to determine the cardinality of the set

$$\left| S \setminus \bigcup_{k=1}^{m} S_k \right|.$$

We assume that we can determine the number of elements having one of the properties, two of the properties and so forth.

It is instructive to work out special cases for a small number of properties. For example, let us consider the case of $m = 2$ properties, see Fig. 11.4.

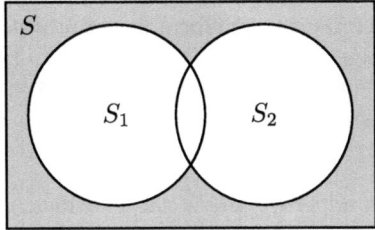

Figure 11.4: The shaded part illustrates the set $S \setminus (S_1 \cup S_2)$, whose elements we would like to count

Since $|S_1 \cup S_2| = |S_1| + |S_2| - |S_1 \cap S_2|$, we have

$$|S \setminus (S_1 \cup S_2)| = |S| - |S_1 \cup S_2| = |S| - |S_1| - |S_2| + |S_1 \cap S_2|.$$

The latter formula is known as the inclusion-exclusion formula for two properties.

Let us give some insights into the inclusion-exclusion formula. We can estimate the number of elements in $S \setminus (S_1 \cup S_2)$ by $|S|$, but this is clearly an overestimate. If we subtract from $|S|$ the number of elements in S_1 and the number of elements in S_2, then we get an underestimate, since the elements in the intersection $S_1 \cap S_2$ were counted both in S_1 and in S_2. As we have shown earlier, the inclusion-exclusion formula

$$|S \setminus (S_1 \cup S_2)| = |S| - |S_1| - |S_2| + |S_1 \cap S_2|$$

gives the correct number of elements of the set.

It might not be immediately obvious why the inclusion-exclusion formula can be attractive. Therefore, we offer a simple example that we will generalize later.

Example 11.35. Suppose that we want to determine the number of integers k in the range $1 \leqslant k \leqslant n$ for $n = 20 = 2^2 5$ that are not divisible by the prime factors 2 and 5. Let $S = \{1, 2, \ldots, 20\}$, P_1 be the property of being divisible by 2, and P_2 the property of being divisible by 5. Then

$$S_1 = \{x \in S \mid x \text{ is divisible by } 2\} = \{2x \mid 1 \leqslant x \leqslant 10\}$$

is a set with $|S_1| = 10$ elements and

$$S_2 = \{x \in S \mid x \text{ is divisible by } 5\} = \{5x \mid 1 \leq x \leq 4\}$$

is a set with $|S_2| = 4$ elements. The numbers that are divisible by both 2 and 5 are given by $|S_1 \cap S_2| = |\{10, 20\}| = 2$.

By the inclusion-exclusion formula, the number $|S \backslash (S_1 \cup S_2)|$ of elements in S that are not divisible by either 2 or 5 is given by

$$|S \backslash (S_1 \cup S_2)| = |S| - |S_1| - |S_2| + |S_1 \cap S_2| = 20 - 10 - 4 + 2 = 8.$$

The appeal of the inclusion-exclusion formula is that it is easy to count the number of elements in the sets S_1, S_2, and $S_1 \cap S_2$. Of course, we could have determined the number of elements of the set $S \backslash (S_1 \cup S_2) = \{1, 3, 7, 9, 11, 13, 17, 19\}$ by simply forming the set, but for larger n, this will not be an attractive solution.

We will now state and prove the general inclusion-exclusion principle. It will be useful to introduce some notation that will allow us to state the result in a compact form. Let I denote a subset of $\{1, 2, \ldots, m\}$. Then we denote by S_I the set

$$S_I = \begin{cases} S & \text{if } I = \emptyset, \\ S_{i_1} \cap S_{i_2} \cap \cdots \cap S_{i_k} & \text{if } I = \{i_1, i_2, \ldots, i_k\}. \end{cases}$$

Theorem 11.36 (Inclusion-Exclusion Formula). *Let S be a finite set with m pairwise different subsets $S_1, S_2, \ldots, S_m$. Then*

$$\left| S \backslash \bigcup_{k=1}^{m} S_k \right| = \sum_{I \subseteq \{1, 2, \ldots, m\}} (-1)^{|I|} |S_I|.$$

Proof. For a subset T of S, we denote by I_T the indicator function of T, meaning that $I_T(x) = 1$ for $x \in T$ and $I_T(x) = 0$ for $x \in S \backslash T$. Then

$$\left| S \backslash \bigcup_{k=1}^{m} S_k \right| = \sum_{x \in S} \prod_{k=1}^{m} (1 - I_{S_k}(x)).$$

We can expand the right-hand side in the form

$$\left| S \backslash \bigcup_{k=1}^{m} S_k \right| = \sum_{x \in S} \sum_{I \subseteq \{1, \ldots, m\}} (-1)^{|I|} \prod_{k \in I} I_{S_k}(x) = \sum_{I \subseteq \{1, \ldots, m\}} (-1)^{|I|} |S_I|,$$

where we used the fact that $\prod_{k \in I} I_{S_k}(x)$ is the characteristic function of the set $S_I = \bigcap_{k \in I} S_k$. □

Example 11.37. Let us find the number of integers in the range from 1 to 100 that are neither a perfect square nor a perfect cube nor a perfect 4th

11.6 The Inclusion-Exclusion Principle

power. Let $S = \{1, 2, \ldots, 100\}$. We denote by S_1 the subset of perfect squares of S, by S_2 the subset of perfect cubes, and by S_3 the subset of perfect fourth powers. Then $|S_1| = |\{1^2, 2^2, \ldots, 10^2\}| = 10$, $|S_2| = |\{1^3, 2^3, 3^3, 4^3\}| = 4$, and $|S_3| = \{1^4, 2^4, 3^4\}| = 3$. Furthermore,

$$|S_1 \cap S_2| = 2, \quad |S_1 \cap S_3| = 3, \quad |S_2 \cap S_3| = 1, \quad \text{and} \quad |S_1 \cap S_2 \cap S_3| = 1.$$

The inclusion-exclusion formula for $m = 3$ yields

$$\begin{aligned}
|S \setminus (S_1 \cup S_2 \cup S_3)| \\
&= |S| - |S_1| - |S_2| - |S_3| + |S_1 \cap S_2| \\
&\quad + |S_1 \cap S_3| + |S_2 \cap S_3| - |S_1 \cap S_2 \cap S_3| \\
&= 100 - 10 - 4 - 3 + 2 + 3 + 1 - 1 = 88.
\end{aligned}$$

Thus, 88 numbers from 1 to 100 are neither a perfect square nor a perfect cube nor a perfect fourth power.

It is worth noting a particular special case of the inclusion-exclusion formula.

Corollary 11.38. *Let S be a finite set and $S_1, \ldots, S_m$ be pairwise distinct subsets of S. If the cardinality C_k of the intersection $S_{i_1} \cap S_{i_2} \cap \cdots \cap S_{i_k}$ depends only on k and not on the particular indices, then the inclusion-exclusion formula simplifies to*

$$\left| S \setminus \bigcup_{k=1}^{m} S_k \right| = \sum_{k=0}^{m} (-1)^k \binom{m}{k} C_k.$$

We will now present some applications. We begin by counting the number of surjective maps from a finite set to another.

Proposition 11.39. *The number of surjective mappings from a finite set A with n elements to a finite set B with m elements is given by*

$$\sum_{k=0}^{m} (-1)^k \binom{m}{k} (m-k)^n$$

when $n \geq m$.

Proof. Let S denote the set of all maps from A to $B = \{b_1, \ldots, b_m\}$. Let S_k denote the subset of maps in S that do not have the element b_k in their range, where k is in the range $1 \leq k \leq m$. For an index set $I \subseteq \{1, 2, \ldots, m\}$ of cardinality k, we have

$$|S_I| = (m-k)^n,$$

since there are $m-k$ remaining elements from B that can be selected for each of the n arguments in the set of functions in S_I. By Corollary 11.38, the number of surjective functions is given by

$$\sum_{k=0}^{m} (-1)^k \binom{m}{k} (m-k)^n,$$

as claimed. □

An important consequence is a formula for the Stirling numbers of the second kind.

Corollary 11.40. *A Stirling number of the second kind can be expressed as follows:*
$$\left\{ {n \atop m} \right\} = \frac{1}{m!} \sum_{k=0}^{m} (-1)^k \binom{m}{k} (m-k)^n.$$

Proof. A surjective map from a set A with n elements to a set with m elements induces a partition on A with m nonempty parts via its preimages. There are $\left\{ {n \atop m} \right\}$ ways to partition A in this way, and $m!$ ways to assign the function values to the partitions, so
$$m! \left\{ {n \atop m} \right\} = \sum_{k=0}^{m} (-1)^k \binom{m}{k} (m-k)^n,$$
which implies the claim. $\square$

Suppose that an office with n office workers is organizing a Secret Santa gift exchange. So everyone is drawing a name of a coworker that she is supposed to surprise with a gift. It is arranged such that no one will be able to draw her own name. In how many different ways is it possible to give the gifts? The next proposition answers this question in a more abstract setting.

Let S be a finite set. Recall that a permutation π on S is a bijection on S. The permutation π is called a **derangement** if and only if it is fixedpoint-free, so $\pi(x) \neq x$ holds for all $x \in S$.

Proposition 11.41. *The number of derangements on a set T of n elements is given by*
$$n! \sum_{k=0}^{n} \frac{(-1)^k}{k!}.$$

Proof. Let $T = \{t_1, t_2, \ldots, t_n\}$. Let S denote the set of all permutations from the set T to itself. We denote by T_k the subset of T consisting of permutations π such that $\pi(t_k) = t_k$. For $I = \{i_1, \ldots, i_k\} \subseteq \{1, \ldots, n\}$, the set $T_I = T_{i_1} \cap \cdots \cap T_{i_k}$ has cardinality $(n-k)!$, since the permutations in T_I fix k elements and permute the remaining $n-k$ elements in an arbitrary fashion. By Corollary 11.38, the number of permutations that do not fix any elements is given by
$$\sum_{k=0}^{n} (-1)^k \binom{n}{k} (n-k)! = \sum_{k=0}^{n} (-1)^k \frac{n!}{k!} = n! \sum_{k=0}^{n} \frac{(-1)^k}{k!},$$
which proves the claim. $\square$

For a positive integer n, **Euler's totient function** $\varphi(n)$ is defined as the number of integers in the set $\{1, 2, \ldots, n\}$ that are relatively prime to n. For instance, $\varphi(4) = 2$, since 1 and 3 are relatively prime to 4.

11.6 The Inclusion-Exclusion Principle

Theorem 11.42. *Suppose that $p_1, p_2, \ldots, p_m$ are all prime factors of n. Then*

$$\varphi(n) = n \prod_{k=1}^{m} \left(1 - \frac{1}{p_k}\right).$$

Proof. We prove the claim using the inclusion-exclusion formula. Let $S = \{1, 2, \ldots, n\}$. We denote by $S_k = \{jp_k \mid 1 \leq j \leq n/p_k\}$ the subset of S consisting of the multiples of the prime p_k. For a subset $\{i_1, i_2, \ldots, i_\ell\}$ of $\{1, 2, \ldots, m\}$, the subset $S_{i_1} \cap S_{i_2} \cap \cdots \cap S_{i_\ell}$ of S consists of all multiples of $p_{i_1} p_{i_2} \cdots p_{i_\ell}$. Therefore, it is a set of cardinality $|S_{i_1} \cap S_{i_2} \cap \cdots \cap S_{i_\ell}| = n/p_{i_1} p_{i_2} \cdots p_{i_\ell}$. By the inclusion-exclusion formula,

$$\begin{aligned}
\varphi(n) &= \left| S \setminus \bigcup_{k=1}^{m} S_k \right| = \sum_{I \subseteq \{1,2,\ldots,m\}} (-1)^{|I|} |S_I| \\
&= n \sum_{I \subseteq \{1,2,\ldots,m\}} (-1)^{|I|} \prod_{k \in I} \frac{1}{p_k} \\
&= n \sum_{I \subseteq \{1,2,\ldots,m\}} \prod_{k \in I} \left(-\frac{1}{p_k}\right)
\end{aligned}$$

Since the identity

$$\prod_{k=1}^{m} (1 + x_k) = \sum_{I \subseteq \{1,2,\ldots,m\}} \prod_{k \in I} x_k$$

holds for all real numbers $x_1, \ldots, x_m$ by Exercise 11.48, substituting $x_k = -1/p_k$ allows one to deduce the claim from the previous equation for $\varphi(n)$. □

Exercises

11.46. How many bit strings of length 8 start with 1 or end with 00?

11.47. Deduce the formula for the inclusion-exclusion principle for $m = 3$,

$$\left| S \setminus \bigcup_{k=1}^{3} S_k \right| = |S| - |S_1| - |S_2| - |S_3| + |S_1 \cap S_2| + |S_1 \cap S_3| + |S_2 \cap S_3| - |S_1 \cap S_2 \cap S_3|,$$

from the inclusion-exclusion principle for $m = 2$,

$$|S \setminus (S_1 \cup S_2)| = |S| - |S_1| - |S_2| + |S_1 \cap S_2|.$$

11.48. Prove that

$$\prod_{k=1}^{m} (1 + x_k) = \sum_{I \subseteq \{1,2,\ldots,m\}} \prod_{k \in I} x_k$$

holds for all real numbers $x_1, \ldots, x_m$.

11.49. Use the inclusion-exclusion formula to find the number of integral solutions to $x_1 + x_2 + x_3 + x_4 = 25$ such that x_k is in the range $0 \leq x_k \leq 10$ for $k \in \{1, 2, 3, 4\}$.

11.50. Suppose that you are given a book with pages numbered from 1 to 500. Use the inclusion-exclusion formula to determine the number of pages that contain a 1 in the page number.

11.7 Pigeonhole Principle

Recall that the pigeonhole principle states the self-evident fact that it is impossible to place n pigeons into fewer than n pigeonholes without occupying at least one pigeonhole more than once. Despite the simplicity of this principle, some applications never fail to astound people.

Example 11.43. The number of hairs on a human head can reach up to 240,000. Since 2014, the population of the metropolitan area of Aggieland exceeds 240,000 people. Therefore, the pigeonhole principle implies that there exist now two people in Aggieland that have the same number of hairs on their head.

Example 11.44. Let n be an integer $n \geq 2$. Any set P of n positive integers contains at least two elements such that their difference is divisible by $n - 1$. Indeed, let us form the congruence classes of the n integers in P modulo $n-1$. There exist $n-1$ different congruence classes, so there must exist two elements a and b in P that are in the same equivalence class, meaning that $a \equiv b$ (mod $n - 1$). Therefore, $a - b$ is divisible by $n - 1$.

Example 11.45. Given five points on a sphere, there must be a closed hemisphere that contains four of them, see Fig. 11.5. Indeed, if we arbitrarily pick two points on the sphere, there must be at least one great circle C passing through these points (and in the case of antipodal points, there are even infinitely many). Now one of the two closed hemispheres defined by the great circle C contains at least two of the remaining three points by the pigeonhole principle, so this hemisphere with border C contains at least four points.

Theorem 11.46 (Strong Pigeonhole Principle). *Let n, k be positive integers. If n pigeons sit in k pigeonholes and $n > k$, then at least one pigeonhole must contain at least $\lceil n/k \rceil$ pigeons.*

Proof. Seeking a contradiction, let us suppose that every pigeonhole contains less than $\lceil n/k \rceil$ pigeons. Since there are at most $\lceil n/k \rceil - 1$ pigeons in a pigeonhole, the total number of pigeons is at most $k \left(\lceil \frac{n}{k} \rceil - 1 \right)$, which is bounded by

$$k \left(\left\lceil \frac{n}{k} \right\rceil - 1 \right) < k \left(\left(\frac{n}{k} + 1 \right) - 1 \right) = k \frac{n}{k} = n.$$

However, this means that there are fewer than n pigeons total, contrary to our assumption. □

11.7 Pigeonhole Principle

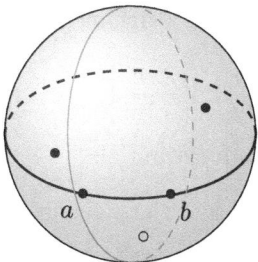

Figure 11.5: There are five points on the sphere. A great circle through the points a and b is shown, which divides the sphere into two hemispheres, an upper and a lower hemisphere. Since there are three remaining points, at least two must be in the same (upper or lower) hemisphere. The points in the upper hemisphere are depicted by a bullet • and the point in the lower hemisphere is depicted by a circle ∘

Example 11.47. Suppose that we are given a string of length 500 over the alphabet $\{a, b, \ldots, z\}$ of lower case letters. Then at least one letter must occur

$$\left\lceil \frac{500}{26} \right\rceil = \lceil 19.2307 \cdots \rceil = 20$$

or more times.

Example 11.48. Ernie is rolling four dice 100 times. Each time, he records the sum of face values of the dice. For instance, if Ernie rolls ⚃⚄⚁⚅ then he records the sum $4 + 5 + 2 + 6 = 17$. Evidently, all sums could be the same, but this is unlikely to occur. Bert wants to know how many sums are guaranteed to be the same (even in the most unfavorable situation)?

Ernie enthusiastically attempts to write down all possible sequences of 100 die rolls. Bert is flabbergasted and interrupts him. Bert points out that there are $6^4 = 1296$ different combinations of face values of four dice, and Ernie cannot possibly write down all sequences of 100 rolls, as 1296^{100} is more than 100 duocentillion. Bert points out that it will take too long to get an answer to his question.

Ernie informs Bert that he already knows the answer. He explains to Bert that the sum of the face values of the four dice ranges from 4 to 24, so there are 21 different sum values. Ernie further elaborates that one sum value must occur at least

$$\left\lceil \frac{100}{21} \right\rceil = 5$$

times by the Strong Pigeonhole Principle. Ernie chuckles and goes back to writing down die roll sequences.

Example 11.49. The Isaac Stern Auditorium in Carnegie Hall can fit 2840 people. In a sold out performance, there must be at least five people having the exact same first and last initial. Indeed, there are $26 \times 26 = 676$ different

initials. By the Strong Pigeonhole Principle, any audience attending a concert must have at least
$$\left\lceil \frac{2840}{676} \right\rceil = 5$$
people with the same first and last initials.

You can find more examples of the pigeonhole principle in Chap. 14.

Exercises

11.51. Mr. Bocks has a drawer that contains 12 pairs of socks. They all have a different color, so they do not match up well. During a power outage, he is not able to see the colors. How many socks should he grab from the drawer to ensure that he gets a pair with matching colors?

11.52. Mr. Bocks bought new socks. His drawer still contains 12 pairs of socks. However, he now has six pairs of white socks, two pairs of beige socks, one pair of red socks, and three pairs of blue socks. There is again a power outage, and he is not able to see the colors of his socks. How many socks should he grab from the drawer to ensure that he gets a pair with matching colors?

11.53. Choose seven different integers from the set $U = \{1, 2, \ldots, 12\}$. Show that at least two of the seven integers need to sum to 13.

11.54. Color each point in the Euclidean plane $\mathbf{R}^2$ either red or blue. For any coloring, show that there are two points of the same color that are at distance 1 from each other.

11.55. (a) Show that if five points are placed anywhere inside a closed equilateral triangle of side length 1, then at least two of the five points will be no farther apart than $1/2$. (b) How many points are guaranteed to be within a Euclidean distance of $1/2$ if 17 points are placed inside the triangle?

11.56. Suppose that 101 Dalmatians are locked in a square room that has an area of 400ft^2. Use the pigeonhole principle to show that at any point in time there are at least two Dalmatians that are less than 3 feet apart.

11.57. Show that every positive integer n has a nonzero integer multiple whose digits consist entirely of 0s and 1s.

11.58. Let T be a subset with $n + 1$ elements of the set $S = \{1, 2, \ldots, 2n\}$. Show that
(a) there are two elements in T that are coprime,
(b) there are two elements in T such that one divides the other.

11.59. Suppose that n people attend a party in a posh new restaurant. Show that there are two people at the party that have the same number of friends at the party. You can assume that the friends' relationship is symmetric and irreflexive and that at least two people attend the party. [Hint: People without friends are a bit suspicious and people that are friends with everyone are a bit overbearing. Investigate them!]

11.60. In this exercise, we formulate a more general version of the pigeonhole principle.
(a) Let $n_1, n_2, \ldots, n_k$ be positive integers. Show that if n pigeons with $n \geqslant n_1 + n_2 + \cdots + n_k - k + 1$ are distributed among k pigeonholes, then there exists a pigeonhole j with $1 \leqslant j \leqslant k$ that contains at least n_j pigeons.
(b) Derive the pigeonhole principle as a consequence of the generalized pigeonhole principle given in part (a).
(c) Suppose that you have a glass jar full of fasteners of three different types. You want to reattach a fence picket. You will either need two long screws of type S_1 or three medium length screws of type S_2 or four short length screws of type S_3. You want to attach the fence picket with just one type of screws for a uniform look. How many screws do you need to grab from the jar to guarantee that you have either two screws of type S_1, three screws of type S_2, or four screws of type S_3.

11.8 Notes

Counting is at the heart of enumerative combinatorics. It is discussed in almost every book on discrete mathematics or enumerative combinatorics. Aigner discusses the basics of counting in [4] and more advanced techniques in [2] and [3]. Graham, Knuth, and Patashnik [30] systematically develop many facts about counting coefficients. Jukna [43] gives more applications of counting. Moll [62] contains a wealth of material on counting coefficients. The seminal work by Stanley [72, 73] is a good place to start if you are interested in more advanced techniques of enumerative combinatorics. There are even entire books dedicated to counting, see for example Beeler [8]. The book by Benjamin and Quinn [9] is entirely dedicated to combinatorial proofs.

Chapter 12

Generating Functions

A generating function is a device somewhat similar to a bag. Instead of carrying many little objects detachedly, which could be embarrassing, we put them all in a bag, and then we have only one object to carry, the bag.

— George Pólya, *Mathematics and Plausible Reasoning, Volume 1*

Some counting problems are difficult to solve by a direct approach. For example, we often want to count the number of elements in a set that have a certain property. Euler and Laplace introduced generating functions that can often help. At first sight it might appear as a mere change of representation, but solutions using generating functions can be surprisingly effective. In the next chapter, we will see how generating functions can help solving recurrence relations.

12.1 The Basic Concept

Given a sequence of real numbers $a = (a_0, a_1, a_2, a_3, \ldots)$, the **ordinary generating function** of a, or **generating function** for short, is given by the power series

$$A(z) = \sum_{k=0}^{\infty} a_k z^k = a_0 + a_1 z + a_2 z^2 + a_3 z^3 + \cdots.$$

Informally, one can view a power series as a generalization of a polynomial with an infinite number of terms. A power series is a polynomial if all but a finite number of the coefficients a_k are equal to zero.

We will denote by $[z^k]$ the operator that extracts the k-th coefficient from the generating function,

$$[z^k]A(z) = a_k.$$

What is the benefit of using a generating function when it apparently encodes the same information as the sequence? One advantage of generating functions is that they allow us to do algebraic manipulations. Viewing the sequences in this way can give us considerable insight, as we will see. Let us have a look at an example to give you a first taste of this tool.

Example 12.1. A six-sided normal die has the face values

⚀,⚁,⚂,⚃,⚄,⚅.

So each value from 1 to 6 occurs precisely once, and all other values do not occur. The generating function of the normal die is

$$D(z) = z + z^2 + z^3 + z^4 + z^5 + z^6.$$

The exponents represent the face values. As each face value occurs only once, $[z^k]D(z) = 1$ for all values k in the range $1 \leq k \leq 6$. Suppose that we want to know in how many different ways a pair of normal dice will yield a sum of 7. Since the problem is so small, we could simply list all possible cases. So here are the only combinations of the pair of dice that sum to 7:

⚀⚅, ⚁⚄, ⚂⚃, ⚃⚂, ⚄⚁, ⚅⚀.

However, it might be easy to overlook a combination, so this approach does not scale well to larger problems. Let's see how we can use generating functions to solve the problem. For each pair (a, b) of values of the dice, the product $z^a z^b = z^{a+b}$ yields the sum $a + b$ of the values in the exponent. Therefore, if we form the product of the generating polynomials of the two dice, $D(z)^2 = D(z)D(z)$, then the coefficient of $z^{a+b} = z^7$ of the resulting polynomial counts the number of pairs that sum to 7. In this case, we have

$$D(z)^2 = z^2 + 2z^3 + 3z^4 + 4z^5 + 5z^6 + 6z^7 + 5z^8 + 4z^9 + 3z^{10} + 2z^{11} + z^{12}.$$

Therefore, we have $[z^7]D(z)^2 = 6$ different pairs of die values that sum to 7, confirming that our aforementioned enumeration of such pairs is complete.

Example 12.2. Sam has a pair of six-sided crazy dice. The first die has the following six sides

⚀,⚁,⚁,⚂,⚂,⚃,

so it has the 2 and 3 repeated twice, and the largest value is a 4. The second die has the following six sides

⚀,⚁,⚂,⚃,⚄,⚅.

The generating functions of the face values of the first and second die are given by

$$C(z) = z + 2z^2 + 2z^3 + z^4 \quad \text{and} \quad W(z) = z + z^3 + z^4 + z^5 + z^6 + z^8.$$

12.2 Operations on Generating Functions

The coefficient $[z^k]C(z)$ counts the number of times the value k occurs on the first die. For instance $[z^3]C(z) = 2$, since the value 3 occurs twice. Multiplying the two generating functions yields

$$C(z)W(z) = z^2 + 2z^3 + 3z^4 + 4z^5 + 5z^6 + 6z^7 + 5z^8 + 4z^9 + 3z^{10} + 2z^{11} + z^{12},$$

which is exactly the same distribution of values as for a pair of standard dice. In other words, even though each die is a bit wacky, their sums behave exactly in the same way as the sum of a pair of normal dice. It is quite remarkable that one can obtain a fair sum of dice values from dice that are not all fair. The generating function allowed us to quickly check this fact without much hassle.

Exercises

12.1. Show that apart from the pair of normal dice and the pair of crazy dice there cannot exist any other pair of dice such that the product of their generating functions is equal to

$$P(z) = z^2 + 2z^3 + 3z^4 + 4z^5 + 5z^6 + 6z^7 + 5z^8 + 4z^9 + 3z^{10} + 2z^{11} + z^{12},$$

assuming that (a) the face values are at least 1 and (b) each die has 6 faces. [Hint: Factor the polynomial $P(z)$ over the rational numbers and inspect all possible combinations of factors.]

12.2. Let us use the notations of Example 12.2. Find all terms z^a and z^b in $C(z)$ and $W(z)$, respectively, such that $z^{a+b} = z^6$. Explain in your own words why there must be five such pairs.

12.3. Determine the number of ways to obtain a sum of 12 when rolling 3 six-sided normal (fair) dice using generating functions.

12.4. Grandpa Dell found 20 collectible baseball cards that he wants to divide among his three grandchildren. Since the oldest grandson Albert just helped with the dishes, he wants to give him an even number of baseball cards, so that Albert receives at least eight but not more than 14 baseball cards. The younger grandchildren Bella and Clara should each receive an odd number of baseball cards; they should get at least three and at most nine cards, but not necessarily the same number of cards. In how many different ways can grandpa Dell distribute the 20 cards subject to these rules?

12.2 Operations on Generating Functions

Let $A(z)$ and $B(z)$ denote the power series

$$A(z) = \sum_{k=0}^{\infty} a_k z^k \quad \text{and} \quad B(z) = \sum_{k=0}^{\infty} b_k z^k,$$

where a_k and b_k are real numbers for all nonnegative integer indices k.

The **sum** $A(z) + B(z)$ is defined as

$$A(z) + B(z) = \sum_{k=0}^{\infty} (a_k + b_k) z^k.$$

The **product** $A(z)B(z)$ of two power series is defined as

$$A(z)B(z) = \sum_{k=0}^{\infty} \left(\sum_{\ell=0}^{k} a_{k-\ell} b_\ell \right) z^k.$$

We denote by $\mathbf{R}[[z]]$ the set of all formal power series

$$\mathbf{R}[[z]] = \left\{ \sum_{k=0}^{\infty} a_k z^k \,\bigg|\, a_k \in \mathbf{R} \text{ for all } k \in \mathbf{N}_0 \right\}.$$

A formal power series does not need to be convergent. A power series that is not convergent cannot be used as a function. However, we can still algebraically manipulate the formal power series, since for instance, sum and product are defined for all formal power series, convergent or not. We can often prove remarkable facts about the sequences using formal power series, even though they might lack any convergence properties.

We say that a power series $B(z)$ is a **multiplicative inverse** of the power series $A(z)$ if and only if their product satisfies

$$A(z)B(z) = 1,$$

so all terms in the product vanish except the constant term. If a multiplicative inverse $B(z)$ of the power series $A(z)$ in $\mathbf{R}[[z]]$ exists, then we can express $A(z)$ also in the form

$$A(z) = \frac{1}{B(z)}.$$

This form is particularly useful when $B(z)$ is a polynomial. Among other things, it will allow us to find a closed-form expression for the Fibonacci numbers.

Not all formal power series have a multiplicative inverse. The next proposition gives a simple criterion for the existence of a multiplicative inverse.

Proposition 12.3. *A formal power series $A(z) = \sum_{k=0}^{\infty} a_k z^k$ has a multiplicative inverse if and only if $a_0 \neq 0$.*

Proof. Suppose that $B(z) = \sum_{k=0}^{\infty} b_k z^k$ is a multiplicative inverse of $A(z)$. Then we must have

$$[z^0] A(z) B(z) = a_0 b_0 = 1,$$

which implies that $a_0 \neq 0$.

12.2 Operations on Generating Functions

Conversely, suppose that $A(z)$ is a formal power series with nonzero constant coefficient $[z^0]A(z) = a_0 \neq 0$. Then the putative inverse power series $B(z) = \sum_{k=0}^{\infty} b_k z^k$ must satisfy $b_0 = 1/a_0$. Let us now consider the coefficients b_k with nonzero index. Suppose that we have already determined the coefficients $b_0, \ldots, b_{k-1}$. Then we can define b_k by the expression

$$b_k = -\frac{1}{a_0} \sum_{\ell=1}^{k} a_\ell b_{k-\ell}.$$

Indeed, all coefficients in the sum on the right-hand side are defined, and $a_0 \neq 0$, so forming the quotient $-1/a_0$ is valid. If we multiply both sides by $-a_0$, then we obtain

$$-a_0 b_k = \sum_{\ell=1}^{k} a_\ell b_{k-\ell}.$$

Adding $a_0 b_k$ on both sides yields

$$\sum_{\ell=0}^{k} a_\ell b_{k-\ell} = 0,$$

as it should. It follows by induction that all coefficients b_k are defined and satisfy $A(z)B(z) = 1$. □

The next example shows how to determine the multiplicative inverse of the constant 1 sequence.

Example 12.4. Let $A(z)$ denote the generating function of the constant sequence $a = (1, 1, 1, \ldots)$, so

$$A(z) = \sum_{k=0}^{\infty} z^k.$$

Then

$$zA(z) = \sum_{k=0}^{\infty} z^{k+1} = \sum_{k=1}^{\infty} z^k = A(z) - 1.$$

It follows that

$$1 = A(z) - zA(z) = A(z)(1 - z).$$

Therefore $B(z) = (1 - z)$ is the multiplicative inverse of $A(z)$. For this reason, we often write

$$\frac{1}{1-z} = \sum_{k=0}^{\infty} z^k.$$

In this case, we even have convergence for all z satisfying $|z| < 1$, since this is a geometric series.

The next example illustrates the product of two formal power series.

Example 12.5. Let $A(z)$ denote the generating function of the constant sequence $a = (1, 1, 1, \ldots)$, so
$$A(z) = \sum_{k=0}^{\infty} z^k.$$
Then
$$A(z)A(z) = \sum_{k=0}^{\infty} \left(\sum_{\ell=0}^{k} 1 \cdot 1 \right) z^k = \sum_{k=0}^{\infty} (k+1) z^k.$$
In view of the previous example, we have
$$\frac{1}{(1-z)^2} = \sum_{k=0}^{\infty} (k+1) z^k.$$
Thus, $(1-z)^{-2}$ is the generating function of the sequence $(1, 2, 3, \ldots)$.

We can generalize the previous example as follows.

Proposition 12.6. *Let $A(z)$ denote the generating function of the sequence $(a_0, a_1, a_2, \ldots)$. Then*
$$\frac{1}{1-z} A(z)$$
is the generating function of the associated sequence of partial sums
$$(a_0, a_0 + a_1, a_0 + a_1 + a_2, a_0 + a_1 + a_2 + a_3, \ldots).$$

Proof. By definition, the product of $1/(1-z)$ and $A(z)$ yields
$$\frac{1}{1-z} A(z) = \left(\sum_{k=0}^{\infty} z^k \right) \left(\sum_{\ell=0}^{\infty} a_\ell z^\ell \right)$$
$$= \sum_{k=0}^{\infty} \left(\sum_{\ell=0}^{k} 1 \cdot a_\ell \right) z^k$$
$$= \sum_{k=0}^{\infty} \left(\sum_{\ell=0}^{k} a_\ell \right) z^k.$$

The right-hand side is the generating function of
$$(a_0, a_0 + a_1, a_0 + a_1 + a_2, a_0 + a_1 + a_2 + a_3, \ldots),$$
as claimed. $\square$

Another useful operation is the multiplication of coefficients by powers of a real number r.

Proposition 12.7. *Let $A(z)$ denote the generating function of the sequence $(a_0, a_1, a_2, a_3, \ldots)$ and let r denote a real number. Then $A(rz)$ is the generating function of the sequence $(a_0, ra_1, r^2 a_2, r^3 a_3, \ldots)$.*

12.2 Operations on Generating Functions

Proof. If we expand the formal power series $A(rz)$, then we get

$$A(rz) = \sum_{k=0}^{\infty} a_k(rz)^k = \sum_{k=0}^{\infty} a_k r^k z^k,$$

which is the generating function of $(a_0, ra_1, r^2 a_2, r^3 a_3, \ldots)$, as claimed. □

Suppose that the sequence $(a_0, a_1, a_2, \ldots)$ has the generating function $A(z)$, then the **right-shifted sequence**

$$(\underbrace{0, \ldots, 0}_{k \text{ zeros}}, a_0, a_1, a_2, \ldots)$$

has the generating function $z^k A(z)$.

Similarly, the **left-shifted sequence** $(a_k, a_{k+1}, a_{k+2}, \ldots)$ has the generating function

$$\frac{A(z) - a_{k-1} z^{k-1} - \cdots - a_1 z - a_0}{z^k}.$$

Example 12.8. The generating function of the sequence $(1, 2, 3, \ldots)$ is given by

$$A(z) = \frac{1}{(1-z)^2}.$$

The generating function of the left-shifted sequence is given by

$$\frac{A(z) - 1}{z} = \frac{1 - (1-z)^2}{z(1-z)^2} = \frac{2}{(1-z)^2} - \frac{z}{(1-z)^2}.$$

This equation asserts that the doubled sequence $(2, 4, 6, \ldots)$ minus the right-shifted sequence $(0, 1, 2, \ldots)$ is equal to the left-shifted sequence $(2, 3, 4, \ldots)$.

Given a formal power series

$$A(z) = \sum_{k=0}^{\infty} a_k z^k,$$

we can define its **formal derivative** by

$$\frac{d}{dz} A(z) = \sum_{k=1}^{\infty} k a_k z^{k-1} = \sum_{k=0}^{\infty} (k+1) a_{k+1} z^k.$$

In other words, if a sequence $(a_0, a_1, a_2, \ldots)$ has the generating function $A(z)$, then the sequence

$$b = (b_0, b_1, b_2, \ldots) = (a_1, 2a_2, 3a_3, \ldots)$$

has the generating function $\frac{d}{dz} A(z)$.

Let $A(z)$ denote the generating function of a sequence

$$(a_0, a_1, a_2, \ldots)$$

Then $\frac{d}{dz}A(z)$ is the generating function of the sequence $(a_1, 2a_2, 3a_3, \ldots)$ that is multiplied by a proportionality factor. However, this also **shifts** the sequence to the left. We can compensate for the left shift of the differential operator by multiplying it with z. Thus, $(z\frac{d}{dz})A(z)$ is the generating function of the sequence

$$(0a_0, 1a_1, 2a_2, 3a_3, \ldots)$$

that multiplies each term by its index.

Let D denote the differential operation $D = \frac{d}{dz}$.

Proposition 12.9. *Let $A(z)$ denote the generating function of the sequence $(a_0, a_1, a_2, \ldots)$. Let $P(x)$ denote a polynomial with real coefficients. Then the sequence $(P(0)a_0, P(1)a_1, P(2)a_2, \ldots)$ has the generating function $P(zD)A(z)$. In particular, $zDA(z)$ is the generating function of $(0a_0, 1a_1, 2a_2, 3a_3, \ldots)$.*

Proof. The generating function of $(a_0, a_1, a_2, \ldots)$ is given by $A(z) = \sum_{k=0}^{\infty} a_k z^k$. Then

$$zDA(z) = z\sum_{k=0}^{\infty}(k+1)a_{k+1}z^k = \sum_{k=0}^{\infty} ka_k z^k$$

is the generating function of the sequence $(na_n)_{n \geq 0}$. It follows that

$$(zD)^m A(z) = \sum_{k=0}^{\infty} k^m a_k z^k$$

is the generating function of the sequence $(n^m a_n)_{n \geq 0}$.

Consequently, if we are given a polynomial $P(x) = \sum_{k=0}^{m} c_k x^k$, then

$$P(zD)A(z) = \sum_{k=0}^{m} c_k (zD)^k A(z)$$

is the generating function of the sequence

$$\left(\sum_{k=0}^{m} c_k n^k a_n \right)_{n \geq 0} = (P(n)a_n)_{n \geq 0},$$

as claimed. $\square$

Given a formal power series

$$A(z) = \sum_{k=0}^{\infty} a_k z^k,$$

12.2 Operations on Generating Functions

we can define its **formal integral** by

$$\int_0^z A(x)\,dx = \sum_{k=0}^{\infty} \frac{a_k}{k+1} z^{k+1} = \sum_{k=1}^{\infty} \frac{a_{k-1}}{k} z^k$$

In other words, if a sequence $(a_0, a_1, a_2, \ldots)$ has the generating function $A(z)$, then the sequence

$$b = (b_0, b_1, b_2, b_3, \ldots) = \left(0, \frac{a_0}{1}, \frac{a_1}{2}, \frac{a_2}{3}, \ldots\right)$$

has the generating function $\int_0^z A(x)\,dx$. This shifts the sequence to the right and divides it by a proportionality factor such that

$$b_k = \frac{a_{k-1}}{k}$$

holds for all $k \geqslant 1$ and $b_0 = 0$.

Example 12.10. If we formally integrate the generating function

$$\frac{1}{1-z} = \sum_{k=0}^{\infty} z^k$$

of the constant one sequence $(1, 1, 1, \ldots)$, then we obtain the generating function

$$\ln\left(\frac{1}{1-z}\right) = \sum_{k=1}^{\infty} \frac{z^k}{k}.$$

of the sequence $(0, 1, 1/2, 1/3, \ldots)$.

If we are given a formal power series $A(z) = \sum_{k=0}^{\infty} a_k z^k$ and a positive integer m, then the **power** $A(z)^m$ is given by

$$\sum_{k=0}^{\infty} \left(\sum_{k_1 + k_2 + \cdots + k_m = k} a_{k_1} a_{k_2} \cdots a_{k_m} \right) z^k.$$

The inner sum ranges over all m-tuples $(k_1, k_2, \ldots, k_m)$ of nonnegative integers such that $k_1 + k_2 + \cdots + k_m = k$.

As an application, we record the following important consequence.

Proposition 12.11. *If m is a positive integer, then*

$$\frac{1}{(1-z)^m} = \sum_{k=0}^{\infty} \binom{m+k-1}{k} z^k.$$

Proof. Since

$$A(z) = \frac{1}{1-z} = \sum_{k=0}^{\infty} z^k,$$

is the generating function of the constant one sequence, its m-th power is given by

$$A(z)^m = \sum_{k=0}^{\infty} \left(\sum_{k_1+k_2+\cdots+k_m=k} a_{k_1} a_{k_2} \cdots a_{k_m} \right) z^k$$

$$= \sum_{k=0}^{\infty} \left(\sum_{k_1+k_2+\cdots+k_m=k} 1 \right) z^k$$

By Corollary 11.26, the number of nonnegative integer solutions to the equation $k_1 + k_2 + \cdots + k_m = k$ is given by $\binom{m+k-1}{k}$. Therefore, we can conclude that

$$A(z)^m = \sum_{k=0}^{\infty} \binom{m+k-1}{k} z^k,$$

as claimed. □

Exercises

12.5. Let $A(z) = 1/(1-z) = \sum_{k=0}^{\infty} z^k$ be the generating function of the constant one sequence $(1, 1, 1, \ldots)$. Then $A(z)^2$ is the generating function of the sequence $(1, 2, 3, 4, \ldots)$. Determine the power series of $A(z)^3$ using the product of $A(z)^2$ and $A(z)$. The coefficients of the formal power series of $A(z)^3 = (1-z)^{-3}$ should look familiar. Identify them.

12.6. Let $A(z)$ be the generating function of the sequence $(a_0, a_1, a_2, a_3, \ldots)$. Determine the generating functions of the sequences

$$(a_0, \underbrace{0, \ldots, 0}_{m \text{ zeros}}, a_1, \underbrace{0, \ldots, 0}_{m \text{ zeros}}, a_2, \underbrace{0, \ldots, 0}_{m \text{ zeros}}, a_3 \ldots)$$

and

$$(a_0, \underbrace{0, \ldots, 0}_{m \text{ zeros}}, ra_1, \underbrace{0, \ldots, 0}_{m \text{ zeros}}, r^2 a_2, \underbrace{0, \ldots, 0}_{m \text{ zeros}}, r^3 a_3 \ldots),$$

where r is a real number.

12.7. Determine the generating function of the sequence

$$(1, 0, 1, 0, 1, 0, \cdots)$$

in closed form (that is, as a rational function) and find its multiplicative inverse.

12.8. Determine the generating function of the sequence

$$(1, -1, 1, -1, 1, -1, \cdots)$$

in closed form (that is, the generating function should be given as a rational function).

12.3 Elementary Generating Functions

12.9. Let $A(z) = \sum_{k=0}^{\infty} a_k z^k$ denote the generating function of the sequence $(a_0, a_1, a_2, \ldots)$. Determine the generating function of the sequences

$$(a_0, 0, a_2, 0, a_4, 0, \ldots) \quad \text{and} \quad (0, a_1, 0, a_3, 0, a_5, \ldots)$$

as a linear combination of terms of the form $A(rz)$, where r is a real number.

12.10. Suppose that $A(z)$ denotes the generating function of the sequence $(a_0, a_1, a_2, \ldots)$. Determine the generating function of the sequence

$$(a_0, a_1 - a_0, a_2 - a_1, a_3 - a_2, \ldots).$$

12.3 Elementary Generating Functions

We will need a little dictionary that collects the generating functions for a few well-known sequences. We have already derived some, so it will be good to collect them in one place for better reference.

Example 12.12. We have seen that the constant sequence $(1, 1, 1, 1, \ldots)$ has the generating function

$$\frac{1}{1-z} = \sum_{k=0}^{\infty} z^k.$$

The left-hand side can be interpreted as a convenient shorthand.

Example 12.13. The sequence $(0, 1, 2, 3, 4, \ldots)$ has the generating function

$$\frac{z}{(1-z)^2} = z \sum_{k=0}^{\infty} (k+1) z^k = \sum_{k=0}^{\infty} k z^k.$$

We obtained this result from right-shifting the sequence $(1, 2, 3, 4, \ldots)$ that has the generating function $1/(1-z)^2$, as we have seen in the previous section.

Example 12.14. The sequence $s = (0, 1, 4, 9, 16, \ldots)$ of squares has the generating function

$$\frac{z(1+z)}{(1-z)^3} = \sum_{k=0}^{\infty} k^2 z^k.$$

Indeed, the sequence $(0, 1, 2, 3, 4, \ldots)$ has the generating function $z/(1-z)^2$. Taking the formal derivative yields the sequence $(1, 4, 9, 16, \ldots)$ with generating function

$$\frac{d}{dz} \frac{z}{(1-z)^2} = \frac{1+z}{(1-z)^3}.$$

Right-shifting yields the generating function of the sequence s.

Example 12.15. The sequence $(0, \frac{1}{1}, \frac{1}{2}, \frac{1}{3}, \cdots)$ of reciprocals of n has generating function

$$\ln\left(\frac{1}{1-z}\right) = \sum_{k=1}^{\infty} \frac{z^k}{k}.$$

Example 12.16. Let n be a nonnegative integer. The binomial sequence $s = \left(\binom{n}{0}, \binom{n}{1}, \binom{n}{2}, \ldots\right)$ has the generating function

$$S(z) = \sum_{k=0}^{\infty} \binom{n}{k} z^k.$$

Since the binomial coefficient $\binom{n}{k} = 0$ when k exceeds n, the power series $S(z)$ is equal to the polynomial

$$S(z) = \sum_{k=0}^{n} \binom{n}{k} z^k.$$

By the binomial theorem, we have

$$S(z) = (1+z)^n.$$

This fact is particularly useful. Many binomial coefficient identities can be derived using the fact that $(1+z)^n$ is the generating function of the binomial coefficient sequence.

There exists a useful generalization of the previous example to noninteger exponents n.

Example 12.17. Let x be a real number. The function $f(z) = (1+z)^x$ can be differentiated arbitrarily often. The first derivative of $f(z)$ is given by $f^{(1)}(z) = x(1+z)^{x-1}$, the second derivative by $f^{(2)}(z) = x(x-1)(1+z)^{x-2}$, and in general, the k-th derivative is given by $f^{(k)}(z) = x^{\underline{k}}(1+z)^{x-k}$. Therefore, the Taylor series of $f(z)$ about $z=0$ is given by

$$(1+z)^x = \sum_{k=0}^{\infty} \frac{f^{(k)}(0)}{k!} z^k = \sum_{k=0}^{\infty} \frac{x^{\underline{k}}}{k!} z^k.$$

The falling power of x divided by $k!$ equals the generalized binomial coefficient $\frac{x^{\underline{k}}}{k!} = \binom{x}{k}$. Therefore, we have

$$(1+z)^x = \sum_{k=0}^{\infty} \binom{x}{k} z^k.$$

This **binomial series** has in general infinitely many nonzero terms, but it converges for all z satisfying $|z| < 1$. The binomial series allows one to derive the formal power series for expressions such as $\sqrt{1+z} = (1+z)^{1/2}$.

We can use results from calculus to expand our repertoire of generating functions.

Example 12.18. The sequence $(\frac{1}{0!}, \frac{1}{1!}, \frac{1}{2!}, \frac{1}{3!}, \frac{1}{4!}, \cdots)$ of reciprocals of the factorials has the **exponential function**

$$\exp(x) = \sum_{k=0}^{\infty} \frac{x^k}{k!}$$

as a generating function.

12.3 Elementary Generating Functions

Example 12.19. The sequence $(1, 0, -\frac{1}{2!}, 0, \frac{1}{4!}, 0, \ldots)$ has the **cosine function**

$$\cos(x) = 1 - \frac{x^2}{2!} + \frac{x^4}{4!} - \frac{x^6}{6!} + \frac{x^8}{8!} + \cdots = \sum_{k=0}^{\infty} (-1)^k \frac{x^{2k}}{(2k)!}$$

as a generating function.

Example 12.20. The sequence $(0, 1, 0, -\frac{1}{3!}, 0, \frac{1}{5!}, 0, \ldots)$ has the **sine function**

$$\sin(x) = x - \frac{x^3}{3!} + \frac{x^5}{5!} - \frac{x^7}{7!} + \frac{x^9}{9!} + \cdots = \sum_{k=0}^{\infty} (-1)^k \frac{x^{2k+1}}{(2k+1)!}$$

as a generating function.

Example 12.21. The sequence $(0, 1, 0, \frac{1}{3}, 0, \frac{2}{15}, 0, \frac{17}{315}, 0, \frac{62}{2835}, \ldots)$ has the generating function

$$\tan(x) = x + \frac{1}{3}x^3 + \frac{2}{15}x^5 + \frac{17}{315}x^7 + \cdots$$

Exercises

12.11. The generating function of the sequence $(1, 2, 4, 8, 16, \ldots)$ of powers of two is given by

$$A(z) = \sum_{k=0}^{\infty} 2^k z^k.$$

Find a closed form of the generating function $A(z)$ that does not use a power series.

12.12. Let a and $d > 0$ be real numbers. Find a closed form of the generating function of the arithmetic progression

$$(a, a+d, a+2d, a+3d, \ldots).$$

The closed form is an expression for the generating function that does not use a power series.

12.13. Let a and $r \neq 0$ be real numbers. Find a closed form of the generating function of the geometric progression

$$(a, ar, ar^2, ar^3, \ldots).$$

The closed form is an expression for the generating function that does not use a power series.

12.14. Use the binomial series to prove that the generating function of the central binomial coefficients $\binom{2n}{n}$ is given by

$$\frac{1}{\sqrt{1+4z}} = \sum_{k=0}^{\infty} \binom{2k}{k} z^k.$$

12.15. Let n be a nonnegative integer. Use the generating function of the binomial sequence
$$\left(\binom{n}{0}, \binom{n}{1}, \binom{n}{2}, \ldots, \binom{n}{n}\right)$$
to prove that
$$\sum_{k=0}^{n} \binom{n}{k} = 2^n.$$

12.16. Let n be a nonnegative integer. Use the generating function of the binomial sequence
$$\left(\binom{n}{0}, \binom{n}{1}, \binom{n}{2}, \ldots, \binom{n}{n}\right)$$
to prove that
$$\sum_{k=0}^{n} (-1)^k \binom{n}{k} = 0.$$

12.17. Let n be a nonnegative integer. Use the generating function of the binomial sequence
$$\left(\binom{n}{0}, \binom{n}{1}, \binom{n}{2}, \ldots, \binom{n}{n}\right)$$
to prove that
$$\sum_{k=0}^{n} 2^k \binom{n}{k} = 3^n.$$

12.18. Deduce Vandermonde's identity
$$\binom{m+n}{k} = \sum_{i=0}^{k} \binom{m}{i} \binom{n}{k-i}$$
from $[z^k](1+z)^{m+n} = [z^k](1+z)^m(1+z)^n$.

12.19. Show that the sum of the Fibonacci numbers is given by
$$f_0 + f_1 + f_2 + \cdots + f_n = f_{n+2} - 1$$
using generating functions. You can use the fact that the Fibonacci numbers have the generating function
$$\sum_{k=0}^{\infty} f_k z^k = \frac{z}{1 - z - z^2},$$
as we will show in the next chapter.

12.20. Use generating functions to show that the sum of the first n Harmonic numbers is given by
$$H_1 + H_2 + \cdots + H_n = (n+1)H_n - n.$$

12.4 Giving Change

Let us conclude this chapter with a classical application of generating functions. Suppose that you want to determine in how many different ways one can give change for an amount of c cents using pennies, nickels, dimes, and quarters. In other words, we need to find the number of nonnegative integer solutions to the equation

$$c = p + 5n + 10d + 25q,$$

where p denote the number of pennies, n the number of nickels, d the number of dimes, and q the number of quarters. Counting the number of such solutions can be a bit tedious if we approach it directly.

Let us reformulate the problem using generating functions. The generating function listing the different possibilities for the number of pennies is given by

$$P(z) = \frac{1}{1-z} = \sum_{k=0}^{\infty} z^k.$$

The generating functions for the number of nickels, dimes, and quarters are respectively given by

$$N(z) = \frac{1}{1-z^5}, \quad D(z) = \frac{1}{1-z^{10}}, \quad Q(z) = \frac{1}{1-z^{25}}.$$

In principle, the answer to the counting problem is

$$[z^c] P(z) N(z) D(z) Q(z).$$

But how can we extract the coefficient of z^c in

$$\frac{1}{1-z} \frac{1}{1-z^5} \frac{1}{1-z^{10}} \frac{1}{1-z^{25}}.$$

We do not want to expand the terms directly, as this leads to a mess when c is large. Instead, we follow a slightly more principled approach and solve simpler problems by restricting the available denominations.

If we only have pennies available, then the problem is very simple. The number of ways to give change for the amount of c cents when only pennies are available is given by

$$[z^c] P(z) = [z^c] \frac{1}{1-z} = [z^c] \sum_{k=0}^{\infty} z^k = 1.$$

This is obvious, since there are no other options than giving c pennies.

If we have pennies and nickels available, then the problem becomes a bit more interesting. Let us write the product $P(z)N(z)$ in the form

$$P(z)N(z) = \frac{1}{1-z} \frac{1}{1-z^5} = \sum_{k=0}^{\infty} n_k z^k.$$

Then

$$P(z) = (1 - z^5)P(z)N(z) = (1 - z^5)\sum_{k=0}^{\infty} n_k z^k.$$

Applying $[z^k]$ to both sides yields $1 = n_k - n_{k-5}$ or $n_k = n_{k-5} + 1$ when $k \geq 5$, and $n_k = 1$ when k is in the range $0 \leq k < 5$. Therefore, $n_k = \lfloor k/5 \rfloor + 1$ for all $k \geq 0$.

Example 12.22. The number of ways to give change for $c = 11$ cents using pennies and nickels is given by $\lfloor 11/5 \rfloor + 1 = 3$. This makes sense, as you have three choices: either use two nickels, merely one nickel, or no nickels at all. Once you have decided which of the three options you would like to choose, then there is no other option than giving the remainder in pennies, and there is just one way to do that. By the multiplication principle, there are 3×1 solutions. The solutions are explicitly given by (a) ⑤,⑤,①, (b) ⑤,①,①,①,①,①,①, and (c) ①,①,①,①,①,①,①,①,①,①,①.

If we have pennies, nickels, and dimes available, then the solutions become quite a bit more varied, but our approach remains the same. We simply try to reduce the problem to smaller cases. We write the product $P(z)N(z)D(z)$ in the form

$$P(z)N(z)D(z) = \frac{1}{1-z}\frac{1}{1-z^5}\frac{1}{1-z^{10}} = \sum_{k=0}^{\infty} d_k z^k.$$

Thus, the coefficient d_k denotes the number of ways to give change to k cents using pennies, nickels, and dimes. Multiplying by $1 - z^{10}$ yields

$$P(z)N(z) = (1 - z^{10})\sum_{k=0}^{\infty} d_k z^k.$$

Applying $[z^k]$ to both sides yields $n_k = d_k - d_{k-10}$ or $d_k = d_{k-10} + n_k$ when $k \geq 10$, and $d_k = n_k$ when k is in the range $0 \leq k < 10$.

Finally, if we have pennies, nickels, dimes, and quarters available, then the number of ways q_k to give change for k cents has the generating function

$$P(z)N(z)D(z)Q(z) = \sum_{k=0}^{\infty} q_k z^k$$

By now it should be entirely routine. We multiply by $1 - z^{25}$ to obtain

$$P(z)N(z)D(z) = (1 - z^{25})\sum_{k=0}^{\infty} q_k z^k.$$

Applying $[z^k]$ to both sides yields $d_k = q_k - q_{k-25}$ or $q_k = d_k + q_{k-25}$ when $k \geq 25$, and $q_k = d_k$ when $0 \leq k < 25$.

12.4 Giving Change

Example 12.23. Suppose that we want to know the number of ways to give change for 75 cents using pennies, nickels, dimes, and quarters. In other words, we would like to calculate q_{75}. We have

$$q_{75} = d_{75} + q_{50} = d_{75} + d_{50} + q_{25} = d_{75} + d_{50} + d_{25} + d_0.$$

We can expand the d_k's in terms of n_k terms

$$d_{75} = n_{75} + n_{65} + n_{55} + n_{45} + n_{35} + n_{25} + n_{15} + n_5$$
$$d_{50} = n_{50} + n_{40} + n_{30} + n_{20} + n_{10} + n_0$$
$$d_{25} = n_{25} + n_{15} + n_5$$
$$d_0 = n_0$$

Since $n_k = \lfloor k/5 \rfloor + 1$, we get

$$d_{75} = 16 + 14 + 12 + 10 + 8 + 6 + 4 + 2 = 72$$
$$d_{50} = 11 + 9 + 7 + 5 + 3 + 1 = 36$$
$$d_{25} = 6 + 4 + 2 = 12$$
$$d_0 = 1$$

Therefore, we can conclude that there are

$$q_{75} = d_{75} + d_{50} + d_{25} + d_0 = 72 + 36 + 12 + 1 = 121$$

ways to make change for 75 cents using pennies, nickels, dimes, and quarters.

Exercises

12.21. In how many different ways can you make change for 89 cents using pennies, nickels, dimes, and quarters?

12.22. Let q_k denote the number of ways to give change for k cents using pennies, nickels, dimes, and quarters. Find a structure in the generating function of (q_k) which implies that if k is a nonnegative integer such that $k \equiv 0 \pmod 5$, then $q_k = q_{k+1} = q_{k+2} = q_{k+3} = q_{k+4}$.

12.23. The country of Binoria uses 1, 2, 4, and 8 cent coins. In how many different ways can you give change for 75 cents using the Binorian denominations?

Chapter 13

Recurrence Relations

> *Running away, then returning. My present was feeding on my past, and my future was waiting for the recursive loop to complete.*
>
> — Steve Saroff, *Paper Targets*
>
> *I wish my wish would not be granted!*
>
> — Douglas Hofstadter, *Gödel, Escher, Bach*

In this chapter, we show how generating functions can be used to solve recurrence relations. We first introduce the basic terminology and then give an example that illustrates the method for a particularly simple recurrence relation. We obtain a closed form for the coefficients of the Fibonacci sequence. Before generalizing this result, we review the partial fraction decomposition. We then show how to find closed-form solutions for linear homogeneous recurrence relations. We illustrate how generating function can even help when the recurrence relations are nonlinear using the recurrence for the Catalan numbers as an example.

13.1 Recurrence Relations

A sequence satisfies a recursive formula if we can describe the next term of a sequence using preceding terms of the sequence. Familiar examples are the interest earned on a bank account, modeling the growth of a population, and estimating the running time of a recursive algorithm.

In many applications, it is not difficult to find a recursive formula that determines a sequence. For instance, for all positive integers n, let p_n denote the number of ways to write the number $n+2$ as an ordered sum in which each term is either 2 or 3. By an ordered sum, we mean that we will count $2+3$ and $3+2$ separately. For example, $p_6 = 4$, since we can express $6+2=8$ in four different ways as a sum of 2 or 3, namely

$$8 = 2+2+2+2 = 2+3+3 = 3+2+3 = 3+3+2.$$

In the same vein, we get

$$\begin{aligned}
p_0 &= 1, \text{ since } 2 = 2, \\
p_1 &= 1, \text{ since } 3 = 3, \\
p_2 &= 1, \text{ since } 4 = 2 + 2, \\
p_3 &= 2, \text{ since } 5 = 2 + 3 = 3 + 2, \\
p_4 &= 2, \text{ since } 6 = 2 + 2 + 2 = 3 + 3, \\
p_5 &= 3, \text{ since } 7 = 2 + 2 + 3 = 2 + 3 + 2 = 3 + 2 + 2.
\end{aligned}$$

We could continue to list subsequent terms, but enumerating all choices gets tedious for larger n. There is an easier way to calculate the numbers p_n. When calculating p_n, we can distinguish the number of sums that start with 2 and the number of sums that start with 3. If the first term is 2, then there are p_{n-2} ways to write the remaining terms, and if the first term is 3, then there are p_{n-3} ways to write the remaining terms. Thus, we have

$$p_n = p_{n-2} + p_{n-3}$$

for all integers $n > 2$. Using the initial terms, $p_0 = p_1 = p_2 = 1$ and this recurrence relation, we can quickly calculate the first few terms

$$(p_0, p_1, p_2, p_3, p_4, \ldots) = (1, 1, 1, 2, 2, 3, 4, 5, 7, 9, 12, 16, \ldots).$$

This sequence is called the **Padovan sequence**, which was named after the architect Richard Padovan, who popularized it in works about mathematical properties of architectural proportions.

The recurrence relation is good for computations as long as n does not become too large. If you would like to know more about the behavior of p_n for large n, then a formula for p_n that does not depend on previous terms would be preferable. In this chapter, we will see how to derive such closed-form expressions for certain recurrence relations. Before we get into this topic, we need to settle some terminology.

Let $a = (a_0, a_1, a_2, \ldots)$ be a sequence of real numbers. We say that the sequence satisfies a **recurrence relation** of finite history if and only if there exists a nonnegative integer d and a function $f \colon \mathbf{R}^d \to \mathbf{R}$ such that

$$a_n = f(a_{n-1}, \ldots, a_{n-d})$$

holds for all $n \geq d$. We call d the **degree** of the recurrence relation.

We say that a satisfies a **linear recurrence relation** of order d with constant coefficients if and only if there exist real numbers $c_1, c_2, \ldots, c_d$ with $c_d \neq 0$ and a function $g \colon \mathbf{N}_0 \to \mathbf{R}$ such that

$$a_n + c_1 a_{n-1} + c_2 a_{n-2} + \cdots + c_d a_{n-d} = g(n)$$

holds for all $n \geq d$. The recurrence is called **homogeneous** if and only if $g(n) = 0$ holds for all $n \geq d$. If $g(n)$ is not identically 0, then the linear recurrence is called **inhomogeneous**.

In general, an infinite number of sequences satisfy a homogeneous linear recurrence relation. One needs to specify the initial conditions $a_0, \ldots, a_{d-1}$ and the recurrence relation to specify the sequence a.

13.1 Recurrence Relations

Example 13.1. The Fibonacci sequence satisfies the recurrence relation
$$f_n - f_{n-1} - f_{n-2} = 0$$
for all $n \geqslant 2$ and has the initial conditions $f_0 = 0$ and $f_1 = 1$. This is a homogeneous linear recurrence relation of degree 2.

Example 13.2. The sequence of squares satisfies the recurrence relation
$$h_n - h_{n-1} = 2n - 1$$
for all $n \geqslant 1$. This is an inhomogeneous linear recurrence relation of degree 1.

The goal of this chapter is to show how generating functions can help to find closed-form solutions to recurrence relations. By a closed form, we mean an expression that explicitly gives the value of the coefficients of the sequence. For instance, the recurrence given in the previous example has the closed-form solution
$$h_n = n^2$$
for all positive integers n. The recurrence relation might be a perfectly reasonable way to calculate the value of h_n, but a closed-form solution might give additional information such as the asymptotic growth of the coefficients h_n.

Exercises

13.1. The **Tribonacci numbers** are given by $t_0 = 0$, $t_1 = 0$, $t_2 = 1$, and the recurrence relation
$$t_n = t_{n-1} + t_{n-2} + t_{n-3}$$
for all integers $n \geqslant 3$. Find the first 21 terms $(t_0, t_1, t_2, t_3, \ldots, t_{20})$ of the Tribonacci sequence.

13.2. Write a program that on input of an integer n calculate the n-th Tribonacci number t_n. Use this program to find t_{302}.

13.3. Give a recursive formulation of the sequence $(m_k)_{k \geqslant 0}$, where $m_k = \lfloor k/3 \rfloor$. Give the initial condition and the recurrence relation.

13.4. Find a recursive formulation of the sequence $(p_k)_{k \geqslant 0}$ of the powers of 2,
$$(p_0, p_1, p_2, p_3, p_4, \ldots) = (1, 2, 4, 8, 16, \ldots).$$
Give the initial condition and the recurrence relation.

13.5. The **centered hexagonal number** is a figurate number h_n that counts a hexagon and its neighbors on a hexagonal grid as follows. The number $h_1 = 1$ counts a single central hexagon. The number $h_2 = 1 + 6 = 7$ counts the central hexagon and its six neighbors on the hexagonal grid. The number $h_3 = 19$ counts the central hexagon, its six direct neighbors, and the neighbors of those neighbors. The next number simply includes another layer of neighbors. Figure 13.1 illustrates the counting numbers $h_1, h_2, h_3,$ and h_4.

We also define $h_0 = 0$. Give the initial conditions and recurrence relation for the sequence of central hexagonal counting numbers $(h_n)_{n \geqslant 0}$.

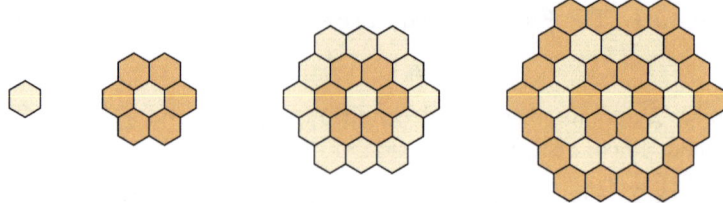

Figure 13.1: The counting number h_n counts how many fields are occupied by a discrete hexagon on the hexagonal grid that is centered about a single hexagon and has side length n. The aforementioned four figures illustrate from left to right the discrete hexagons that are associated with the central hexagonal counting numbers $h_1 = 1$, $h_2 = 7$, $h_3 = 19$, and $h_4 = 37$.

13.6. There is a simple non-recursive expression for the centered hexagonal counting numbers h_n. This exercise guides you to discover and prove it yourself.
(a) Form the sequence of the partial sums of the centered hexagonal numbers and numerically evaluate the first 10 terms
$$(h_0, h_0 + h_1, h_0 + h_1 + h_2, \ldots, h_0 + h_1 + \ldots + h_9).$$
(b) Guess a simple formula for the n-th partial sum s_n of centered hexagonal numbers,
$$s_n = \sum_{k=0}^{n} h_k,$$
based on your observations in part (a).
(c) Express h_n for $n \geqslant 1$ as a difference of two successive terms of the sequence given in parts (a) and (b).
(d) Prove the formula that you have found for h_n in part (c) by strong induction.

13.7. Find a recursive formulation of the sequence $(n!!)_{n \geqslant 0}$ of double factorials. The double factorial is defined as
$$n!! = \prod_{k=0}^{\lceil \frac{n}{2} \rceil - 1} (n - 2k),$$
and the empty product in particular as $0!! = 1$. Give the initial conditions and the recurrence relation.

13.8. In general, there are many different recursive formulations for a given sequence. For example, prove that the sequence $(q_n)_{n \geqslant 0}$ satisfying the recurrence relation $q_n = q_{n-1} - q_{n-5}$ for integers $n \geqslant 5$ and the initial conditions $q_0 = 1$, $q_1 = 1$, $q_2 = 1$, $q_3 = 2$, and $q_4 = 2$ coincides with the Padovan sequence $(p_n)_{n \geqslant 0}$ satisfying the recurrence relation $p_n = p_{n-2} + p_{n-3}$ for $n \geqslant 3$ and the initial conditions $p_0 = 1$, $p_1 = 1$, and $p_2 = 1$. In other words, the Padovan sequence $(p_n)_{n \geqslant 0}$ satisfies the same recurrence equation as the sequence $(q_n)_{n \geqslant 0}$.

13.9. The Fibonacci sequence $(f_n)_{n \geq 0}$ satisfies the recurrence relation $f_n - f_{n-1} - f_{n-2} = 0$ for $n \geq 2$ and the initial conditions $f_0 = 0$ and $f_1 = 1$. Form the quotients $r_n = f_{n+1}/f_n$ of subsequent Fibonacci numbers.
(a) Find the numerical values of the first 10 well-defined quotients $r_1, r_2, \ldots, r_{10}$, rounded to 3 digits.
(b) Prove that the quotients satisfy $r_n = 1 + 1/r_{n-1}$.
(c) Let $\phi = (1+\sqrt{5})/2 \approx 1.618$ denote the golden ratio. Show that $\phi = 1+1/\phi$.
(d) Use parts (b) and (c) to show that $|r_n - \phi| \leq \frac{1}{\phi}|r_{n-1} - \phi|$.
(e) Use part (d) to show that $\lim_{n \to \infty} |r_n - \phi| = 0$ and deduce that

$$\lim_{n \to \infty} \frac{f_{n+1}}{f_n} = \lim_{n \to \infty} r_n = \phi.$$

13.2 A Motivating Example

The recursive formulation of a sequence can be convenient when calculating a few terms, but the dependency of a term on previous terms can be a nuisance when analyzing the behavior of the sequence. Therefore, we often strive to find a non-recursive formulation for the value of the coefficients of a sequence given by a recurrence. The problem is that guessing such a formula can be difficult. Generating functions are often a good tool to find a non-recursive formulation of the sequence coefficients. In this section, we illustrate how generating functions can help with this task.

Suppose that you are given a sequence $(h_0, h_1, h_2, \ldots)$ such that its coefficients are defined by the recurrence relation

$$h_n = \begin{cases} 0 & \text{when } n = 0, \\ h_{n-1} + 2n - 1 & \text{when } n \geq 1. \end{cases}$$

You will hardly need any special methods to solve such a simple recurrence, especially after reading the previous section. However, let us pretend for the moment that we do not know the answer. Our goal is to find a closed-form solution using the tool of generating functions.

The ordinary generating function of the sequence $(h_0, h_1, h_2, \ldots)$ is given by the formal power series

$$H(z) = \sum_{k=0}^{\infty} h_k z^k.$$

Using the definition of the coefficients h_n, we can rewrite this formal power series in the form

$$H(z) = h_0 + \sum_{k=1}^{\infty} (h_{k-1} + 2k - 1) z^k.$$

Keeping in mind that $h_0 = 0$, the distributive law yields

$$H(z) = \sum_{k=1}^{\infty} h_{k-1} z^k + \sum_{k=1}^{\infty} 2k z^k + \sum_{k=1}^{\infty} (-1) z^k. \tag{13.1}$$

The three sums of the right-hand side are all simple variations of generating functions that we are familiar with. The first sum is a shifted version of $H(z)$, namely

$$\sum_{k=1}^{\infty} h_{k-1} z^k = \sum_{k=0}^{\infty} h_k z^{k+1} = zH(z).$$

The second sum is a scaled version of the generating function of the sequence $(0, 1, 2, 3, 4, \ldots)$ from Example 12.13,

$$\sum_{k=1}^{\infty} 2k z^k = 2 \sum_{k=0}^{\infty} k z^k = \frac{2z}{(1-z)^2}.$$

The last sum is the negation of the generating function of the constant one sequence from Example 12.12 shifted to the right,

$$\sum_{k=1}^{\infty} (-1) z^k = -\frac{z}{1-z}.$$

If we substitute these results back into the Eq. (13.1), then we get

$$H(z) = zH(z) + \frac{2z}{(1-z)^2} - \frac{z}{1-z}.$$

Subtracting $zH(z)$ from both sides yields

$$H(z) - zH(z) = (1-z)H(z) = \frac{2z}{(1-z)^2} - \frac{z}{1-z}.$$

Dividing by $1-z$ and bringing to a common denominator gives us

$$H(z) = \frac{2z - z(1-z)}{(1-z)^3} = \frac{z(z+1)}{(1-z)^3}.$$

We recognize this generating function from Example 12.14. It is simply the generating function of the sequence of squares $(0, 1, 4, 9, 16, \ldots)$. In other words, we have

$$H(z) = \sum_{k=0}^{\infty} k^2 z^k.$$

Comparing coefficients, we can conclude that

$$h_n = n^2$$

holds for all $n \geq 0$.

The beauty of this approach is that we can solve much more difficult recurrence relations in essentially the same way. In general, the solution is a linear combination of elementary generating functions with known coefficient sequences. The advantage of generating functions is that we can use familiar algebraic methods to find the linear combinations. In the next section, we give an example that illustrates this approach.

Exercises

13.10. (a) Find a closed form of the generating function of the sequence $(h_n)_{n=0}^{\infty}$ given by the recurrence relation

$$h_0 = 1 \quad \text{and} \quad h_n = 2h_{n-1} + 1 \text{ when } n \geq 1.$$

(b) Find a closed form for the coefficients.

13.11. (a) Find a closed form of the generating function of the sequence $(h_n)_{n=0}^{\infty}$ given by the recurrence relation

$$h_0 = 1 \quad \text{and} \quad h_n = h_{n-1} + n + 3 \text{ when } n \geq 1.$$

(b) Find a closed form for the coefficients.

13.3 Fibonacci Sequence

Recall that the first few terms of the Fibonacci sequence (with the term $f_0 = 0$ included) are given by

$$(f_0, f_1, f_2, f_3, f_4, f_5, f_6, f_7 \ldots) = (0, 1, 1, 2, 3, 5, 8, 13, \ldots).$$

This sequence is defined by the recurrence relation

$$f_n = \begin{cases} 0 & \text{when } n = 0, \\ 1 & \text{when } n = 1, \\ f_{n-1} + f_{n-2} & \text{when } n \geq 2. \end{cases}$$

If we denote the ordinary generating function of the sequence $(f_n)_{n=0}^{\infty}$ by $F(z)$, then we have

$$F(z) = \sum_{k=0}^{\infty} f_k z^k.$$

Using the definition of the recurrence, we obtain

$$F(z) = \sum_{k=0}^{\infty} f_k z^k = z + \sum_{k=2}^{\infty} (f_{k-1} + f_{k-2}) z^k.$$

If we use the distributive law and the fact that $f_0 = 0$, then we can rewrite this in the form

$$F(z) = z + \sum_{k=1}^{\infty} f_{k-1} z^k + \sum_{k=2}^{\infty} f_{k-2} z^k$$
$$= z + zF(z) + z^2 F(z).$$

Subtracting $zF(z) + z^2 F(z)$ from both sides, it follows that

$$F(z) - zF(z) - z^2 F(z) = F(z)(1 - z - z^2) = z.$$

Dividing both sides by $1-z-z^2$, we can conclude that the ordinary generating function of the Fibonacci sequence is given by

$$F(z) = \frac{z}{1-z-z^2}.$$

We can do even better and find a closed form solution for the Fibonacci numbers f_n with $n \geqslant 0$. For this purpose, we factor the polynomial $1-z-z^2$ and write it in the form

$$1-z-z^2 = (1-\rho_1 z)(1-\rho_2 z),$$

where ρ_1 is the golden ratio and ρ_2 its conjugate root,

$$\rho_1 = \frac{1+\sqrt{5}}{2} \quad \text{and} \quad \rho_2 = \frac{1-\sqrt{5}}{2}.$$

We claim that there exist real numbers c and d such that

$$F(z) = \frac{z}{1-z-z^2} = \frac{c}{1-\rho_1 z} + \frac{d}{1-\rho_2 z}. \tag{13.2}$$

Indeed, bringing the fractions to the common denominator $1-z-z^2$ yields

$$\frac{z}{1-z-z^2} = \frac{c(1-\rho_2 z)}{1-z-z^2} + \frac{d(1-\rho_1 z)}{1-z-z^2}.$$

Therefore, $c+d-(\rho_2 c + \rho_1 d)z = z$. Comparing coefficients shows that the real numbers c and d must satisfy the system of linear equations

$$c + d = 0,$$
$$-\rho_2 c - \rho_1 d = 1.$$

It follows that $\rho_2 d - \rho_1 d = (\rho_2 - \rho_1)d = 1$, so $d = -1/\sqrt{5}$ and $c = 1/\sqrt{5}$. Therefore, we found the coefficients c and d of the representation (13.2).

The reason why we chose the representation (13.2) is that the terms

$$\frac{c}{1-\rho_1 z} \quad \text{and} \quad \frac{d}{1-\rho_2 z}$$

are easy to expand into power series with explicitly known coefficients, since they are just given by geometric series

$$\frac{c}{1-\rho_1 z} = c \sum_{k=0}^{\infty} \rho_1^k z^k \quad \text{and} \quad \frac{d}{1-\rho_2 z} = d \sum_{k=0}^{\infty} \rho_2^k z^k.$$

So let us now substitute the calculated values for the variables ρ_1, ρ_2, c, and d into (13.2). Using the aforementioned power series expansion yields the remarkable generating function

$$F(z) = \sum_{k=0}^{\infty} \frac{1}{\sqrt{5}} \left(\left(\frac{1+\sqrt{5}}{2} \right)^k - \left(\frac{1-\sqrt{5}}{2} \right)^k \right) z^k.$$

13.3 Fibonacci Sequence

In conclusion, the Fibonacci number f_k is given by

$$f_k = \frac{1}{\sqrt{5}}\left(\left(\frac{1+\sqrt{5}}{2}\right)^k - \left(\frac{1-\sqrt{5}}{2}\right)^k\right)$$

for all nonnegative integers k. The latter expression is known as **Binet's formula** for the Fibonacci coefficients.

Exercises

13.12. Show that the n-th Fibonacci number is asymptotically equal to

$$f_n \sim \frac{1}{\sqrt{5}}\left(\frac{1+\sqrt{5}}{2}\right)^n.$$

13.13. Let us denote by $[x]$ the nearest integer function given by $[x] = \lfloor x + 1/2 \rfloor$. Show that

$$f_n = \left[\frac{1}{\sqrt{5}}\left(\frac{1+\sqrt{5}}{2}\right)^n\right]$$

holds for all $n \geqslant 0$.

13.14. Derive a closed form for the generating function for the sequence $(f_{2n})_{n=0}^{\infty}$ of even-indexed Fibonacci numbers. [Hint: Use the fact that the generating function of the sequence $(f_n)_{n=0}^{\infty}$ of Fibonacci numbers is given by $F(z) = z/(1-z-z^2)$.]

13.15. Let f_n denote the Fibonacci numbers given by $f_0 = 0$, $f_1 = 1$ and $f_n = f_{n-1} + f_{n-2}$ for $n \geqslant 2$. Determine the value of the infinite sum

$$\sum_{n=0}^{\infty} \frac{f_n}{2^n}.$$

[Hint: Use generating functions.]

13.4 Partial Fractions

We can generalize the method to find a closed-form solution of the Fibonacci recurrence to a much wider class of recurrence relations. We will need an essential tool: the partial fraction decomposition of rational functions. You might have encountered the partial fraction decomposition in calculus, when calculating integrals of rational functions. In this section, we review the partial fraction decomposition, but we formulate it in a form that will be convenient for solving recurrence relations.

Suppose that we are given a rational function $p(z)/q(z)$, where $p(z)$ and $q(z)$ are polynomials with real or complex coefficients and $p(z)$ has smaller degree than $q(z)$. Suppose that $q(z)$ factors into linear terms

$$q(z) = (a_1 z + b_1)^{d_1}(a_2 z + b_2)^{d_2} \cdots (a_k z + b_k)^{d_k}, \qquad (13.3)$$

where a_i and b_i are complex numbers, $a_i \neq 0$, and the roots $r_i = -b_i/a_i$ of the linear terms are pairwise distinct. In other words, the root $r_i = -b_i/a_i$ occurs with multiplicity d_i in the polynomial $q(z)$.

The partial fraction decomposition expresses the rational function $p(z)/q(z)$ as a linear combination of the partial fractions

$$\frac{1}{(a_j z + b_j)^{m_j}},$$

where m_j is an integer in the range $1 \leq m_j \leq d_j$ and the integer j is an integer in the range $1 \leq j \leq k$.

The precise statement of the partial fraction decomposition is given in the next proposition. We adapt the inductive argument given in Walter [78] to the slightly more general form of the partial fraction decomposition that we use here.

Proposition 13.3. *Let $p(z)/q(z)$ be a rational function such that $\deg p < \deg q$ and $q(z)$ is a polynomial with the factorization (13.3). Then there exist complex numbers $C_{i,j}$ such that*

$$\frac{p(z)}{q(z)} = \sum_{i=1}^{k} \left(\frac{C_{i,1}}{a_i z + b_i} + \frac{C_{i,2}}{(a_i z + b_i)^2} + \cdots + \frac{C_{i,d_i}}{(a_i z + b_i)^{d_i}} \right).$$

This decomposition of the rational function $p(z)/q(z)$ is unique.

Proof. We will prove the claim about the existence of the partial fraction decomposition by strong induction on the degree $n = \deg q(z)$ of the polynomial $q(z)$ in the denominator.

Induction Basis When the polynomial $q(z)$ has degree $n = 1$, then $p(z)$ must have degree 0, so it must be a constant. Therefore, $p(z)/q(z)$ is of the claimed form.

Induction Step Suppose that every proper rational function with denominator polynomial of degree less than n has a decomposition of the claimed form.

13.4 Partial Fractions

Suppose further that we are given a proper rational function $p(z)/q(z)$ with $n = \deg q$. Let $r_1 = -b_1/a_1$ denote a root of the polynomial $q(z)$ of multiplicity $d_1 \geqslant 1$. This means that we can factor $q(z)$ in the form

$$q(z) = (a_1 z + b_1)^{d_1} a(z),$$

where $a(z)$ is a polynomial of degree $n - d_1$ satisfying $a(r_1) \neq 0$. We will now show that $p(z)/q(z)$ has a partial fraction decomposition of the claimed form.

Let us first consider the related rational function

$$\frac{p(z)}{a(z)} - \frac{p(r_1)}{a(r_1)} = \frac{p(z)a(r_1) - p(r_1)a(z)}{a(z)a(r_1)}.$$

Since r_1 is a root of the numerator, we can rewrite this expression in the form

$$\frac{p(z)}{a(z)} - \frac{p(r_1)}{a(r_1)} = \frac{(a_1 z + b_1)b(z)}{a(z)},$$

where $b(z)$ is a polynomial of degree $\deg b(z) \leqslant n - 2$. It follows that

$$\frac{p(z)}{(a_1 z + b_1)^{d_1} a(z)} - \frac{p(r_1)}{(a_1 z + b_1)^{d_1} a(r_1)} = \frac{b(z)}{(a_1 z + b_1)^{d_1 - 1} a(z)}.$$

Since $\deg b \leqslant n - 2$ and $\deg\left((a_1 z + b_1)^{d_1 - 1} a(z)\right) \leqslant n - 1$, we can express the right-hand side by induction hypothesis in the claimed form. Adding $\frac{a(r_1)^{-1} p(r_1)}{(a_1 z + b_1)^{d_1}}$ to both sides of the equation shows that

$$\frac{p(z)}{q(z)} = \frac{p(z)}{(a_1 z + b_1)^{d_1} a(z)}$$

can be expressed in the claimed form as well.

Therefore, the existence of the partial fraction decomposition follows by strong induction on $n = \deg q$.

We will now prove the uniqueness of the partial fraction decomposition. Seeking a contradiction, let us assume that there exist two different decompositions of $p(z)/q(z)$, say

$$\frac{p(z)}{q(z)} = \sum_{i=1}^{k} \sum_{j=1}^{d_i} \frac{C_{i,j}}{(a_i z + b_i)^j} = \sum_{i=1}^{k} \sum_{j=1}^{d_i} \frac{D_{i,j}}{(a_i z + b_i)^j}.$$

Subtracting one decomposition from the other, we obtain

$$\sum_{i=1}^{k} \sum_{j=1}^{d_i} \frac{C_{i,j} - D_{i,j}}{(a_i z + b_i)^j} = 0.$$

Since the decompositions are different, there must exist some index μ such that $C_{\mu,j} = D_{\mu,j}$ does not hold for all indices j. Let c_μ denote the largest index such that $C_{\mu,c_\mu} \neq D_{\mu,c_\mu}$. If we multiply both sides of the previous equation by $(a_\mu z + b_\mu)^{c_\mu}$ and let $z \to r_\mu = -b_\mu/a_\mu$, then we get

$$\lim_{z \to r_\mu} \sum_{i=1}^{k} \sum_{j=1}^{d_i} \frac{(C_{i,j} - D_{i,j})(a_\mu z + b_\mu)^{c_\mu}}{(a_i z + b_i)^j} = 0.$$

Let us now examine the behavior of the individual terms of this expression. Since $(a_\mu z + b_\mu)^{c_\mu}/(a_i z + b_i)^j \to 0$ as $z \to r_\mu$ for all $i \neq \mu$, and $(a_i z + b_\mu)^{c_\mu}/(a_i z + b_\mu)^j \to 0$ as $z \to r_\mu$ for all $j < c_\mu$, we can conclude that

$$\lim_{z \to r_\mu} \sum_{i=1}^{k} \sum_{j=1}^{d_i} \frac{(C_{i,j} - D_{i,j})(a_\mu z + b_\mu)^{c_\mu}}{(a_i z + b_i)^j} = C_{\mu,c_\mu} - D_{\mu,c_\mu} = 0,$$

so $C_{\mu,c_\mu} = D_{\mu,c_\mu}$, which contradicts the choice $C_{\mu,c_\mu} \neq D_{\mu,c_\mu}$.

Therefore, $C_{i,j}$ must equal $D_{i,j}$ for all indices i and j, so the partial fraction decomposition is unique. □

The previous theorem states the form of the partial fraction decomposition of a rational function $p(z)/q(z)$. After factoring the polynomial $q(z)$ of the denominator, the overall form of the decomposition is known. Finding the particular coefficients $C_{i,j}$ amounts to solving a system of linear equations. The next example is rather typical and illustrates the standard approach to find the partial fraction decomposition.

Example 13.4. Suppose that $p(z) = z^2 + 5$ and $q(z) = (z-1)(z-3)^2$. By the partial fraction decomposition, there must exist complex constants such that

$$\frac{z^2 + 5}{(z-1)(z-3)^2} = \frac{C_{1,1}}{z-1} + \frac{C_{2,1}}{(z-3)} + \frac{C_{2,2}}{(z-3)^2}.$$

Multiplying both sides by $q(z)$ yields

$$z^2 + 5 = C_{1,1}(z-3)^2 + C_{2,1}(z-1)(z-3) + C_{2,2}(z-1).$$

Expanding the terms on the right-hand side yields

$$z^2 + 5 = C_{1,1}(z^2 - 6z + 9) + C_{2,1}(z^2 - 4z + 3) + C_{2,2}(z-1).$$

If we compare coefficients on both sides of this equation, then we obtain the following system of linear equations,

$$9C_{1,1} + 3C_{2,1} - C_{2,2} = 5$$
$$-6C_{1,1} - 4C_{2,1} + C_{2,2} = 0$$
$$C_{1,1} + C_{2,1} = 1$$

Solving this system of equations yields $C_{1,1} = 3/2$, $C_{2,1} = -1/2$, and $C_{2,2} = 7$. In other words, the partial fraction decomposition of $p(z)/q(z)$ is given by

$$\frac{z^2 + 5}{(z-1)(z-3)^2} = \frac{3}{2(z-1)} - \frac{1}{2(z-3)} + \frac{7}{(z-3)^2}.$$

13.5 Reciprocal Polynomials

In general, the factorization of the polynomial $q(z)$ completely determines the form of the partial fraction decomposition. Finding the coefficients of this partial fraction decomposition can be done by comparing coefficients and solving the resulting linear system of equations, as we have shown in this example.

Exercises

13.16. Find the partial fraction decomposition of
$$\frac{2z+1}{z^2+z}.$$

13.17. Find the partial fraction decomposition of
$$\frac{3z^2+7z+1}{z(z+1)^2}.$$

13.18. Find the partial fraction decomposition of
$$\frac{2z+36}{z^2+4}.$$

13.19. Find the partial fraction decomposition of
$$\frac{z+1}{1-4z+3z^2}.$$

13.20. Find the partial fraction decomposition of
$$\frac{1}{1-x^2}.$$

13.5 Reciprocal Polynomials

We need to establish some terminology that will help us to formulate our proofs. For a given polynomial, we can find a new polynomial by reversing the order of the coefficients. In this section, we give a few basic facts about such polynomials.

For a polynomial $q(z) = q_0 + q_1 z + q_2 z^2 + \cdots + q_d z^d$, its **reciprocal polynomial** or **reflected polynomial** $g^R(z)$ is defined as
$$g^R(z) = z^d g(1/z).$$
The polynomial g^R is called the reflected polynomial, since it is of the form $g^R(z) = g_d + g_{d-1} z + \cdots + g_0 z^d$, with the coefficients of $g(z)$ in reverse order.

Lemma 13.5. *Suppose that $c(z)$ is a polynomial of the form $c(z) = 1 + c_1 z + \cdots + c_d z^d$ with $c_d \neq 0$. Then $c(z)$ can be written in the form*
$$c(z) = \prod_{k=1}^{m}(1-a_k z)^{d_k}$$
for some pairwise distinct complex numbers a_k and positive integers d_k such that $d_1 + d_2 + \cdots + d_m = d$.

Proof. By the fundamental theorem of algebra, we can factor the reciprocal polynomial $c^R(z)$ of the polynomial $c(z)$ into linear factors

$$c^R(z) = \prod_{k=1}^{m} (z - a_k)^{d_k}$$

for some complex numbers a_k that are pairwise distinct and positive integers d_k such that $d_1 + d_2 + \cdots + d_m = d$. Since the reciprocal polynomial of $c^R(z)$ is given by $c(z)$, we have

$$c(z) = z^d c^R(1/z) = z^d \prod_{k=1}^{m} \left(\frac{1}{z} - a_k \right)^{d_k} = \prod_{k=1}^{m} (1 - a_k z)^{d_k},$$

as claimed. □

Example 13.6. For the Fibonacci sequence, the polynomial $c(z) = 1 - z - z^2$. Factoring the reciprocal polynomial $c^R(z) = z^2 c(1/z)$ yields

$$c^R(z) = z^2 - z - 1 = \left(z - \frac{1 + \sqrt{5}}{2} \right) \left(z - \frac{1 - \sqrt{5}}{2} \right).$$

Therefore, we recover the factorization

$$c(z) = \left(1 - \frac{1 + \sqrt{5}}{2} z \right) \left(1 - \frac{1 - \sqrt{5}}{2} z \right)$$

that we have used in Sect. 13.3.

Exercises

13.21. Show that a nonzero complex number r is a root of the polynomial $c(z)$ if and only if r^{-1} is a root of the reciprocal polynomial $c^R(z)$.

13.22. Let $c(z) = 1 - 8z + 15z^2$. Find complex numbers a_1 and a_2 such that

$$c(z) = (1 - a_1 z)(1 - a_2 z)$$

by factoring the reciprocal polynomial $c^R(z)$.

13.23. Let $c(z) = 1 + 2z + 3z^2$. Find complex numbers a_1 and a_2 such that

$$c(z) = (1 - a_1 z)(1 - a_2 z)$$

by factoring the reciprocal polynomial $c^R(z)$.

13.24. Let $c(z) = 1 - 6z + 11z^2 - 6z^3$. Find complex numbers a_1, a_2 and a_3 such that

$$c(z) = (1 - a_1 z)(1 - a_2 z)(1 - a_3 z)$$

by factoring the reciprocal polynomial $c^R(z)$.

13.25. A polynomial $c(z)$ is called a **palindromic polynomial** if and only if it coincides with its reciprocal polynomial, $c(z) = c^R(z)$. Let $c_n(z) = (z-1)^n$. Determine the set of all positive integers n such that $c_n(z)$ is palindromic.

13.6 Linear Homogeneous Recurrence Relations

Suppose that we are given a sequence $(g_0, g_1, g_2, \ldots)$ satisfying a linear homogeneous recurrence relation of order d with constant coefficients, that is, there exist coefficients $c_1, c_2, \ldots, c_d$ such that

$$g_n + c_1 g_{n-1} + \cdots + c_d g_{n-d} = 0 \tag{13.4}$$

holds for all $n \geq d$. What can we say about the generating function of the sequence? The next proposition gives an answer to this question.

Proposition 13.7. *Suppose that a sequence $(g_0, g_1, g_2, \ldots)$ satisfies the linear homogeneous recurrence relation (13.4) of order d. Then its generating function $G(z) = \sum_{k=0}^{\infty} g_k z^k$ is a rational function*

$$G(z) = \frac{p(z)}{c(z)},$$

where $p(z)$ is a polynomial of degree less than d and $c(z)$ is the polynomial of degree d given by

$$c(z) = 1 + c_1 z + c_2 z^2 + \cdots + c_d z^d,$$

which is formed by the coefficients of the recurrence relation (13.4).

Proof. We are going to show that the coefficient $[z^n] c(z) G(z) = 0$ for all $n \geq d$. Since this implies that $c(z) G(z) = p(z)$ for some polynomial $p(z)$ of degree less than d, this will prove our claim.

Let us first consider the generating function $G(z)$ and its shifted and scaled versions $c_k z^k G(z)$ for $1 \leq k \leq d$. We have

$$
\begin{array}{rlllllll}
G(z) & = & g_0 & + & g_1 z & + & g_2 z^2 & + \cdots + & g_n z^n & + \cdots \\
c_1 z G(z) & = & & & c_1 g_0 z & + & c_1 g_1 z^2 & + \cdots + & c_1 g_{n-1} z^n & + \cdots \\
c_2 z^2 G(z) & = & & & & & c_2 g_0 z^2 & + \cdots + & c_2 g_{n-2} z^n & + \cdots \\
\vdots & & & & & & \ddots & & \vdots & \vdots \\
c_d z^d G(z) & = & & & & & & \cdots & c_d g_{n-d} z^n & \cdots
\end{array}
$$

Since the sum of these equations is equal to $c(z) G(z)$, we get

$$[z^n] c(z) G(z) = g_n + c_1 g_{n-1} + \cdots + c_d g_{n-d} = 0$$

for all integers n satisfying $n \geq d$.

Thus, $c(z) G(z) = p(z)$ for some polynomial $p(z)$ of degree less than d, which implies that the generating function $G(z) = p(z)/c(z)$, as claimed. $\square$

The previous proposition shows that a sequence satisfying a linear homogeneous recurrence relation has a fairly simple generating function, namely a rational function. We will now exploit this fact to express the coefficients of the sequence in closed form.

Let us first take a look at some key examples that will guide the way.

Example 13.8. If the generating function $G(z)$ is a rational function of the form
$$\frac{p(z)}{c(z)} = \frac{1}{1-az},$$
then $g_k = a^k$ for all $k \geq 0$. This follows from Example 12.12.

Example 13.9. If the generating function $G(z)$ is a rational function of the form
$$\frac{p(z)}{c(z)} = \frac{1}{(1-az)^m},$$
for some integer $m \geq 1$, then $g_k = \binom{k+m-1}{k} a^k$ for all $k \geq 0$. Indeed, this follows from Proposition 12.11. Note that for $m = 1$, we recover the previous example as a special case.

The generating function $G(z)$ of a homogeneous linear recurrence of degree d might not be of the form given by the previous examples. However, it can always be written as a linear combination of terms that are of the form given by the previous two examples. This means that we can find an explicit form for the coefficients g_k of the generating function of *every* homogeneous linear recurrence with constant coefficients! The next proposition shows how to find the closed-form expression for the coefficients.

Proposition 13.10. *Suppose that a sequence* $(g_0, g_1, g_2, \ldots)$ *satisfies the linear homogeneous recurrence relation (13.4) of order d. Then its generating function* $G(z) = \sum_{k=0}^{\infty} g_k z^k$ *is a rational function*
$$G(z) = \frac{p(z)}{c(z)},$$
where $p(z)$ is a polynomial of degree less than d and $c(z)$ is the polynomial of degree d given by
$$c(z) = 1 + c_1 z + c_2 z^2 + \cdots + c_d z^d,$$
which is formed by the coefficients of the recurrence relation (13.4). There exist pairwise distinct complex numbers a_i such that the polynomial $c(z)$ factors into the form
$$c(z) = \prod_{k=1}^{m} (1 - a_k z)^{d_k}.$$
Then there exist complex numbers $C_{i,j}$ such that
$$\frac{p(z)}{c(z)} = \sum_{k=1}^{m} \left(\frac{C_{k,1}}{1-a_k z} + \frac{C_{k,2}}{(1-a_k z)^2} + \cdots + \frac{C_{k,d_k}}{(1-a_k z)^{d_k}} \right).$$

Note that each term in the sum is of a known form, see Examples 13.8 and 13.9.

13.6 Linear Homogeneous Recurrence Relations

Proof. It follows from the previous proposition that the generating function $G(z)$ is a rational function of the claimed form. Since the constant coefficient of the denominator polynomial $c(z)$ is equal to 1, we can factor $c(z)$ in the form

$$c(z) = \prod_{k=1}^{m} (1 - a_k z)^{d_k}$$

by Lemma 13.5. By the partial fraction decomposition, there exist complex numbers $C_{i,j}$ such that

$$\frac{p(z)}{c(z)} = \sum_{k=1}^{m} \left(\frac{C_{k,1}}{1 - a_k z} + \frac{C_{k,2}}{(1 - a_k z)^2} + \cdots + \frac{C_{k,d_k}}{(1 - a_k z)^{d_k}} \right),$$

as claimed. □

Let us have a look at an example.

Example 13.11. The Pell sequence $(p_0, p_1, p_2, p_3, \ldots)$ has the initial conditions $p_0 = 1$ and $p_1 = 2$ and satisfies the linear homogeneous recurrence relation of order 2 given by

$$p_n - 2p_{n-1} - p_{n-2} = 0$$

for all $n \geq 2$. The first few terms of the Pell sequence are given by

$$(1, 2, 5, 12, 29, 70, \ldots).$$

The generating function $G(z)$ of the Pell sequence satisfies

$$[z^n](1 - 2z - z^2)G(z) = 0$$

for all $n \geq 2$. Furthermore, the initial conditions imply that

$$[z^0](1 - 2z - z^2)G(z) = 1 \quad \text{and} \quad [z^1](1 - 2z - z^2)G(z) = 0.$$

Therefore,

$$G(z) = \frac{1}{1 - 2z - z^2}.$$

The reciprocal polynomial of the denominator polynomial has the factorization $c^R(z) = z^2 - 2z - 1 = (z - (1 + \sqrt{2}))(z - (1 - \sqrt{2}))$. The polynomial $c^R(z)$ has the two distinct roots $a_1 = 1 + \sqrt{2}$ and $a_2 = 1 - \sqrt{2}$. Therefore, $G(z)$ is of the form

$$G(z) = \frac{C_{1,1}}{1 - (1 + \sqrt{2})z} + \frac{C_{2,1}}{1 - (1 - \sqrt{2})z}.$$

By performing the partial fraction decomposition, we get

$$G(z) = \frac{1}{4} \left(\frac{2 + \sqrt{2}}{1 - (1 + \sqrt{2})z} + \frac{2 - \sqrt{2}}{1 - (1 - \sqrt{2})z} \right).$$

Therefore, the closed form of the coefficient p_n is given by

$$p_n = [z^n]G(z) = \frac{1}{4} \left((2 + \sqrt{2})(1 + \sqrt{2})^n + (2 - \sqrt{2})(1 - \sqrt{2})^n \right)$$

for all $n \geq 0$.

Exercises

13.26. Suppose that a sequence $(a_0, a_1, a_2, \ldots)$ of real numbers satisfies the recurrence relation $a_n - 5a_{n-1} + 6a_{n-2} = 0$ for all $n \geq 2$.
(a) What is the order of the linear recurrence relation?
(b) Express the generating function of the sequence as a rational function.
(c) Find a generic closed-form solution for this recurrence relation.
(d) Find the terms $a_0, a_1, \ldots, a_5$ of this sequence when the initial conditions are given by $a_0 = 2$ and $a_1 = 5$.
(e) Find the closed-form solution when $a_0 = 2$ and $a_1 = 5$.

13.27. Suppose that a sequence $(a_0, a_1, a_2, \ldots)$ of real numbers satisfies the recurrence relation $a_n + 7a_{n-1} + 10a_{n-2} = 0$ for all $n \geq 2$.
(a) Express the generating function of the sequence as a rational function.
(b) Find a generic closed-form solution for this recurrence relation.
(c) Find the terms $a_0, a_1, \ldots, a_4$ of this sequence when the initial conditions are given by $a_0 = 1$ and $a_1 = 4$.
(d) Find the closed-form solution when $a_0 = 1$ and $a_1 = 4$.

13.28. The Pell sequence can be used to approximate $\sqrt{2}$. Show that
$$\frac{p_{n-1} + p_n}{p_n} \sim \sqrt{2}.$$

13.7 Characteristic Polynomials

We can also find the closed-form solution of a linear homogeneous recurrence relation using a different approach. We seemingly eschew generating functions and obtain the coefficients as linear combinations of the roots of a polynomial. However, it is simply a reformulation of the results from the previous section.

Let $(g_0, g_1, g_2, \ldots)$ be a sequence satisfying a linear homogenous recurrence relation of order d with constant coefficients

$$g_n + c_1 g_{n-1} + c_2 g_{n-2} + \cdots + c_{d-1} g_{n-d+1} + c_d g_{n-d} = 0 \tag{13.5}$$

for all integers n such that $n \geq d$. We define the **characteristic polynomial** $\chi(z)$ of this recurrence relation by

$$\chi(z) = z^d + c_1 z^{d-1} + c_2 z^{d-2} + \cdots + c_{d-1} z + c_d.$$

The closed form of the recurrence relation is essentially obtained by a linear combination of the roots of the characteristic polynomial.

Proposition 13.12. *Let $(g_0, g_1, g_2, \ldots)$ be a sequence satisfying the homogeneous linear recurrence relation (13.5) of order d. If the characteristic polynomial $\chi(z)$ of this recurrence has the factorization*

$$\chi(z) = \prod_{k=1}^{m} (z - a_i)^{d_k},$$

13.7 Characteristic Polynomials

where the root a_k occurs with multiplicity d_k, then

$$g_n = \sum_{k=1}^{m} p_k(n) a_k^n,$$

where $p_k(n)$ is a polynomial of degree $d_k - 1$. In particular, if each root a_k has multiplicity $d_k = 1$, then g_n is a linear combination of the n-th powers of the roots a_n.

Proof. By Proposition 13.10, the generating function of the sequence is a rational function of the form

$$\frac{p(z)}{c(z)} = \sum_{k=1}^{m} \left(\frac{C_{k,1}}{1 - a_k z} + \frac{C_{k,2}}{(1 - a_k z)^2} + \cdots + \frac{C_{k,d_k}}{(1 - a_k z)^{d_k}} \right).$$

It follows from Example 13.9 that the n-th coefficient of $p(z)/c(z)$ is of the form

$$[z^n]\frac{p(z)}{q(z)} = \sum_{k=1}^{m} \left(C_{k,1} \binom{n}{n} a_k^n + C_{k,2} \binom{n+1}{n} a_k^n + \cdots + C_{k,d_k} \binom{n + d_k - 1}{n} a_k^n \right).$$

The claim follows, since

$$p_k(n) = C_{k,1} \binom{n}{n} + C_{k,2} \binom{n+1}{n} + \cdots + C_{k,d_k} \binom{n + d_k - 1}{n}$$

is a polynomial of degree $d_k - 1$ in n. □

The following special case is particularly easy to understand.

Corollary 13.13. *We keep the notations of the previous proposition. If the roots of the characteristic polynomial are pairwise distinct, then g_n is the linear combination of the n-th powers of the roots of the characteristic polynomial $\chi(z)$. In other words, there exist complex coefficients C_k such that*

$$g_n = \sum_{k=1}^{m} C_k a_k^n$$

for all $n \geq 0$.

Proof. This follows from the previous proposition by considering the special case when all roots of the characteristic polynomial have multiplicity 1, so $d_1 = d_2 = \cdots = d_k = 1$. □

Let us have a look at some examples that illustrate the previous corollary and proposition.

Example 13.14. Let $(g_0, g_1, g_2, \ldots)$ be a sequence of real numbers such that $g_0 = 1$ and $g_1 = 2$ and $g_n - 10g_{n-1} + 21g_{n-2} = 0$ for all $n \geqslant 2$. This sequence has the characteristic function

$$\chi(z) = z^2 - 10z + 21 = (z-3)(z-7).$$

By Corollary 13.13, the closed form for the coefficients g_n must be of the form

$$g_n = C_1 3^n + C_2 7^n$$

for some complex numbers C_1 and C_2. We can find the value of the constants C_1 and C_2 in this generic solution using the two known initial conditions

$$g_0 = 1 = C_1 + C_2,$$
$$g_1 = 2 = C_1 3 + C_2 7.$$

Solving this system of linear equations yields $C_1 = 5/4$ and $C_2 = -1/4$. Therefore, we can conclude that

$$g_n = \frac{5}{4} 3^n - \frac{1}{4} 7^n$$

is the closed-form solution to the recurrence relation.

Example 13.15. Let $(g_0, g_1, g_2, \ldots)$ be a sequence of real numbers such that $g_0 = 1$ and $g_1 = 2$ and $g_n - 6g_{n-1} + 9g_{n-2} = 0$ for all $n \geqslant 2$. This sequence has the characteristic function

$$\chi(z) = z^2 - 6z + 9 = (z-3)^2.$$

By Proposition 13.12, the closed form for the coefficients g_n must be of the form

$$g_n = D_{1,1} 3^n + D_{1,2} n 3^n$$

for some complex coefficients $D_{1,1}$ and $D_{1,2}$. If we substitute the initial conditions into this generic solution, we get the system of equations

$$g_0 = 1 = D_{1,1},$$
$$g_1 = 2 = D_{1,1} 3 + D_{1,2} 3.$$

Solving this system of equations yields $D_{1,1} = 1$ and $D_{1,2} = -1/3$. Therefore, we can conclude that

$$g_n = 3^n - n 3^{n-1}$$

is the closed-form solution to the recurrence relation.

13.8 Inhomogeneous Linear Recurrence Relations

Exercises

13.29. Let $G(z) = p(z)/c(z)$ denote the generating function of a sequence satisfying a linear homogeneous recurrence relation of order d. How is the characteristic polynomial $\chi(z)$ of the recurrence relation related to the polynomials $p(z)$ and $c(z)$?

13.30. Use the characteristic equation to find a closed form solution for the coefficients of the sequence $(g_0, g_1, g_2, \ldots)$ satisfying $g_0 = 2$, $g_1 = 1$, and $g_n - 7g_{n-1} + 12g_{n-2} = 0$ for all $n \geqslant 2$.

13.31. Use the characteristic equation to find a closed-form solution for the coefficients of the sequence $(g_0, g_1, g_2, \ldots)$ satisfying $g_0 = 2$, $g_1 = 2$, and $g_n - 2g_{n-1} - 15g_{n-2} = 0$ for all $n \geqslant 2$.

13.32. Use the characteristic equation to find a closed-form solution for the coefficients of the sequence $(g_0, g_1, g_2, \ldots)$ satisfying $g_0 = 1$, $g_1 = 1$, and $g_n - 10g_{n-1} + 25g_{n-2} = 0$ for all $n \geqslant 2$.

13.33. Use the characteristic equation to find a closed-form solution for the coefficients of the sequence $(g_0, g_1, g_2, \ldots)$ satisfying $g_0 = 1$, $g_1 = 5$, and $g_n - 14g_{n-1} + 49g_{n-2} = 0$ for all $n \geqslant 2$.

13.8 Inhomogeneous Linear Recurrence Relations

A sequence $(g_0, g_1, g_2, \ldots)$ satisfies an inhomogeneous linear recurrence relation of order d with constant coefficients if and only if there exist constants $c_1, c_2, \ldots, c_d$ with $c_d \neq 0$ and a nonzero function $f \colon \mathbf{N}_0 \to \mathbf{R}$ such that

$$g_n + c_1 g_{n-1} + c_2 g_{n-2} + \cdots + c_d g_{n-d} = f(n) \tag{13.6}$$

holds for all $n \geqslant d$. The associated homogeneous linear recurrence relation is given by

$$g_n + c_1 g_{n-1} + c_2 g_{n-2} + \cdots + c_d g_{n-d} = 0. \tag{13.7}$$

Proposition 13.16. *Two solutions to an inhomogeneous linear recurrence relation of order d with constant coefficients differ by a solution to the associated homogeneous recurrence relation.*

Proof. Let $(g_0, g_1, g_2, \ldots)$ and $(g'_0, g'_1, g'_2, \ldots)$ be two solutions to the inhomogeneous linear recurrence relation (13.6), so

$$g_n + c_1 g_{n-1} + c_2 g_{n-2} + \cdots + c_d g_{n-d} = f(n),$$
$$g'_n + c_1 g'_{n-1} + c_2 g'_{n-2} + \cdots + c_d g'_{n-d} = f(n),$$

hold for all $n \geqslant d$. Subtracting the two equations yields

$$(g_n - g'_n) + c_1(g_{n-1} - g'_{n-1}) + c_2(g_{n-2} - g'_{n-2}) + \cdots + c_d(g_{n-d} - g'_{n-d})$$
$$= f(n) - f(n) = 0,$$

so the sequence $(h_0, h_1, h_2, \ldots)$ of differences $h_n = g_n - g'_n$ is a solution to the associated homogeneous recurrence relation (13.7). □

In the previous sections, we showed how homogeneous linear recurrence relation with constant coefficients can be solved. If you know one particular solution s to an inhomogeneous recurrence relation, then the previous proposition implies that you can obtain all other solutions by adding solutions to the associated homogeneous linear recurrence relation to your solution s.

The main question is now: How can we find one solution to an inhomogeneous recurrence relation? The answer to this question depends on the nature of the function $f(n)$ in the inhomogeneous part of the recurrence. Fortunately, there are some types of functions $f(n)$ for which it is possible to solve inhomogeneous linear recurrence relations by reducing it to a linear homogeneous recurrence relation.

Let us have a look at an example that illustrates the approach.

Example 13.17. Let us consider the inhomogeneous recurrence relation

$$p_n - 2p_{n-1} - p_{n-2} = f(n)$$

with $f(n) = \frac{5}{4}3^n - \frac{1}{4}7^n$. The associated homogeneous recurrence relation

$$p_n - 2p_{n-1} - p_{n-2} = 0$$

is the Pell sequence recurrence. The inhomogeneous part $f(n) = \frac{5}{4}3^n - \frac{1}{4}7^n$ is a function that satisfies a recurrence relation itself, namely

$$f(n) - 10f(n-1) + 21f(n-2) = 0$$

for all $n \geqslant 2$, as we have seen in Example 13.14.

In this case, we can homogenize the inhomogeneous recurrence relation, meaning that we can derive a homogeneous recurrence relation for the sequence. Indeed, we express the three terms $f(n)$, $-10f(n-1)$, and $21f(n-2)$ using the Pell recurrence

$$\begin{aligned} p_n - 2p_{n-1} - p_{n-2} &= f(n) \\ -10(p_{n-1} - 2p_{n-2} - p_{n-3}) &= -10f(n-1) \\ 21(p_{n-2} - 2p_{n-3} - p_{n-4}) &= 21f(n-2) \end{aligned}$$

Adding the three equations yields the homogeneous recurrence relation

$$p_n - 12p_{n-1} + 40p_{n-2} - 32p_{n-3} - 21p_{n-4} = \\ f(n) - 10f(n-1) + 21f(n-2) = 0.$$

Since this is a *homogeneous* linear recurrence relation, we know at least in principle how to find a closed-form solution.

Let us use the characteristic polynomials to find the form of the closed solution. The characteristic polynomial of the Pell sequence recurrence (which

13.8 Inhomogeneous Linear Recurrence Relations

is the associated homogeneous recurrence relation) is given by $\chi_p(z) = z^2 - 2z - 1$ and the characteristic function of the inhomogeneous part is $\chi_f(z) = z^2 - 10z + 21$. A short calculation shows that the characteristic polynomial $\chi(z)$ of the sequence $(p_0, p_1, p_2, \ldots)$ is given by the product

$$\chi(z) = z^4 - 12z^3 + 40z^2 - 32z - 21 = \chi_p(z)\chi_f(z)$$

Therefore, $\chi(z) = (z - (1 + \sqrt{2}))(z - (1 - \sqrt{2}))(z - 3)(z - 7)$, so there exist complex numbers C_1, C_2, C_3, and C_4 such that

$$p_n = C_1(1 + \sqrt{2})^n + C_2(1 - \sqrt{2})^n + C_3 3^n + C_4 7^n.$$

Proposition 13.18. *Let* $f\colon \mathbf{N}_0 \to \mathbf{R}$ *be a function satisfying a linear homogeneous recurrence relation with characteristic polynomial* $\chi_f(z)$. *Let* $g = (g_0, g_1, g_2, \ldots)$ *be any sequence satisfying the inhomogeneous linear recurrence relation*

$$g_n + c_1 g_{n-1} + c_2 g_{n-2} + \cdots + c_d g_{n-d} = f(n)$$

with constant coefficients $c_1, c_2, \ldots, c_d$ *and* $c_d \neq 0$ *such that the associated homogeneous linear recurrence relation has the characteristic polynomial* $\chi_c(z)$. *Then the sequence* g *satisfies a linear homogeneous recurrence relation with characteristic polynomial* $\chi_c(z)\chi_f(z)$.

Proof. Let $c(z)$ denote the reciprocal polynomial of the characteristic polynomial $\chi_c(z)$, so $c(z) = \chi_c^R(z)$. By assumption, the generating function $G(z) = \sum_{k=0}^{\infty} g_k z^k$ of the sequence g satisfies

$$[z^n]c(z)G(z) = f(n)$$

for all $n \geq \chi_c(z)$.

Similarly, let $d(z)$ denote the reciprocal polynomial of the characteristic polynomial $\chi_f(z)$, so $d(z) = \chi_f^R(z)$. Therefore, the generating function $F(z) = \sum_{k=0}^{\infty} f(k) z^k$ of the function f satisfies

$$[z^n]d(z)F(z) = 0$$

for all $n \geq \deg \chi_f(z)$.

The generating function $c(z)G(z)$ is almost the same as the generating function $F(z)$, except for a few initial terms. Specifically, the difference between the generating functions $c(z)G(z)$ and $F(z)$ is a polynomial $t(z)$ of degree less than $\deg \chi_c(z)$, namely the tail polynomial $t(z)$ given by

$$t(z) = c(z)G(z) - F(z).$$

Since $c(z)G(z) = t(z) + F(z)$, we can conclude that

$$d(z)c(z)G(z) = d(z)t(z) + d(z)F(z).$$

Therefore,

$$[z^n]d(z)c(z)G(z) = [z^n]d(z)t(z) + [z^n]d(z)F(z).$$

If $n \geq \deg_f(z) + \deg_c(z)$, then both terms on the right-hand side are equal to 0, so
$$[z^n]d(z)c(z)G(z) = 0.$$
In other words, $G(z)$ satisfies a linear homogeneous recurrence relation with characteristic polynomial $\chi_f(z)\chi_c(z) = d^R(z)c^R(z)$, as claimed. □

Example 13.19. The Tower of Hanoi is a puzzle consisting of three rods and n discs of different diameters that can slide on the rods. Initially, all discs are stacked on the leftmost rod in decreasing order of diameter with the largest disc at the bottom and the smallest of the n discs at the top. The goal is to transfer the stack of discs from the leftmost rod to the rightmost rod. One is allowed to move one disc at a time, removing the upper disc of one stack and placing it on top of another stack at a different rod. It is not allowed to place a larger disc on top of a smaller disc.

Let t_n denote the minimal number of moves that are needed to transfer n discs from one rod to another. The minimal number of moves is given by the recurrence relation $t_n = 2t_{n-1} + 1$, as the top $n-1$ discs are moved to the middle rod in t_{n-1} moves, then the largest disc is moved from the leftmost to the rightmost rod in a single move, and finally, the $n-1$ discs from the middle rod are moved to the rightmost rod in t_{n-1} moves.

We would like to find the closed form for the inhomogeneous recurrence relation
$$t_n - 2t_{n-1} = 1.$$
The associated homogeneous recurrence $t_n - 2t_{n-1} = 0$ has the characteristic polynomial $z - 2$, and the inhomogeneous part has the characteristic polynomial $z - 1$. Therefore, the sequence $(h_0, h_1, h_2, \ldots)$ of the minimal number of Hanoi tower moves has the characteristic polynomial $\chi(z) = (z-2)(z-1)$. Consequently, a closed form solution for t_n must be of the form
$$t_n = C_1 2^n + C_2 1^n = C_1 2^n + C_2$$
for some constants C_1 and C_2. Since the initial conditions are $t_0 = 0$ and $t_1 = 1$, we obtain the following system of linear equations
$$0 = C_1 + C_2$$
$$1 = C_1 2^1 + C_2$$
Solving this system of equations yields $C_1 = 1$ and $C_2 = -1$, so the closed form solution is given by
$$t_n = 2^n - 1$$
for all $n \geq 0$.

One can extend Proposition 13.18 to other types of functions $f(n)$, but all known results have some limitations on $f(n)$. One can always use generating functions to solve an inhomogeneous recurrence relations, as in Sect. 13.2.

Exercises

13.34. Solve the recurrence $t_n - 3t_{n-1} = 2$ for $n \geq 1$ with initial condition $t_0 = 1$ using characteristic polynomials. As an aside, the number t_n gives the number of triangles in the Sierpinski graph.

13.35. Let f_n denote the n-th Fibonacci numbers, so $f_0 = 0$, $f_1 = 1$, and $f_n = f_{n-1} + f_{n-2}$ for $n \geq 2$. Solve the recurrence $t_n - 2t_{n-1} = f_n$ for $n \geq 1$ with initial condition $t_0 = 1$ using characteristic polynomials.

13.36. Solve the recurrence relation given by $t_n - t_{n-1} = n$ for $n \geq 1$ and $t_0 = 0$ using generating functions.

13.9 Catalan Numbers

In this section, we will study the Catalan numbers that occur in surprisingly many combinatorial counting problems. We will introduce them as numbers C_n that count certain paths in the upper real plane.

Specifically, we will count paths that connect the origin $(0,0)$ with the point (n,n), where n is a nonnegative integer. The path travels along integer coordinates by going either one step up or one step to the right. In other words, from a point with coordinates (a,b), one can travel up

$$U(a,b) = (a, b+1)$$

or right

$$R(a,b) = (a+1, b).$$

Additionally, we require that the path never goes below the line $y = x$. This means that the coordinates of all points (a,b) on the path must satisfy

$$0 \leq a \leq b \leq n.$$

The **Catalan number** C_n is defined as the number of different paths from $(0,0)$ to (n,n) that use n moves U and n moves R and never go below the line $y = x$.

Example 13.20. The Catalan number $C_3 = 5$, since there are five ways to connect the origin $(0,0)$ to $(3,3)$ using three moves U and three moves R in any order, as long as the path does not go below the line $y = x$. Indeed, we have the following five paths.

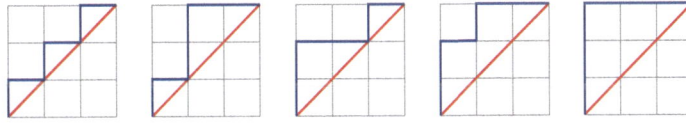

The paths are completely specified by their so-called Dyck words that describe the moves. The five paths correspond to the Dyck words URURUR, URUURR, UURRUR, UURURR, and UUURRR.

The basic principle of constructing the paths should now be apparent. Let us have a look at a few more examples. We have $C_0 = 1$, since there is only one way to construct a path of length 0. Similarly, we have $C_1 = 1$, since every Dyck word must start with an up move U and end with a right move R. Finally, we have $C_2 = 2$, since the only two valid paths from $(0,0)$ to $(2,2)$ are given by the Dyck words URUR and UURR. Larger examples are a bit tedious without a systematic way of enumerating the number of paths.

We can find a recursion for the Catalan number C_n by splitting the path into two parts. Let (m, m) denote the first coordinate where the path meets the line $y = x$ for the first time when starting from $(0, 0)$. Then we get an initial path from $(0, 0)$ to (m, m), and a second path from (m, m) to (n, n). The initial path segment is a bit more constrained, since it does not meet the line $y = x$ at any point (a, a) for any integer a in the range $0 < a < m$. Figure 13.2 illustrates the main idea behind this approach to counting the paths that define a Catalan number. This figure might be helpful when reading the explanation in the following text.

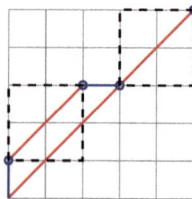

Figure 13.2: Counting Catalan numbers that meet the line $y = x$ for the first time at $(m, m) = (3, 3)$. The C_{m-1} initial paths are in the first dashed box on the lower left, and the second segment of the path is within the upper right dashed box. The possible initial paths start with U and end with R, and within the dashed box they follow either URUR or UURR. The possible second paths are in this example URUR or UURR

The initial path segment must go up U, then follow some Dyck word, before it ends in a right move R and then meets the coordinate (m, m). All we need to do is count the number of paths from $(0, 1)$ to $(m - 1, m)$ that do not go below the line $y = x + 1$. The offset by 1 is a result of the fact that the initial path will not meet the line $y = x$ before (m, m). Evidently, there are C_{m-1} such initial paths, since this is simply a smaller Catalan counting problem in a shifted grid.

Counting the number of second paths is much simpler. Indeed, the number of paths from (m, m) to (n, n) is given by C_{n-m}, since it follows the same constraints as our original problem, except that we start from (m, m) instead of $(0, 0)$.

Let us now put our observations together. The number of paths from $(0, 0)$ that meet the line $y = x$ for the first time at (m, m) for some integer m in the range $1 \leqslant m \leqslant n$ is given by $C_{m-1}C_{n-m}$. Therefore, the Catalan number C_n

13.9 Catalan Numbers

for $n \geq 1$ satisfies the recurrence

$$C_n = C_0 C_{n-1} + C_1 C_{n-2} + C_2 C_{n-3} + \cdots + C_{n-1} C_0.$$

Let us verify our previous computations of the Catalan numbers using this recurrence formula. We have

$$C_0 = 1,$$
$$C_1 = C_0 C_0 = 1 \cdot 1 = 1,$$
$$C_2 = C_0 C_1 + C_1 C_0 = 1 \cdot 1 + 1 \cdot 1 = 2,$$
$$C_3 = C_0 C_2 + C_1 C_1 + C_2 C_0 = 1 \cdot 2 + 1 \cdot 1 + 2 \cdot 1 = 5.$$

In principle, we can use this recurrence formula to calculate the Catalan numbers for arbitrary n. However, we can do even better and use generating functions to find an expression for Catalan numbers in terms of binomial coefficients.

In the next lemma, we find a closed-form expression for the generating function

$$C(z) = \sum_{n=0}^{\infty} C_n z^n$$

of the Catalan numbers.

Lemma 13.21. *The generating function of the Catalan numbers is given by*

$$C(z) = \frac{1 - \sqrt{1 - 4z}}{2z}.$$

Proof. We can express the Catalan number recurrence in the form

$$C_{n+1} = C_0 C_n + C_1 C_{n-1} + \cdots + C_n C_0 = \sum_{m=0}^{n} C_m C_{n-m}$$

for all nonnegative integers n. Multiplying both sides by z^n and summing over all nonnegative integers n yields

$$\frac{C(z) - 1}{z} - C(z)^2.$$

It follows that the generating function satisfies the quadratic equation

$$zC(z)^2 - C(z) + 1 = 0.$$

Solving for $C(z)$ yields by the quadratic formula

$$C(z) = \frac{1 \pm \sqrt{1 - 4z}}{2z}.$$

We will have to find out which choice of sign is appropriate. By the Binomial series, we have

$$\sqrt{1 - 4z} = \sum_{k=0}^{\infty} \binom{\frac{1}{2}}{k} (-1)^k 4^k z^k = 1 - 2z + - \cdots.$$

If we were to take the positive sign in the quadratic formula, we would obtain
$$\frac{1+\sqrt{1-4z}}{2z} = \frac{1+(1-2z+\cdots)}{2z} = \frac{1}{z} - 1 + \cdots.$$
However, this is not a formal power series, so we can rule out this choice. We can conclude that
$$C(z) = \frac{1-\sqrt{1-4z}}{2z},$$
as claimed. □

Proposition 13.22. *The Catalan number C_n is given by*
$$C_n = \frac{1}{n+1}\binom{2n}{n}$$
for all nonnegative integers n.

Proof. The Binomial series yields
$$\sqrt{1-4z} = \sum_{k=0}^{\infty} \binom{\frac{1}{2}}{k}(-1)^k 4^k z^k.$$
Therefore, the generating function $C(z)$ of the Catalan numbers satisfies
$$C(z) = \frac{1-\sqrt{1-4z}}{2z}$$
$$= -\frac{1}{2}\sum_{k=1}^{\infty}\binom{\frac{1}{2}}{k}(-1)^k 4^k z^{k-1}.$$
It follows that the n-th Catalan number is given by
$$C_n = [z^n]C(z) = -\frac{1}{2}\binom{\frac{1}{2}}{n+1}(-1)^{n+1}4^{n+1}.$$
In other words, we have
$$C_n = -\frac{1}{2}\frac{(1/2)(-1/2)(-3/2)\cdots((-2n+1)/2)}{(n+1)!}(-1)^{n+1}4^{n+1}.$$
We first notice that all the signs cancel. If we collect the $n+1$ factors of $1/2$, then we get
$$C_n = \frac{1}{2}\frac{1\cdot 1\cdot 3\cdot 5\cdots(2n-1)}{(n+1)!}2^{n+1}$$
$$= \frac{1\cdot 3\cdot 5\cdots(2n-1)}{(n+1)!}2^n$$

13.10 Notes

If we multiply the numerator and denominator by $n!$ and keep in mind that $n!2^n = 2 \cdot 4 \cdot 6 \cdots 2n$, then we can rewrite this expression in the form

$$C_n = \frac{(2n)!}{(n+1)!n!} = \frac{1}{n+1}\binom{2n}{n},$$

which proves our claim. $\square$

Exercises

13.37. Show that the Catalan number C_n is also given by

$$C_n = \binom{2n}{n} - \binom{2n}{n+1}.$$

13.38. Suppose that there is a group of $2n$ people that all have different heights. We want to take a group photo. For the photo, we would like to have two rows of n people each. The photographer requests that the height in each row increases from left to right, and that every person in the second row is taller than the person positioned directly in front. Show that there are

$$C_n = \frac{1}{n+1}\binom{2n}{n}$$

ways to arrange the $2n$ people for the photo.

13.39. A stack is a useful data structure in computer science. There are two operations that can manipulate a stack. A *push(m)* operation that will place an element m on the top of the stack and a *pop* operation that will retrieve and remove an element from the top of the stack. There is one restriction, though. A pop operation should never be issued to an empty stack. In how many different ways can you issue n push and n pop operations (in any order) such that an initially empty stack will be again empty after these $2n$ operations, and none of the intermediate pop operation will be issued to an empty stack.

13.10 Notes

Generating functions are an extremely useful tool in the solution of recurrence relations. You can find a more in-depth treatment of this method in books on enumerative combinatorics. We recommend the books by Aigner [3], Graham, Knuth, and Patashnik [30] and Stanley [72] for further reading.

The Catalan numbers were introduced by Euler, who discussed and solved the number of triangulations of a regular $(n+2)$-gon. They were rediscovered by the Belgian mathematician Eugène Catalan. The Catalan numbers occur frequently in combinatorial problems. You can find much more material about Catalan numbers in the delightful books by Koshy [53] and Stanley [74]. Stanley gives more than 200 combinatorial interpretations of Catalan numbers, so there is a myriad of applications where these numbers occur.

Chapter 14

Graphs

> *The four-color conjecture was easy to state and easy to understand, no large amount of technical mathematics is needed to attack it, and errors in proposed proofs are hard to see, even for professionals; what an ideal combination to attract cranks!*
>
> — Underwood Dudley, *Mathematical Cranks*

In this chapter, we will study the basics of undirected graphs. Graphs are widely used to abstractly model networks, circuits, friendship relations, and more. They find many applications in mathematics and computer science. We focus on fundamental examples and basic principles. We introduce common graphs and a few graph constructions, discuss connected graphs and trees. We also include a brief discussion of planar graphs and vertex coloring of graphs, paying tribute to the map coloring problem that inspired a large part of graph theory.

14.1 Undirected Graphs

Recall that a graph is a pair (V, E) of sets such that the E consists of 2-element subsets of V. The elements of V are called vertices or nodes of the graph, and the elements of E are called the edges of the graph. Given a graph G, we will denote by $V(G)$ the set of vertices and by $E(G)$ the set of edges. This notation is particularly useful when dealing with multiple graphs at the same time.

We will focus on graphs that have a finite number of vertices. The number of vertices of a graph is called the **order** of the graph, and the number of edges is called the **size** of the graph.

Graphs have an abundance of applications. For instance, subway maps are often colorful depictions of graphs. They rarely give an accurate representation of distances, but show how to get from one station to another. Figure 14.1 shows a small example of a subway map. The underlying graph distills some

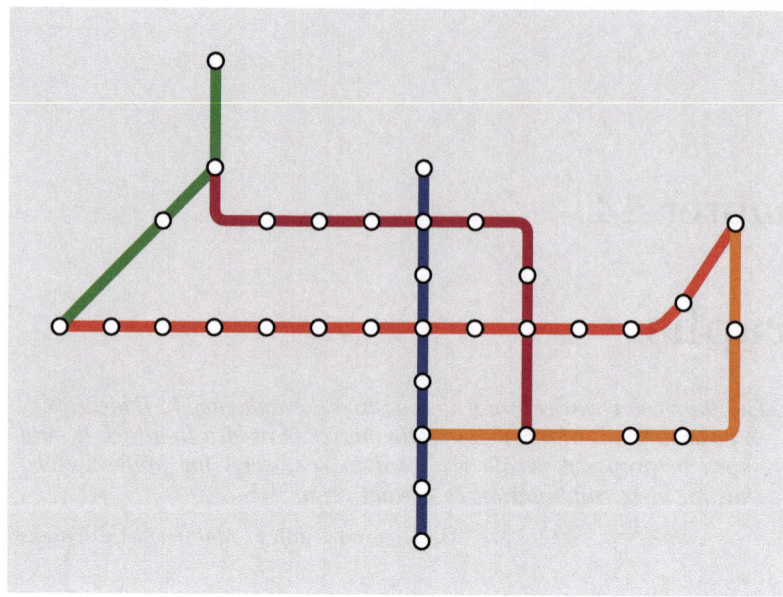

Figure 14.1: The subway map of a fictional city. The circles denote subway stations. We omitted the names of the stations. The different colors correspond to different subway lines

of the salient features of the subway map, for instance, whether one station is next to another.

Recall that in a visual representation of a graph, its vertices are depicted by points in the plane, and its edges are represented by line segments or curves connecting the points of the two vertices representing the edge. For instance, in the subway example, we can represent the subway stations as vertices and edges between any two subway stations that can be reached by a subway without intermediate stops at other stations. Figure 14.2 gives the visual representation of a graph corresponding to the subway map from Fig. 14.1.

Let us have a closer look at an even simpler example.

Example 14.1. Suppose that we are given the graph with four vertices and five edges with the following visual representation

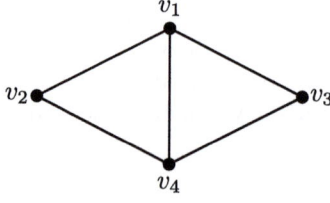

The graph (V, E) corresponding to this visual representation has the set

14.1 Undirected Graphs

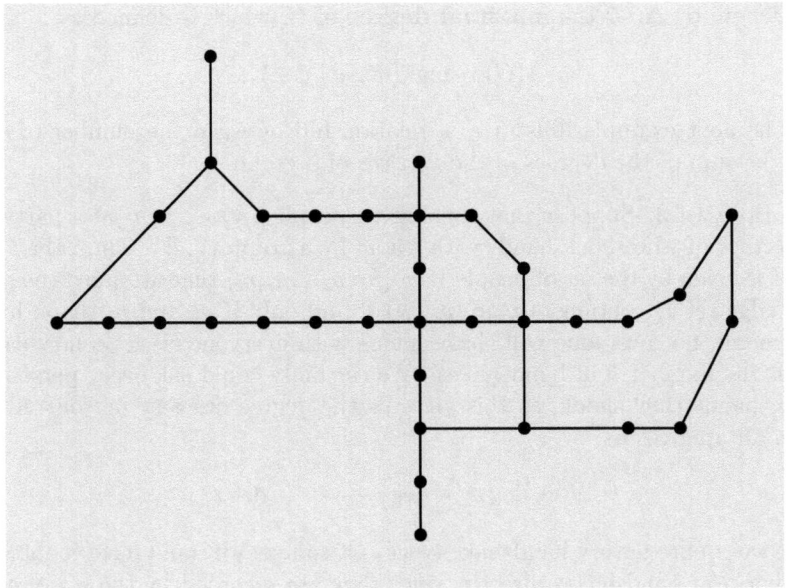

Figure 14.2: The graph corresponding to the subway map. There is an edge between two vertices if and only if they correspond to adjacent subway stops

$V = \{v_1, v_2, v_3, v_4\}$ of vertices and the edge set

$$E = \{\{v_1, v_2\}, \{v_1, v_3\}, \{v_1, v_4\}, \{v_2, v_4\}, \{v_3, v_4\}\}.$$

It is often very helpful to illustrate concepts in graph theory with small examples that have a succinct visual representation.

Two vertices that are connected by an edge are called **adjacent**. A vertex v is called **incident** with an edge e if and only if $v \in e$. Given a vertex v, the **neighborhood** $N(v)$ in a graph is the set of all vertices that are adjacent to v in G. In other words, $N(v) = \{u \in V \mid \{u, v\} \in E\}$. The neighborhood $N(v)$ of v does not contain v itself. The **closed neighborhood** $N[v]$ of v is defined as the neighborhood of v with the node v included, $N[v] = N(v) \cup \{v\}$.

The **degree** $\deg v$ of a node v is the cardinality of its neighborhood, so $\deg v = |N(v)|$.

Example 14.2. We continue our previous example. The node v_1 has the neighborhood $N(v_1) = \{v_2, v_3, v_4\}$. Since the neighborhood of v_1 consists of three nodes, we have $\deg v_1 = 3$. The node v_2 has degree 2, since its neighborhood $N(v_2)$ contains only the two nodes v_1 and v_4.

Given a graph $G = (V, E)$ with finite vertex set V, we denote by $\delta(G)$ the **minimal degree** of the graph G, which is defined as

$$\delta(G) = \min\{\deg v \mid v \in V\}.$$

We denote by $\Delta(G)$ the **maximal degree** of G, which is defined as
$$\Delta(G) = \max\{\deg v \mid v \in V\}.$$

The next example illustrates a relationship between the number of edges and the sum of the degrees of the vertices of a graph.

Example 14.3. Suppose that there are n people $p_1, p_2, \ldots, p_n$ at a party. We keep track of who shook hands with whom by a graph (V, E). Thus, the vertex set V is given by the set of people $V = \{p_1, p_2, \ldots, p_n\}$ that attended the party. The edge set E contains a pair $\{p_k, p_\ell\}$ if and only if p_k and p_ℓ shook hands. In general, not everyone will shake hands with everyone else. So how can we count the number h of handshakes? We certainly could ask every person how many hands they shook, so this gives us the degree of every person. Adding them all up gives us

$$2h = \deg p_1 + \deg p_2 + \cdots + \deg p_n,$$

since we counted every handshake twice. Of course, we could have counted the number h of handshakes directly, since they are recorded by the set of edges, so we have
$$h = |E|.$$
This double counting argument gives us the relation
$$2|E| = \deg p_1 + \deg p_2 + \cdots + \deg p_n$$
between the number of edges and the sum of the degrees in G.

The handshake example applies to arbitrary graphs with a finite number of vertices.

Proposition 14.4 (Handshaking Lemma). *Let $G = (V, E)$ be a graph of order n and size m. Then*
$$2m = \sum_{v \in V} \deg v.$$
In particular, the minimum degree $\delta(G)$ and the maximum degree $\Delta(G)$ satisfy the bound
$$\delta(G) \leqslant \lfloor 2m/n \rfloor \quad \text{and} \quad \Delta(G) \geqslant \lceil 2m/n \rceil.$$

Proof. The same argument as in the previous example shows the equality
$$2m = \sum_{v \in V} \deg v.$$
Since the degree of each of the n vertices is lower-bounded by $\delta(G)$ and upper-bounded by $\Delta(G)$, we have
$$n\delta(G) \leqslant \sum_{v \in V} \deg v = 2m \leqslant n\Delta(G).$$

14.1 Undirected Graphs

Dividing all terms by n yields

$$\delta(G) \leqslant 2m/n \leqslant \Delta(G),$$

which implies $\delta(G) \leqslant \lfloor 2m/n \rfloor$ and $\Delta(G) \geqslant \lceil 2m/n \rceil$. □

Let $G = (V, E)$ be a graph with a finite set V of vertices. The graph G is called **k-regular** if and only if every vertex has degree k. In other words, a graph G is k-regular if and only if $k = \delta(G) = \Delta(G)$.

Corollary 14.5. *Let $G = (V, E)$ be a k-regular graph with n vertices. Then G has $kn/2$ edges.*

Proof. By the Handshaking Lemma, the number m of edges of G satisfies

$$2m = \sum_{v \in V} \deg v = kn,$$

which implies the claim. □

Exercises

14.1. Suppose that an undirected graph G has n vertices. What is the largest number of edges that the graph G can have?

14.2. How many graphs are there that have n vertices?

14.3. Does there exist an undirected graph G with seven vertices $v_1, v_2, \ldots, v_7$ such that
$$\deg v_1 = 1, \deg v_2 = 2, \deg v_3 = 3, \ldots, \deg v_7 = 7?$$
Explain your answer.

14.4. Does there exist an undirected graph G with seven vertices $v_1, v_2, \ldots, v_7$ such that

$\deg v_1 = 0, \deg v_2 = 2, \deg v_3 = 2, \deg v_4 = 4, \deg v_5 = 4, \deg v_6 = 4, \deg v_7 = 6$?

Explain your answer.

14.5. Tony explains to his friend that he constructed a graph with five vertices, where 2 of the vertices have degree 3, and the other three vertices have degree 2. Tony's little brother brags that he constructed a graph with five vertices such that two of the vertices have degree 4, and three of the vertices have degree 3. Fact check the claim by Tony's little brother.

14.6. Suppose that G is a graph with 12 vertices and 30 edges. It contains 2 vertices of degree 4, 1 vertex of degree 5, 4 vertices of degree 6, and 2 vertices of degree 7. All other vertices have the same degree d. What is the degree d?

14.7. Show that a graph cannot have an odd number of vertices that have odd degree.

14.8. Suppose that five couples meet at a party. One of the couples is the host and hostess of the party. The host asks everyone, including his wife, how many people they shook hands with. Curiously, everyone questioned shook hands with a different number of people. Naturally, the host did not ask himself the question. We also know that none of the people shook hands with his or her partner. How many people shook hands with the hostess? (Source: Zeitz)

14.9. Let a and b be adjacent vertices in a graph G.
(a) How many vertices can be in $N(a) \cup N(b)$?
(b) How many vertices can be in $N(a) \cap N(b)$?

14.10. Let G be a graph of order n, where $n \geqslant 2$. Show that there must exist two vertices in G that have the same degree.

14.11. A **triangle** in a graph is a subset of three distinct edges that all pairwise intersect, meaning that they are of the form $\{a,b\}, \{b,c\}, \{c,a\}$. Show that a graph G of order n and size m has at least

$$\frac{m}{n}\left(m - \frac{1}{4}n^2\right)$$

triangles. In particular, this shows that any graph of order n at least 3 and size $m > n^2/4$ must have at least one triangle.

14.2 Common Graphs

In this section, we will discuss a number of families of graphs that occur frequently in applications. These families of graphs are part of the basic vocabulary of graph theory. We will also introduce some basic constructions of graphs that allow us to form more elaborate graphs from such basic primitives.

We begin with a very simple example. The **empty graph** E_n is a graph with n vertices and no edges. Distinct vertices are not adjacent in the empty graph E_n. Since every vertex has degree 0, this is a 0-regular graph. Figure 14.3 shows an example of an empty graph.

• • • • •

Figure 14.3: The empty graph E_5 with five vertices. It does not have any edges

The **path graph** P_n is the graph with vertex set $V = \{v_0, v_1, \ldots, v_{n-1}\}$ and edge set
$$E = \{\{v_k, v_{k+1}\} \mid 0 \leqslant k < n-1\}.$$
The degree of the vertices v_0 and v_{n-1} is 1, and the degree of vertex v_k is 2 when $1 \leqslant k < n-1$. Thus, P_n is not a regular graph when $n > 2$. By contrast,

14.2 Common Graphs

Figure 14.4: The path P_5 with five vertices. There is an edge between successive nodes

the graph P_1 is 0-regular and the graph P_2 is 1-regular. Figure 14.4 shows a typical example of a path graph.

For an integer $n \geq 3$, the **cycle graph** C_n is the graph with vertex set $V = \{v_0, v_1, \ldots, v_{n-1}\}$ and edge set

$$E = \{\{v_k, v_{k+1 \, (\mathrm{mod}\, n)}\} \mid 0 \leq k \leq n-1\}.$$

We can obtain the graph C_n from the path graph P_n by adding the edge $\{v_{n-1}, v_0\}$ connecting the last vertex v_{n-1} to the first vertex v_0. Examples of cycle graphs are shown in Fig. 14.5.

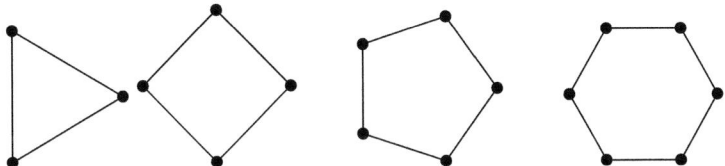

Figure 14.5: The cycle graphs C_3, C_4, C_5, and C_6

The **complete graph** K_n on n vertices is a graph with vertex set $V = \{0, 1, \ldots, n-1\}$ and edge set $E = \{\{a, b\} \mid a, b \in V, a \neq b\}$. The complete graph contains an edge between any two distinct vertices. The complete graph K_n is $(n-1)$-regular, since each vertex is adjacent to each of the other $n-1$ vertices. Figure 14.6 shows some examples of small complete graphs.

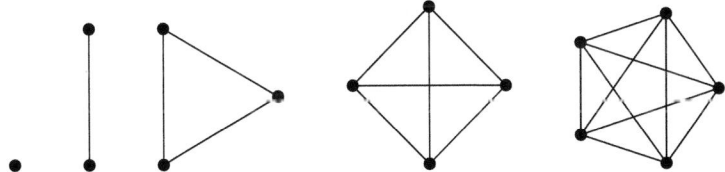

Figure 14.6: The complete graphs K_1, K_2, K_3, K_4, and K_5

A graph $G = (V, E)$ is called **bipartite** if and only if the set of vertices can be partitioned into two disjoint subsets V_1 and V_2 such that every edge joins a vertex in V_1 with a vertex in V_2. In other words, the set of edges is restricted such that no edge connects two vertices in V_1 and no edge connects two vertices in V_2. This implies that the neighbors of a vertex in V_1 are in V_2, and the neighbors of a vertex in V_2 are in V_1. The sets V_1 and V_2 are called the **partite classes** of the bipartite graph. Figure 14.7 illustrates the concept of bipartite graph in the case of cyclic graphs.

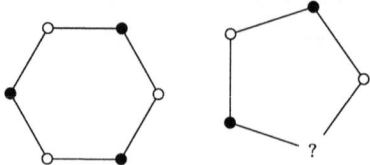

Figure 14.7: The cycle graph C_n is bipartite if and only if n is even. For instance, the figure on the left shows the cycle graph C_6 with vertices in V_1 depicted as ○ and vertices in V_2 depicted as ●, so this is a bipartite graph as each edge is incident to one ○ node and one ● node. The figure on the right shows the cycle graph C_5. No matter how you partition the vertices, there is always one vertex whose neighborhood contains both ○ and ● nodes, which simply means that the graph C_5 is not bipartite

The **complete bipartite graph** $K_{m,n}$ has the vertex set

$$V = \{a_1, a_2, \ldots, a_m\} \cup \{b_1, b_2, \ldots, b_n\}$$

and the edge set $E = \{\{a_k, b_\ell\} \mid 1 \leqslant k \leqslant m, 1 \leqslant \ell \leqslant n\}$. The graph $K_{m,n}$ is indeed bipartite, since every edge is incident with one a-vertex and one b-vertex. The graph $K_{m,n}$ has order $m+n$ and size mn. Figure 14.8 shows the complete bipartite graph $K_{3,5}$.

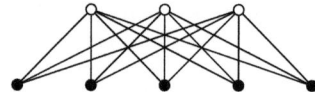

Figure 14.8: The complete bipartite graph $K_{3,5}$

One special case is worth noting. The complete bipartite graph $K_{1,n}$ is also called a **star graph**. Figure 14.9 shows the star graph $K_{1,6}$.

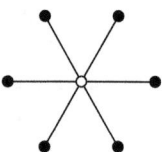

Figure 14.9: The graph $K_{1,6}$ is also known as a star graph

There are various operations that allow us to construct more elaborate graphs from one or more given graphs.

Let T denote the set of all 2-element subsets of V. The **complement** of a graph $G = (V, E)$ is the graph $\overline{G} = (V, T \backslash E)$ that has the same vertex set and the edges that are missing in G as an edge set. In other words, two vertices are adjacent in the complementary graph $\overline{G}$ if and only if they are not adjacent in the graph G. For instance, the complement of the empty graph E_n is the complete graph $K_n = \overline{E_n}$. Another example of a graph and its complement is shown in Fig. 14.10.

14.2 Common Graphs

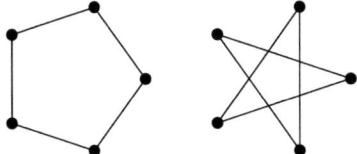

Figure 14.10: The graph C_5 is shown on the left and its complementary graph $\overline{C_5}$ on the right

Suppose that $G_1 = (V_1, E_1)$ and $G_2 = (V_2, E_2)$ are two undirected graphs. The **Cartesian product** $G_1 \square G_2$ is the graph with vertex set $V_1 \times V_2$ such that two vertices (u_1, v_1) and (u_2, v_2) in $V_1 \times V_2$ are adjacent if and only if either (a) $u_1 = u_2$ and $(v_1, v_2) \in E_2$ or (b) $(u_1, u_2) \in E_1$ and $v_1 = v_2$ holds.

For instance, the product of the path graph P_m and the path graph P_n is the grid graph $P_m \square P_n$. An example of the grid graph is shown in Fig. 14.11.

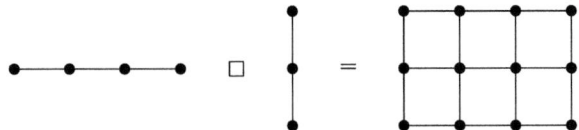

Figure 14.11: The Cartesian product of the path graphs P_4 and P_3 is shown on the right. It is self-explanatory why the Cartesian product $P_4 \square P_3$ is called a grid graph

The n-dimensional **hypercube graph** Q_n is the graph of order 2^n that is obtained by repeatedly forming the Cartesian product of n copies of the complete graph K_2, so

$$Q_n = \underbrace{K_2 \square K_2 \square \cdots \square K_2}_{n \text{ copies of } K_2}.$$

The hypercube Q_n is an n-regular graph of order 2^n and size $n2^{n-1}$. Figure 14.12 shows some examples of hypercube graphs of small dimension.

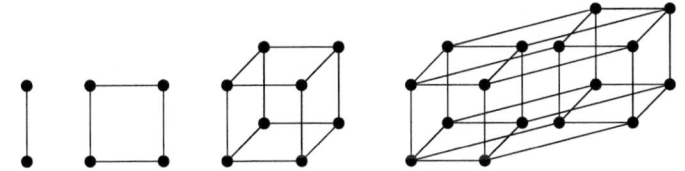

Figure 14.12: The figure shows the hypercubes Q_1, Q_2, Q_3, and Q_4

Given a graph G and a positive integer k, we denote by kG the graph given by the disjoint union of k copies of G. We can also construct kG by forming the Cartesian product of the empty graph E_k with G,

$$kG = E_k \square G.$$

If G has order n and size m, then kG has order kn and size km.

Exercises

14.12. Determine the subset B of positive integers n such that K_n is bipartite,
$$B = \{n \mid K_n \text{ is bipartite}\}.$$
You need to prove your result. In other words, for each n, you need to give an argument whether or not $K_n \in B$ holds.

14.13. Suppose that G is a graph of order n and size m. What is the order and the size of its complementary graph $\overline{G}$?

14.14. Show that the Cartesian product of graphs is associative.

14.15. Suppose that G_1 is a graph of order n_1 with e_1 edges and G_2 is a graph of order n_2 with e_2 edges. Show that the Cartesian product graph $G_1 \square G_2$ is a graph of order $n_1 n_2$ with $n_1 e_2 + n_2 e_1$ edges.

14.16. Prove that the hypercube Q_n is an n-regular graph of order 2^n and size $n2^{n-1}$.

14.3 Connected Graphs

Exploring a graph by walking along the edges in various ways can give valuable insights about the nature of the graph. We introduce the important class of connected graphs and some sufficient conditions that ensure connectivity.

A **walk** from a vertex u to a vertex v in an undirected graph $G = (V, E)$ is an alternating sequence of vertices and edges that begins with the vertex u and ends with the vertex v such that each edge is incident with the vertices preceding and following the edge. In other words, a walk from u to v is a sequence
$$(v_0, e_1, v_1, e_2, v_2, \ldots, e_k, v_k) \tag{14.1}$$
of vertices $v_0, v_1, \ldots, v_k$ in V and edges $e_1, e_2, \ldots, e_k$ in E such that the start vertex v_0 equals u and the end vertex v_k equals v and the edges are of the form $e_m = \{v_{m-1}, v_m\}$ for all m in the range $1 \leq m \leq k$. One should note that an edge or a vertex can occur multiple times in a walk. The number of edges that are traversed on a walk is called the **length** of the walk. For instance, the walk given in (14.1) has length k. Figure 14.13 illustrates a walk in a grid graph.

A **trail** from a vertex u to a vertex v is a walk from u to v that does not contain repeated edges. A trail might still contain a vertex multiple times. A **circuit** is a trail from a vertex u to itself. By definition, a circuit does not contain an edge twice, but may contain repeated vertices (even in addition to the vertex u). Figure 14.14 gives an example of a trail in a grid graph.

A **path** from a vertex u to a vertex v in a graph is a walk without repeated vertices. A path is in particular a trail. A **cycle** is a circuit that does not

14.3 Connected Graphs

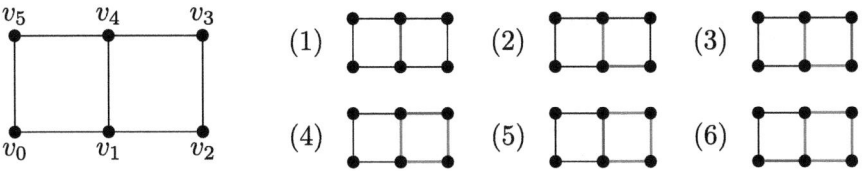

Figure 14.13: The figure on the left shows the grid graph $P_3 \square P_2$. A walk from v_4 to v_0 is given by

$$(v_4, \{v_4, v_1\}, v_1, \{v_1, v_2\}, v_2, \{v_2, v_3\}, v_3, \{v_3, v_4\}, v_4, \{v_4, v_1\}, v_1, \{v_1, v_0\}, v_0).$$

The steps of this walk of length 6 are depicted in the six figures on the right. This walk is not a trail, since it uses the edge $\{v_4, v_1\}$ twice

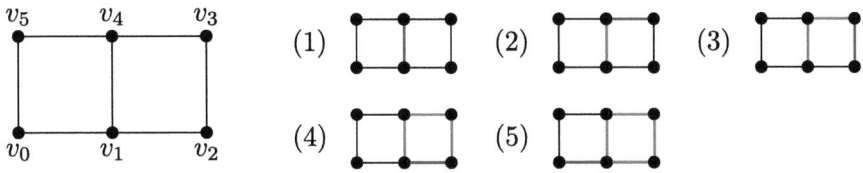

Figure 14.14: The figure on the left shows the grid graph $P_3 \square P_2$. A trail from v_1 to v_0 is given by

$$(v_1, \{v_1, v_4\}, v_4, \{v_4, v_3\}, v_3, \{v_3, v_2\}, v_2, \{v_2, v_1\}, v_1, \{v_1, v_0\}, v_0).$$

This is indeed a trail, since it does not pass any edge twice, but it is not a path, since it visits the vertex v_1 twice

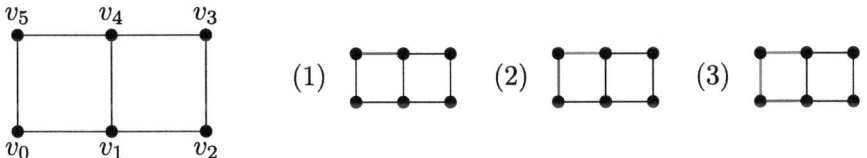

Figure 14.15: The figure on the left shows the grid graph $P_3 \square P_2$. A path from v_4 to v_1 is given by

$$(v_4, \{v_4, v_5\}, v_5, \{v_5, v_0\}, v_0, \{v_0, v_1\}, v_1)$$

contain repeated vertices with the exception that the first vertex is equal to the last vertex. Figure 14.15 gives an example of a path.

A graph $G = (V, E)$ is called **connected** if and only if for any two distinct vertices u and v in V there exists some path from u to v.

The next proposition gives a simple sufficient condition for connectedness.

Proposition 14.6. *Let $G = (V, E)$ be a graph of order n. If every pair of non-adjacent vertices u and v in V satisfy the degree-constraint*

$$\deg u + \deg v \geqslant n - 1,$$

then G is a connected graph.

Proof. Suppose that u and v are distinct vertices in V. We are going to show that there must exist a path between them. We distinguish the cases whether the vertices u and v are adjacent or not.
(a) If u and v are adjacent, then there exists a path of length 1 between them.
(b) If u and v are non-adjacent vertices, then there are $n-2$ vertices in $V\backslash\{u,v\}$. By the degree constraint, the sum of the degrees of u and v is given by $\deg u + \deg v \geqslant n - 1$. Thus, by the pigeonhole principle, there must exist a vertex w in $V\backslash\{u,v\}$ that is adjacent to both vertices u and v. In other words, there exists a path of length 2 between any two non-adjacent vertices.

In conclusion, we have shown that there exists a path between any two vertices of G, so the graph G is connected. □

The next proposition is in a similar spirit, but has a hypothesis that is easier to check.

Proposition 14.7. *Let $G = (V, E)$ be a graph of order n. If $\Delta(G) + \delta(G) \geqslant n - 1$, then G is a connected graph.*

Proof. Let v be a vertex of maximal degree, $\deg v = \Delta(G)$. It suffices to show that for every vertex u in $V\backslash\{v\}$, there exists a path from u to v.

Let u be an arbitrary vertex in $V\backslash\{v\}$. If u is adjacent to v, then there is a path from u to v.

If u is not adjacent to v, then $\deg u + \deg v \geqslant \delta(G) + \Delta(G) \geqslant n - 1$. Since there are $n - 2$ nodes in $V\backslash\{u,v\}$, and the number of edges from $\{u,v\}$ to $V\backslash\{u,v\}$ is at least $n - 1$, there must be at least one vertex w in $V\backslash\{u,v\}$ that is adjacent to both u and v by the pigeonhole principle. Thus, there is a path from u to v.

If u and u' are arbitrary vertices of G, then there is a path from u to the vertex v of maximal degree that we have singled out, and one from this vertex v to u', so there is a path from u to u'. Therefore, the graph G is connected. □

Given a graph $G = (V, E)$ and vertices u and v in V, we write $u \equiv v$ if and only if there is a path from u to v. The relation $\equiv$ is reflexive, symmetric, and transitive. Therefore, $\equiv$ is an equivalence relation on V. We encourage the reader to prove this simple fact, see Exercise 14.18. The equivalence class C_u containing a vertex u is called a component of the graph. We have

$$C_u = \{v \in V \mid u \equiv v\}.$$

We denote by $k(G)$ the **number of components** of the graph G.

Example 14.8. The empty graph E_n with n vertices has $n = k(E_n)$ components. Each component contains just a single vertex.

14.3 Connected Graphs

Suppose that $G = (V, E)$ is an undirected graph and e is an edge of G. We denote by $G - e$ the graph $(V, E\backslash\{e\})$. In other words, $G - e$ has the same vertex set as G and all the edges of G with the exception of e.

We call an edge e of a graph G a **bridge** if and only if $k(G - e) > k(G)$. Evidently, an edge e of a graph G is a bridge if and only if it is not part of any cycle in G.

The **distance** $d(u, v)$ between vertices u and v of a graph is the length of the shortest path between them. We set $d(u, v) = \infty$ if and only if there does not exist a path from u to v.

In a connected graph $G = (V, E)$, the distance is a metric, since the distance is a nonnegative function $V \times V \to \mathbf{R}$ such that for all vertices u, v, w in V it satisfies

M1. the identity of indiscernibles, meaning $d(u, v) = 0$ if and only if $u = v$,

M2. the symmetry condition $d(u, v) = d(v, u)$,

M3. and the triangle inequality, $d(u, v) + d(v, w) \geq d(u, w)$.

The **diameter** $d(G)$ of a graph G is the maximum distance between any two vertices

$$d(G) = \max_{u,v \in V(G)} d(u, v).$$

In a connected graph G, the diameter of G is the maximum length among all shortest paths between any pair of vertices in G.

Exercises

14.17. Show that if a graph $G = (V, E)$ contains a walk from a vertex u to a vertex v, then it contains a path from u to v.

14.18. Let $G = (V, E)$ be an undirected graph. Let $\equiv$ be the relation on the set of vertices V given by $u \equiv v$ if and only if there is path from u to v. Show that $\equiv$ is an equivalence relation.

14.19. Show that a graph G of order n and size m has at least $n - m$ components; in other words, $k(G) \geq n - m$.

14.20. Let $G = (V, E)$ be a graph of order n and size m. Show that if the size of the graph satisfies $m > \binom{n-1}{2}$, then G must be connected.

14.21. Suppose that $G = (V, E)$ is a disconnected graph. Prove that its complementary graph $\overline{G}$ must be connected.

14.22. Suppose that $G = (V, E)$ is a connected graph. Prove or disprove: The complementary graph $\overline{G}$ is disconnected.

14.23. Determine the diameter of
(a) the empty graph E_n,

(b) the path graph P_n,
(c) the cycle graph C_n,
(d) the complete graph K_n,
(e) the complete bipartite graph $K_{m,n}$,
(f) the hypercube Q_n of order 2^n.

You can assume that m and n are positive integers.

14.4 Trees

A **tree** is a connected graph that does not contain any cycles. Typical examples of trees include the star graphs $K_{1,n}$ for $n \geqslant 1$ and the path graphs P_n for $n \geqslant 1$. Another simple example of a tree is illustrated in Fig. 14.16.

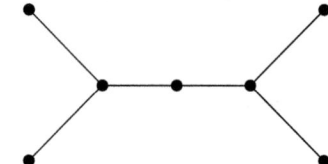

Figure 14.16: A tree of order 7 and size 6

By inspecting the tree in Fig. 14.16, you might notice that there is only one path from one vertex to another. The next proposition shows that this property is shared by all trees.

Proposition 14.9. *In a tree, any two vertices are connected by a unique path.*

Proof. Seeking a contradiction, let us assume that $T = (V, E)$ is a tree that has two vertices u and v, which are connected by two distinct paths A and B. This implies that there exists a cycle consisting of edges from A and B, contradicting the fact that a tree does not contain any cycles. □

In a tree, a vertex of degree 1 is called a **leaf**. A vertex in a tree that is not a leaf is called an **internal vertex** or **internal node**.

Proposition 14.10. *A tree T of order $n \geqslant 2$ has at least two leaves.*

Proof. Let P denote a path of maximal length in T that starts from a vertex u and ends at a vertex v. The nodes u and v cannot be adjacent to any vertex that is not contained in P, since this would allow us to extend the path, contradicting the maximality of the length of P. The vertex u (and similarly the vertex v) is adjacent to precisely one vertex in P, since otherwise we would have a cycle in T. Therefore, u and v are both vertices of degree 1, so they are two distinct leaves. □

Figure 14.16 shows an example of a tree with 7 vertices and $6 = 7-1$ edges. In general, any tree of order n will have size $n-1$, as the next proposition shows.

14.4 Trees

Proposition 14.11. *A tree of order n must have size $m = n - 1$.*

Proof. We prove the claim by induction on the order n.
Induction Basis If the order of the tree is $n = 1$, then there are no edges, so the size of the tree is indeed $n - 1 = 0$.
Induction Step Let us suppose that the claim holds for all trees of order n or less. Suppose that T is a tree of order $n + 1$. Let e be an edge of T. Since the edge e is not part of any cycle, it must be a bridge. Therefore, $T-e$ decomposes into two components T_a and T_b. Since each component is cycle-free, they are both trees. Suppose that the order of T_a is a and the order of T_b is b. The sum of these orders give us the order of T, so

$$n + 1 = a + b.$$

By induction hypothesis, T_a has size $a - 1$ and T_b has size $b - 1$. Since the set of edges of the tree T is given by

$$E(T) = E(T_a) \cup E(T_b) \cup \{e\},$$

the size of T is given by

$$(a - 1) + (b - 1) + 1 = a + b - 1 = n.$$

Therefore, the claim follows by induction on n. □

A graph H is called a **subgraph** of a graph G if and only if its set of vertices satisfies $V(H) \subseteq V(G)$ and its set of edges satisfies $E(H) \subseteq E(G)$. A subgraph H of G is called **spanning** if and only if their vertex sets are the same, $V(H) = V(G)$. A spanning subgraph that happens to be a tree is called a **spanning tree**.

Proposition 14.12. *Every connected graph G has at least one spanning tree.*

Proof. Let $\mathcal{S}$ denote the family of connected spanning subgraphs of G, ordered by inclusion. Choose a minimal element T of $\mathcal{S}$. Then T must be a spanning tree of G. Indeed, if T would contain a cycle, then we could delete an edge e from this cycle and $T - e$ would be a connected spanning graph, contradicting the minimality of T. □

Exercises

14.24. Show that a graph G is a tree if there is a unique path between any two vertices.

14.25. A graph without cycles is called a **forest**. Show that if a forest F of order n has $k(F)$ components, then it has size $n - k(F)$.

14.26. Prove that a connected graph G is a tree if and only if every edge is a bridge.

14.27. Show that a connected graph G of order n and size $n-1$ must be a tree.

14.28. Determine the set consisting of all trees T such that their complement $\overline{T}$ is a tree as well.

14.5 Planar Graphs

A graph G is called **planar** if and only if it can be represented in the plane such that edges do not intersect except in end points. Even though some representations of the graph might have intersections of edges, the definition of a planar graph guarantees that there exists some way of drawing the edges using curves that do not intersect.

Example 14.13. The complete graph K_4 of order 4 is planar. Indeed, the next figure shows on the left-hand side how we drew K_4 earlier with two edges crossing. However, the right-hand side illustrates that there is a way to draw K_4 without crossing edges.

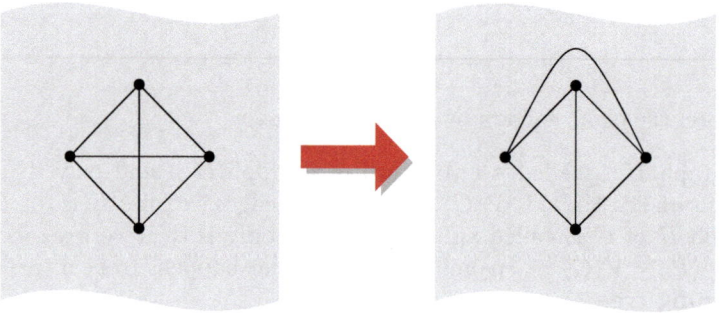

We can conclude that K_4 is a planar graph.

Example 14.14. The complete graph K_5 of order 5 is not planar.

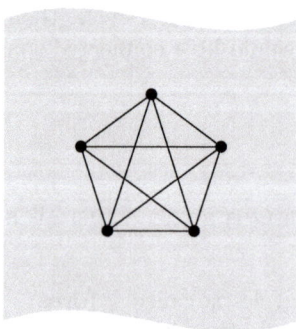

Any representation of K_5 in the plane requires that some edges cross. We will later discover a simple proof of this fact.

14.5 Planar Graphs

If a planar graph is represented in the plane without edges crossing, then it divides the plane into regions that are called **faces**. The faces are maximal open regions of the plane that are bounded by the edges of the graph. There exists always one face that is unbounded, surrounding the graph. The concept of faces is best illustrated with a small example.

Example 14.15. The following planar graph G has five vertices, eight edges, and five faces. The faces F_0, F_1, F_2, F_3, and F_4 are shown in different colors.

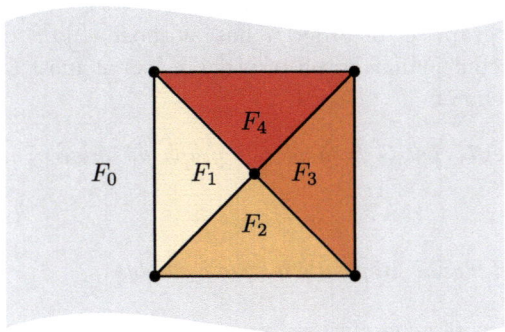

The face F_0 is the unbounded face surrounding the graph, whereas the faces F_1 to F_4 are each bounded by three edges forming a triangle.

A drawing of a planar graph in the plane without crossings is called a **plane graph**. So a plane graph is one particular representation of a planar graph.

Proposition 14.16 (Euler's Formula). *If a connected plane graph G has order n, size m, and f faces, then*

$$n - m + f = 2.$$

Proof. We will argue by induction on the number of faces. If we have a single face, so $f = 1$, then the graph G cannot contain a cycle. In other words, G must be a tree. A tree with n vertices has $m = n - 1$ edges. Therefore, we indeed have

$$n - m + f = n - (n - 1) + 1 = 2,$$

so Euler's formula holds in this case.

Suppose that we have more than one face, $f > 1$, and the result holds for a smaller number of faces. Since we have more than one face, there must be a cycle. Choose an edge $e = \{u, v\}$ in a cycle of G. The cycle encloses a face. Let us create a new graph $G - e$ that is obtained from G by removal of the edge e. The edge e must have separated two faces, so the graph $G - e$ has order n, size $m - 1$ and $f - 1$ faces. By induction hypothesis, $G - e$ satisfies Euler's formula

$$n - (m - 1) + (f - 1) = 2.$$

Simplifying shows that
$$n - m + f = 2$$
holds, which is Euler's formula for the graph G. □

It follows from the previous proposition that any two representations of a planar graph by plane graphs must have the same number of faces.

A **maximal planar graph** is a planar graph to which one cannot add any edge without losing planarity. Maximal planar graphs with at least three vertices have the property that the boundary of every face is a triangle.

In general, a graph G of order n has at most $\binom{n}{2} = \Theta(n^2)$ edges. By contrast, a connected planar graph of order n has at most $O(n)$ edges as the next proposition shows.

Proposition 14.17. *Let G be a planar graph of order n and size m, where $n \geqslant 3$. Then*
$$m \leqslant 3n - 6.$$
If G is a maximal planar graph, then equality holds.

Proof. Without loss of generality, we may assume that G is a maximal planar graph. Indeed, if G is not a maximal planar graph, then we can add edges without changing the order of the graph.

In a plane embedding of a maximal planar graph, every face is bounded by exactly three edges. On the other hand, every edge is on the boundary of two faces. Therefore, the number of (face, edge) pairs satisfies $3f = 2m$. By Euler's formula, $n - m + f = 2$, so $n - m + \frac{2}{3}m = n - \frac{1}{3}m = 2$. Multiplying both sides of this equality by 3, we get $3n - m = 6$ or $m = 3n - 6$. We can conclude that an arbitrary (not necessarily maximal) planar graph satisfies $m \leqslant 3n - 6$. □

Corollary 14.18. *Every planar graph has at least one vertex of degree 5 or less.*

Proof. Seeking a contradiction, let's assume that G is a planar graph of order n and size m such that every vertex has degree 6 or more. By the Handshaking Lemma, we would have $2m \geqslant 6n$, so $m \geqslant 3n$. However, any planar graph satisfies $m \leqslant 3n - 6$, leading to the contradiction $3n \leqslant 3n - 6$. □

Example 14.19. The complete graph K_5 is a connected graph of order $n = 5$ and size $m = \binom{5}{2} = 10$. If K_5 were planar, then it would satisfy $m \leqslant 3n - 6$. However, here we have $10 = m > 3n - 6 = 9$, so K_5 cannot be a planar graph.

Exercises

14.29. Show that if $n \geqslant 5$, then the complete graph K_n is not a planar graph.

14.30. Show that the complete bipartite graph $K_{3,3}$ is not a planar graph.

14.6 Graph Coloring

14.31. Show that the Petersen graph of order 10 and size 15 (shown below) is not a planar graph.

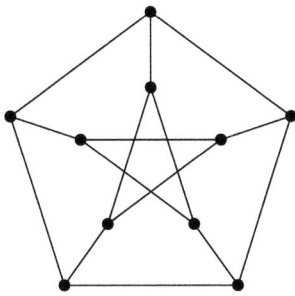

14.32. Let G be a planar graph of order n. Recall that $\Delta(G)$ denotes the maximal degree of G. Let n_k denote the number of vertices of degree k for $0 \leqslant k \leqslant \Delta(G)$. Show that

$$\sum_{k=0}^{5}(6-k)n_k \geqslant \sum_{k=7}^{\Delta(G)}(k-6)n_k + 12.$$

Note that this implies that G must contain a vertex of degree 5 or less.

14.33. Let G be a graph of order $n \geqslant 11$. Show that either G or its complement $\overline{G}$ is not planar.

14.6 Graph Coloring

A political map is supposed to show the governmental boundaries of countries, states, or counties. Good maps use several distinct colors so that one can easily differentiate between countries, states, or counties. Figure 14.17 shows the states of the western half of the United States. The states are colored such that neighboring states do not share the same color, except when they merely share a corner. Remarkably, very few colors are needed to color any map. This was observed in 1852 by the student Francis Guthrie, and his brother related the observation to De Morgan. Soon it was mathematical folklore that any map can be colored with at most four colors, but a proof remained elusive. In fact, it took more than a 100 years until it was proven in 1977 by Haken and Appel using a computer-aided proof. The proof remained a bit controversial, since it was too long to be checked by humans. Meanwhile, Gonthier obtained a shorter computer-aided formal proof.

The search for a proof of the four-color theorem spurred the development of graph theory. One can easily relate the map coloring problem to the coloring of planar graphs. For a given map of states, we can find an associated graph as follows. Each state is represented by a vertex. Two vertices are joined by an edge if and only if the corresponding states share a border. The map coloring problem is now translated in a vertex coloring problem of the corresponding

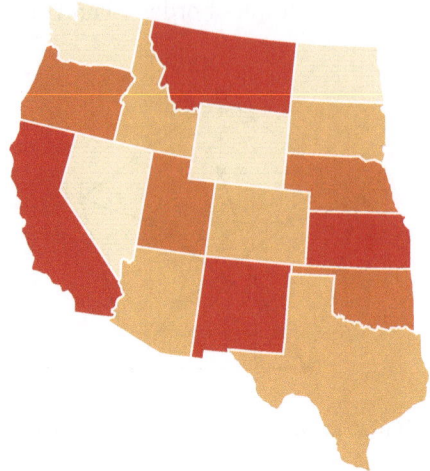

Figure 14.17: A map of the western half of the United States. It is colored with four different colors

planar graph. Here we require the vertices of the graph to have different colors whenever they are adjacent. For instance, Fig. 14.18 shows the graph that is associated with the map given in Fig. 14.17. In this example, we assigned the same colors to the vertices that were used in the map. We will explain the vertex color problem for arbitrary (not necessarily planar) graphs in the next paragraph.

A **vertex coloring** of a graph $G = (V, E)$ is an assignment of colors to the vertices in V such that no two adjacent vertices receive the same color. Since we merely care about the number of distinct colors rather than the particular choice of colors, we might as well choose the elements of an arbitrary finite set C as colors. In other words, we have no qualms using the numbers 1, 2, 3, and 4 as "colors," since we can replace them with actual colors such as green, blue, red, and yellow.

In other words, a vertex coloring of a graph $G = (V, E)$ is a map $c \colon V \to C$ such that $c(u) \neq c(v)$ whenever u and v are adjacent vertices. If u and v are nonadjacent vertices, then there are no restrictions on the choice of colors for u and v.

It is easy to color any graph with a large number of colors. It is a challenge to find a coloring that has the smallest possible number of colors. The **chromatic number** $\chi(G)$ of a graph G is the smallest number of colors that can be used in any vertex coloring of G that assigns different colors to adjacent vertices. Let us have a look at some examples.

Example 14.20. The empty graph E_n on n vertices has the chromatic number $\chi(E_n) = 1$. In fact, any graph G on n vertices with $\chi(G) = 1$ must be the empty graph, since the existence of a single edge in G would require at least two colors in any vertex coloring.

14.6 Graph Coloring

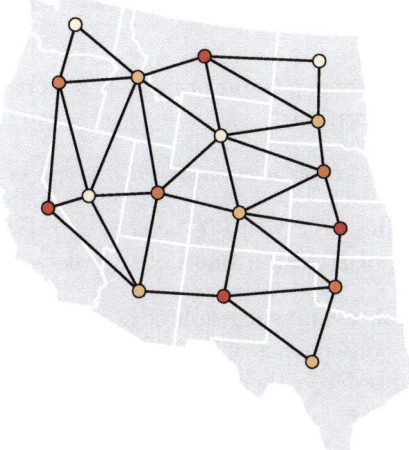

Figure 14.18: The graph corresponding to the map of the western half of the United States. Each state is a vertex, and there is an edge between two states if and only if they share a border (not just a corner). The vertices are colored in the same colors as the map in Fig. 14.17. Any admissible coloring of the vertices uses different colors for adjacent vertices

Example 14.21. The path graph P_n has chromatic number $\chi(P_n) = 2$ for all $n \geqslant 2$. Indeed, we can give the color black to the vertices with even index and the color white to the vertices with odd index, so $\chi(P_n) \leqslant 2$. For instance, the path graph P_5 on five vertices has the vertex coloring

We must use at least two colors in any vertex coloring of P_n with $n \geqslant 2$, since the graph has at least one edge. Therefore, $\chi(P_n) = 2$, as claimed.

Example 14.22. The complete graph K_n on n vertices has chromatic number $\chi(K_n) = n$. Indeed, since there is an edge between any pair of vertices, the colors for the vertices must be pairwise different. Therefore, n different colors are needed in a vertex coloring of K_n.

Example 14.23. The cycle graph C_n with an even number $n \geqslant 2$ of vertices has chromatic number $\chi(C_n) = 2$, since we obtain a valid vertex coloring if we assign vertices with even index the color black and with odd index the color white, so $\chi(C_n) \leqslant 2$. Furthermore, C_n contains an edge, so $\chi(C_n) \geqslant 2$.

The cycle graph C_n with an odd number $n \geqslant 3$ of vertices has chromatic number $\chi(C_n) = 3$. Indeed, we can color C_n with three colors by assigning the vertex with index 0 to be gray, the vertices with even nonzero index the color black, and the vertices with odd indices the color white, whence $\chi(C_n) \leqslant 3$. Since C_n has an edge, we have $\chi(C_n) \geqslant 2$. We cannot color C_n with merely two colors, since C_n is not bipartite when n is odd. Therefore, $\chi(C_n) = 3$ when n is odd and $n \geq 3$.

Let's have a look at some general properties of vertex colorings. We begin with a very simple observation.

Proposition 14.24. *Let H be a subgraph of a graph G. Then $\chi(H) \leq \chi(G)$.*

Proof. Choose a coloring of G with $\chi(G)$ colors and restrict it to $V(H)$. Then this is a valid coloring of H with at most $\chi(G)$ colors, so $\chi(H) \leq \chi(G)$. □

A **clique** in a graph G is a subgraph of G that happens to be a complete graph. An immediate consequence of the previous proposition is that if a graph G contains a clique of order k, then

$$\chi(G) \geq k.$$

The **clique number** $\omega(G)$ of a graph G denotes the size of a clique in G with maximum size. Thus, the chromatic number is lower-bounded by the clique number, $\chi(G) \geq \omega(G)$.

An upper bound on the chromatic number $\chi(G)$ is given in the next proposition.

Proposition 14.25. *Let G be a graph with maximal degree $\Delta(G)$. Then the chromatic number of G satisfies*

$$\chi(G) \leq \Delta(G) + 1.$$

Proof. Suppose that the vertex set of G is of the form $\{v_1, v_2, \ldots, v_n\}$. We choose $C = \{1, 2, \ldots, \Delta(G) + 1\}$ as a set of colors. We are going to assign colors from C to the vertices in the order of the indices.

We assign v_1 the color 1. We can inductively assign colors to the remaining vertices. Suppose that we already have colored the vertices $P_{k-1} = \{v_1, v_2, \ldots, v_{k-1}\}$. Then we can color the vertex v_k with the smallest color in C that is not contained in the colors of its neighboring vertices in $N(v_k) \cap P_{k-1}$. Since v_k has at most $\Delta(G)$ neighbors in $N(v_k) \cap P_{k-1}$, there is always a color available.

This vertex coloring shows that G can be colored with at most $1 + \Delta(G)$ colors. Therefore, we can conclude that $\chi(G) \leq 1 + \Delta(G)$, as claimed. □

A subset I of the set of vertices of a graph G is called **independent** if and only if no two vertices in I are adjacent. A **maximum independent set** is an independent set of largest possible cardinality in a graph. The **independence number** $\alpha(G)$ of a graph is the cardinality of a maximum independent set in the graph G.

If $c\colon V \to C$ is vertex coloring of a graph, then the set V_k of all vertices of color k,

$$V_k = \{v \mid c(v) = k\},$$

14.6 Graph Coloring

is a **color class** of G. Each color class is an independent set. Furthermore, the set $V(G)$ is partitioned into pairwise disjoint color classes

$$V(G) = \bigcup_{k \in C} V_k.$$

Therefore, it is not surprising that we can find lower and upper bounds on the chromatic number of a graph in terms of its independence number.

Proposition 14.26. *Let G be a graph with n vertices. Then the chromatic number $\chi(G)$ is bounded by*

$$\left\lceil \frac{n}{\alpha(G)} \right\rceil \leq \chi(G) \leq n - \alpha(G) + 1.$$

Proof. Let I be a maximal independent set of G. We can color all vertices in I with the same color, say 1. The remaining $n - \alpha(G)$ vertices can be colored with pairwise distinct numbers greater than 1. This is a coloring with $n - \alpha(G) + 1$ colors, which shows that

$$\chi(G) \leq n - \alpha(G) + 1.$$

For the lower bound, consider any vertex coloring c of G with $\chi(G)$ colors in $\{1, 2, \ldots, \chi(G)\}$. Let I_k denote the subset of vertices with color k. Then I_k is an independent set. Furthermore, the set of vertices $V(G)$ is partitioned by the sets

$$V(G) = I_1 \cup I_2 \cup \cdots \cup I_{\chi(G)}.$$

Since $|I_k| \leq \alpha(G)$, we get

$$n \leq \chi(G)\alpha(G).$$

Dividing all sides by $\alpha(G)$ and realizing that $\chi(G)$ is an integer yields the claimed lower bound on $\chi(G)$. $\square$

We will now prove that all planar graphs can be colored with at most five colors. This means that all maps can be colored with at most five colors.

Proposition 14.27. *A planar graph can be colored with at most 5 colors.*

Proof. We argue by induction over the number of vertices. The claim certainly holds for all planar graphs of order 5 or less.

Suppose that the claim holds for all planar graphs of order n. Consider now a planar graph G of order $n + 1$. We are going to show that G can be colored with five colors.

The graph G contains a vertex v of degree $\deg v \leq 5$ by Corollary 14.18. If we delete v from G, then the resulting graph $G - v$ can be colored with at most five colors by induction hypothesis. If $\deg v$ is actually strictly less than 5 or if two neighbors of v are colored the same in this coloring of $G - v$, then there is a color available so that we can extend the coloring of $G - v$ to a coloring of G.

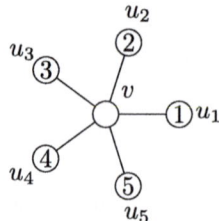

Figure 14.19: One particular embedding of the neighborhood of v in the plane

It remains to deal with the case that v has degree 5 and all five neighboring nodes u_1, u_2, u_3, u_4, u_5 of v have been assigned different colors in the coloring of $G - v$. It will ease our discussion if we assume that u_1, u_2, u_3, u_4, and u_5 are respectively assigned the colors[1] $1, 2, 3, 4$, and 5. Figure 14.19 shows the neighborhood of v in one particular embedding of G in the plane.

Let us denote by $H_{a,b}$ the induced subgraph of $G - v$ consisting of the nodes that are colored with the colors a and b, where $1 \leq a, b \leq 5$. We will use these subgraphs to identify ways to change our coloring of $G - v$ so that we can extend it to a five coloring of G. We begin by considering the subgraph $H_{1,4}$. There is nothing special about the choice of the colors 1 and 4, except that the nodes u_1 and u_4 are not both adjacent to the same face. We will now distinguish whether u_1 and u_4 are in a different component in the induced subgraph $H_{1,4}$ or not.

Case 1. If u_1 and u_4 are contained in different components of $H_{1,4}$, then we can swap the colors of 1 and 4 in the component of $H_{1,4}$ that contains u_1. This leads to another valid coloring of $G - v$ that has the color 1 available at v, see Fig. 14.20. Coloring v with the color 1 leads to the desired coloring of G with five colors.

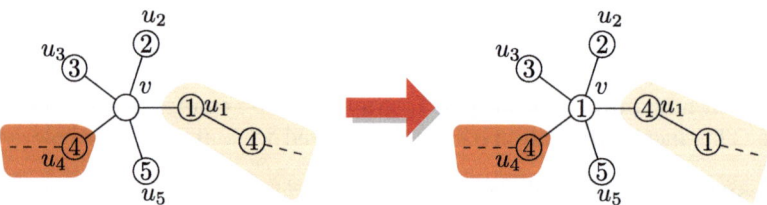

Figure 14.20: If u_1 and u_4 in different components in $H_{1,4}$, then we can swap the colors 1 and 4 in the component that contains u_1. The right-hand side shows the recolored neighborhood of v, which permits us to color v with the color 1

Case 2. If u_1 and u_4 are contained in the same component of $H_{1,4}$, then there must exist a path P from u_1 to u_4 consisting of nodes that are alternatingly colored 1 and 4. The path P together with the path (u_4, v, u_1) forms a

---

[1] Of course, you are welcome to substitute your favorite five colors for the numbers from 1 to 5.

14.6 Graph Coloring

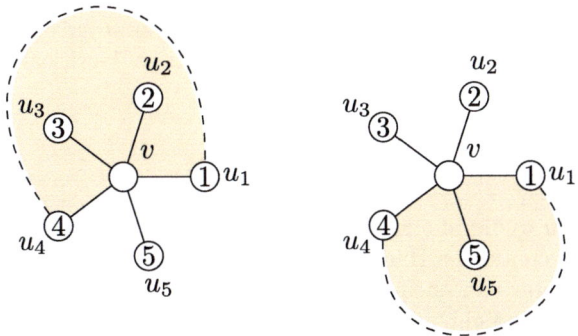

Figure 14.21: If u_1 and u_4 are in the same component of $H_{1,4}$, then there is a cycle of nodes colored either with 1 or 4 that encloses the vertices u_2 and u_3 (shown on the left) or the vertex u_5 (shown on the right). In either case, this implies that there is no path from u_2 to u_5 in $H_{2,5}$. In other words, u_2 and u_5 are in different components in $H_{2,5}$, so we can recolor the component of u_2 in $H_{2,5}$ to make a color available at v

cycle C of G that encloses either u_5 or both u_2 and u_3. Figure 14.21 shows the two possible resulting situations in plane embeddings.

In either case, this implies that in $H_{2,5}$ there is no path from u_2 to u_5, as this path would have to cross the cycle C, which is impossible in a planar graph. Put differently, the vertices u_2 and u_5 lie in different components in $H_{2,5}$ so we can swap the colors 2 and 5 in the component containing u_2, so the color 2 becomes available at v. We can then color v with 2 to obtain the desired five coloring of G.

It follows by induction that any planar graph can be colored with five colors, as claimed. □

Exercises

14.34. The map shown in Fig. 14.17 was colored with four colors. It is not surprising that we were able to do that given that the four color theorem asserts that *any* map can be colored with at most four colors. Yet, some maps can be colored with *fewer* colors. Show that the map given in Fig. 14.17 cannot be colored with less than four colors. [Hint: Look at Nevada and its neighboring states.]

14.35. Show that a graph G of order n that is not a complete graph has chromatic number $\chi(G) < n$.

14.36. Determine the chromatic number $\chi(Q_n)$ of the hypercube graph on 2^n vertices with $n \geq 2$.

14.37. Let G be a graph of size m. Show that

$$\chi(G)(\chi(G) - 1) \leq 2m.$$

14.38. Show that a graph G has chromatic number $\chi(G) = 2$ if and only if G does not contain a cycle with an odd number of vertices.

14.39. A graph G is called a k-**critical graph** if and only if $\chi(G) = k$, but $\chi(H) < k$ for all subgraphs H of G. Show that a k-critical graph satisfies $\delta(G) \geqslant k - 1$.

14.7 Hamiltonian Cycles and Paths

A **Hamiltonian cycle** in a graph G is a cycle that contains every vertex of G. A graph G that contains a Hamiltonian cycle is called a **Hamiltonian graph**. In the first chapter, we encountered Hamiltonian cycles on knight graphs in the form of closed knight's tours.

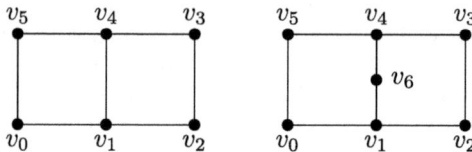

Figure 14.22: The graph on the left is Hamiltonian. Indeed, a Hamiltonian cycle is given by $(v_0, v_1, v_2, v_3, v_4, v_5, v_0)$. The graph on the right is not Hamiltonian

A path in a graph G that contains every vertex of G is called a **Hamiltonian path**. Every Hamiltonian graph contains a Hamiltonian path, as simply deleting one edge of the Hamiltonian cycle yields a Hamiltonian path. On the other hand, there exist graphs that contain a Hamiltonian path, but do not have a Hamiltonian cycle. Indeed, the graph on the right in Fig. 14.22 has the Hamiltonian path $(v_6, v_1, v_0, v_5, v_4, v_3, v_2)$, but does not contain a Hamiltonian cycle.

The next theorem gives a sufficient condition for the existence of a Hamiltonian cycle given that the graph contains sufficiently many edges. However, the edges need to be distributed well-enough. The Norwegian mathematician Øystein Ore formulated a condition that looks at two non-adjacent vertices. Since they are not directly connected by an edge, Ore ensured that they can be reached by connecting them to at least n other vertices. If one of the two nodes has a small degree, then the other node must offer a high degree.

Theorem 14.28 (Øystein Ore)**.** *Let $G = (V, E)$ be a graph with $n \geqslant 3$ vertices that satisfies*

$$\deg u + \deg v \geqslant n \tag{14.2}$$

for all non-adjacent vertices u and v in V. Then G is a Hamiltonian graph.

Proof. Seeking a contradiction, suppose that G is a graph satisfying the Ore condition (14.2) that does not contain a Hamiltonian cycle. We may assume that G contains a maximal number of edges among all graphs with n vertices that satisfy (14.2) and do not have a Hamiltonian cycle.

Then G is extremal in the sense that it contains a Hamiltonian path, say

$$(v_1, v_2, \ldots, v_n),$$

14.7 Hamiltonian Cycles and Paths

but not a Hamiltonian cycle. Indeed, a graph G without even a Hamiltonian path would allow one to add an edge to G without creating a Hamiltonian cycle, contradicting the maximality of the number of edges in G.

Since the graph G is not Hamiltonian, the vertices v_1 and v_n are not adjacent, as this would imply the existence of a Hamiltonian cycle. By the Ore condition,
$$\deg v_1 + \deg v_n \geqslant n.$$
Now consider all $n-1$ pairs of vertices $P = \{(v_{k-1}, v_k) \mid 2 \leqslant k \leqslant n\}$ on the Hamiltonian path. By the Pigeonhole principle, there must exist a pair (v_{k-1}, v_k) in P such that v_{k-1} is adjacent to v_n, and v_k is adjacent to v_1. It follows that the cycle
$$(v_1, v_2, \ldots, v_{k-1}, v_n, v_{n-1}, \ldots, v_k, v_1)$$
is a Hamiltonian cycle, contradicting the assumption that G does not contain a Hamiltonian cycle. $\square$

Ore's theorem allows us to assert the existence of a Hamiltonian cycle in a graph. All we need to check is that the sum of the degrees of pairs of non-adjacent vertices is large. If a graph does not have such a large number of edges, then deciding whether the graph has a Hamiltonian cycle can become a challenge. In particular, the proof that a given graph does *not* have a Hamiltonian cycle can become quite involved.

There exists a remarkable theorem by the Latvian mathematician Emanuel Grinberg that gives a necessary condition for the existence of Hamiltonian cycles for planar graphs. Grinberg's theorem can be used to show that a planar graph does not have a Hamiltonian cycle.

Let G be a plane graph with a Hamiltonian cycle C. An edge of G that does not lie on the Hamiltonian cycle C is called a **chord**. A chord either lies in the interior or exterior of C. Similarly, every face of the plane graph G is either inside or outside of C.

Example 14.29. Consider the plane graph G with 10 vertices and 15 edges shown below. This graph has a Hamiltonian cycle that is shown with thick edges.

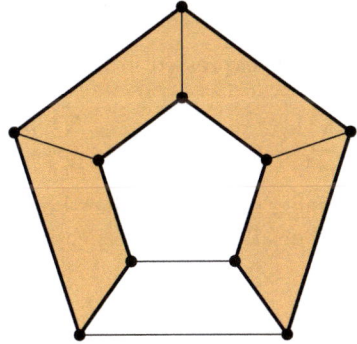

The four interior faces are shaded, and the three outside faces are not shaded. Notice that the unbounded face surrounding this graph is one of the outside faces. The plane graph G has three interior and two exterior chords.

An essential role in Grinberg's theorem are the following numerical characteristics of a plane graph with a Hamiltonian cycle. For a plane graph G with Hamiltonian cycle C, we denote by

- f_k the number of interior faces of G that have k edges on their boundary,
- f'_k the number of exterior faces of G that have k edges on their boundary.

We continue our previous example to illustrate these notions.

Example 14.30 (Continued). In the plane graph G of the previous example, there are four interior faces with four edges on their boundary, but merely one exterior face with four edges on its boundary. The first line of the following equations states these facts more succinctly,

$$f_4 = 4, \quad f'_4 = 1,$$
$$f_5 = 0, \quad f'_5 = 2.$$

The characteristics shown in the second line tell us that there are two exterior faces with five edges on their boundary, but there is no interior face that has fives edges on its boundary.

Grinberg's theorem gives a numerical invariant that plane graphs with a Hamiltonian cycle have to satisfy.

Theorem 14.31 (Grinberg). *A plane graph G of order n with a Hamiltonian cycle C must satisfy*

$$\sum_{k=3}^{n}(k-2)(f_k - f'_k) = 0. \tag{14.3}$$

Proof. Suppose that c chords of G lie in the interior of C and c' chords of G lie in the exterior of C. Then there are $c+1$ faces interior to C and $c'+1$ faces exterior to C. Therefore, we can conclude that

$$\sum_{k=3}^{n} f_k = c+1 \quad \text{and} \quad \sum_{k=3}^{n} f'_k = c'+1.$$

In other words, c and c' can be expressed in the form

$$c = \sum_{k=3}^{n} f_k - 1 \quad \text{and} \quad c' = \sum_{k=3}^{n} f'_k - 1.$$

Each face has a certain number of edges on their boundary. Summing all edges as the summation ranges over all interior (respectively exterior) faces yields

$$\sum_{k=3}^{n} k f_k = n + 2c \quad \text{and} \quad \sum_{k=3}^{n} k f'_k = n + 2c'.$$

14.7 Hamiltonian Cycles and Paths

Indeed, the first sum counts each of the n edges on C once and the interior chords twice. Similarly, the second sum counts each of the n edges on C once and the exterior chords twice. Substituting our expressions for c and c' in the previous equations yields

$$\sum_{k=3}^{n} k f_k = n + 2 \left(\sum_{k=3}^{n} f_k - 1 \right) \quad \text{and} \quad \sum_{k=3}^{n} k f'_k = n + 2 \left(\sum_{k=3}^{n} f'_k - 1 \right).$$

We can rewrite these equations in the form

$$\sum_{k=3}^{n} (k-2) f_k = n - 2 \quad \text{and} \quad \sum_{k=3}^{n} (k-2) f'_k = n - 2.$$

From these two equations, we can deduce

$$\sum_{k=3}^{n} (k-2)(f_k - f'_k) = (n-2) - (n-2) = 0,$$

which proves our claim. $\square$

The equation (14.3) is called the **Grinberg equation**. The next example illustrates how we can use the Grinberg equation to show that a plane graph does not have a Hamiltonian cycle.

Example 14.32. Consider the plane graph G of order 9 and size 12 shown as follows.

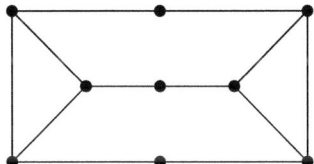

We observe that there are two faces with three edges, and three faces with six edges. Seeking a contradiction, we assume that the graph has a Hamiltonian cycle C. It follows that the plane graph G must satisfy the Grinberg equation

$$1(f_3 - f'_3) + 4(f_6 - f'_6) = 0.$$

The graph contains two faces with three edges on their boundary, $f_3 + f'_3 = 2$. In other words, the possible values of the term $1(f_3 - f'_3)$ are $\{-2, 0, 2\}$. Reducing the Grinberg equation modulo 4 shows that $f_3 - f'_3 \equiv 0 \pmod{4}$. Therefore, we can conclude that one of the triangular faces must be interior and the other exterior, so $f_3 = 1 = f'_3$. Therefore, the Grinberg equation reduces to

$$4(f_6 - f'_6) = 0.$$

We can conclude that $f_6 = f_6'$, so there must be the same number of faces with six edges on their boundary lying in the interior of C as the ones lying in the exterior of C. In particular, there must be an even number of faces with six edges on their boundary. This contradicts the fact that there are $f_6 + f_6' = 3$ such faces in G. This contradiction shows that the plane graph G cannot possess a Hamiltonian cycle C.

Exercises

14.40. Lady Hamilton is having a dinner party. She wants to seat her guests at a big round table such that everyone is seated next to good friends. There are 18 guests and two hosts, so a total of twenty people altogether. She realized that she invited five people belonging to the astronomy society ($a0, \ldots, a4$), five people from the bowling club ($b0, \ldots, b4$), five people from the cat club ($c0, \ldots, c4$), and five mechanics disciples ($d0, \ldots, d4$) that adore her husband's work.

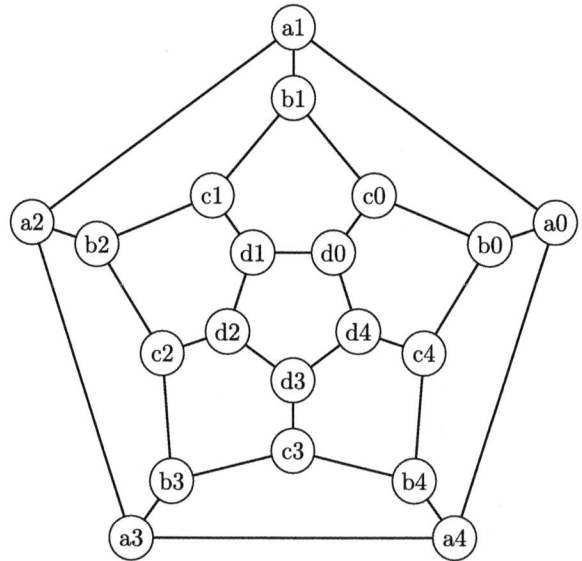

She drew a line between any two people that liked each other. Since this resulted in merely 30 friendships, she was getting a bit worried about her plan. Can Lady Hamilton seat her guests around the table? Give a solution if one exists.

14.41. For what values of n does the hypercube graph Q_n have a Hamiltonian cycle?

14.42. For what values of the parameters does the complete bipartite graph $K_{m,n}$ have a Hamilonian cycle?

14.7 Hamiltonian Cycles and Paths

14.43. Show that the graph on the right in Fig. 14.22 cannot have a Hamiltonian cycle.

14.44. Let $G = (V, E)$ be a Hamiltonian graph. Let k be a positive integer. Show that if k vertices are deleted from G (along with incident edges), then the resulting graph has at most k different components. [By contrast, a non-Hamiltonian graph can possibly decompose into more than k components when k vertices are deleted.]

14.45. We call a graph G **Hamiltonian-connected** if and only if there is a Hamiltonian path between any two vertices of G. Show that a Hamiltonian-connected graph must contain a Hamiltonian cycle.

14.46. A vertex v in a graph G is called a **cut-vertex** if and only if the removal of the vertex v and adjacent edges from G will result in a disconnected graph. Show that a graph G with a cut-vertex cannot be a Hamiltonian graph.

14.47. Show that the Ore condition is sharp in the sense that there exists a graph G with n vertices satisfying

$$\deg u + \deg v \geqslant n - 1$$

for all non-adjacent vertices, but G does not contain a Hamiltonian cycle.

14.48. Deduce with the help of Grinberg's theorem that the plane graph

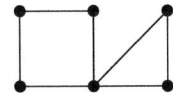

does not have a Hamiltonian cycle.

14.49. Deduce with the help of Grinberg's theorem that the plane graph

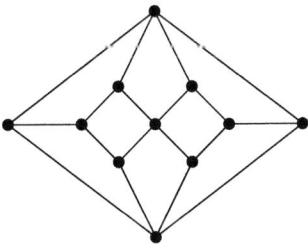

does not have a Hamiltonian cycle.

14.50. Suppose that G is a plane graph such that its unbounded face has a border with 15 edges. All other faces are bordered by 4-cycles, 6-cycles, or 8-cycles. Use Grinberg's Theorem to show that the graph G cannot contain a Hamiltonian cycle.

14.51. Give an example of a plane graph that is not Hamiltonian, but satisfies Grinberg's criterion.

14.52. Find two plane graphs of order 8 and size 12, one with a Hamiltonian cycle and the other without, such that each face in both graphs have degree 4, meaning that there are four edges on the boundary of each face. What does this tell you about Grinberg's criterion?

14.53. Follow in Grinberg's footsteps and show that the plane graph of order 44 given below does not have a Hamiltonian cycle.

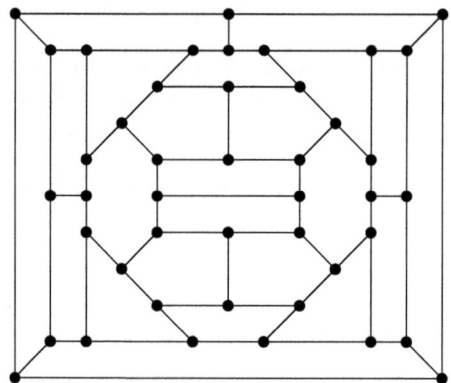

[Hint: After setting up Grinberg's equation, it is a good strategy to find a positive integer M such that reducing the equation modulo M yields few terms.]

Chapter 15

Probability Theory

The 50–50–90 rule: Anytime you have a 50–50 chance of getting something right, there's a 90 percent probability you'll get it wrong.

— Andy Rooney

In this chapter, we will introduce basic notions of probability theory. We discuss sample spaces and probability measures. We introduce conditional probabilities and discuss independent events. Then we explore random variables and expectation. We conclude the chapter with some examples that illustrate the probabilistic method.

15.1 Probability Spaces

Probability theory allows us to deal with uncertainty in a quantitative way. The main object of study in probability theory is experiments with outcomes that are subject to chance. For example, tossing a coin yields either heads or tails as outcomes. Rolling a die yields the face values 1, 2, 3, 4, 5, or 6 as outcomes. These experiments have in common that one cannot predict with certainty the outcome of the experiment.

The set of outcomes of an experiment is called its **sample space**. We will typically denote the sample space by Ω. In this chapter, *we will assume that the sample spaces are discrete*, meaning Ω is either finite or countably infinite. This simplifies the theory a little bit.

A subset of the sample space is called an **event**. For instance, the subset $A = \{1, 3, 5\}$ of the sample space $\Omega = \{1, 2, 3, 4, 5, 6\}$ describes the event that the outcome of rolling a die results in an odd face value. The event $B = \{1, 2, 3, 4\}$ describes the event that the face value of the die is at most 4. Events given by singleton sets are called **elementary events**. For instance, the elementary event $\{6\}$ describes that face value of 6 is the outcome of a die roll.

The sample space Ω can be viewed as an event, as it is a subset of itself, and is referred to as the **certain event**. The empty set $\emptyset$ is also an event, called the **impossible event**. Two events A and B are called disjoint if and only if no outcome is common to both events; in other words, A and B are disjoint if and only if $A \cap B = \emptyset$.

A probability measure is a function that assigns to an event the probability of this event. In more detail, a **probability measure** $\Pr$ on a sample space Ω is a function $\Pr \colon P(\Omega) \to [0, 1]$ satisfying the following two axioms

P1 The certain event satisfies $\Pr[\Omega] = 1$.

P2 If the events $E_1, E_2, \ldots$ in $P(\Omega)$ are pairwise disjoint (meaning that the events E_i and E_k do not have elements in common when the indices i and k are not the same), then

$$\Pr[\bigcup_{k=1}^{\infty} E_k] = \sum_{k=1}^{\infty} \Pr[E_k].$$

The two axioms are there to ensure consistency among these assignments of probabilities. For instance, we will see shortly that if A and B are events such that $A \subseteq B$, then $\Pr[A] \leqslant \Pr[B]$. So the probability measure is monotonic. The largest values must thus be assigned to the sample space, which contains all events. The axiom **P1** ensures that the largest probability is normalized to $\Pr[\Omega] = 1$.

Since a discrete sample space is a countable set, it is a countable union of elementary events, cf. Proposition 3.55. Therefore, all probability measures on a discrete sample space can be obtained by assigning probabilities to elementary events. This is detailed in the next remark.

Remark 15.1. Let Ω be a finite or countably infinite nonempty set. Let $p \colon \Omega \to [0, 1]$ be a function such that

$$\sum_{x \in \Omega} p(x) = 1.$$

For a subset A of Ω, we define

$$\Pr[A] = \sum_{x \in A} p(x).$$

Then $\Pr$ is a probability measure on Ω. In particular, an elementary event $\{x\}$, with $x \in \Omega$, is assigned the probability $p(x)$. Since a discrete sample space is the countable union of elementary events, the sum of the probabilities of elementary events must add up to 1.

For a finite sample space, this method of assigning probabilities to elementary events is particularly straightforward. The next example details the simple instance of rolling a die.

Example 15.2. Let $\Omega = \{1, 2, 3, 4, 5, 6\}$ be the sample space of face values of a die. If we assign the same probabilities to each of the elementary events,

$$\Pr[\{1\}] = \frac{1}{6}, \Pr[\{2\}] = \frac{1}{6}, \ldots, \Pr[\{6\}] = \frac{1}{6},$$

15.1 Probability Spaces

then the compound event $A = \{1, 3, 5\}$ has the probability

$$\Pr[A] = \frac{|A|}{|\Omega|} = \frac{3}{6} = \frac{1}{2}.$$

More generally, an arbitrary subset B of Ω has probability $|B|/|\Omega| = |B|/6$, the number of elements in the compound event B divided by 6.

More generally, we can define a probability measure on any finite sample space by assigning each elementary event the probability $1/|\Omega|$. We will study such uniform probability distributions in the next section.

On a countably infinite sample space, one cannot assign a uniform probability to elementary events. However, one can use convergent series to define probability measures, as the next example shows in a particular instance.

Example 15.3. Let $\Omega = \mathbf{N}_1$ be the set of positive integers. Leonard Euler famously showed that the sum of the squares of the reciprocals of positive integers converges to $\pi^2/6$, so

$$\sum_{k=1}^{\infty} \frac{1}{k^2} = \frac{1}{1^2} + \frac{1}{2^2} + \frac{1}{3^2} + \cdots = \frac{\pi^2}{6}.$$

Therefore, we can obtain a probability measure on $\Omega = \mathbf{N}_1$ by assigning to the elementary event $\{k\}$, where k is a positive integer, the probability

$$\Pr[\{k\}] = \frac{1}{k^2} \frac{6}{\pi^2}.$$

The sum over all positive integers k shows that

$$\Pr[\Omega] = \sum_{k=1}^{\infty} \Pr[\{k\}] = \sum_{k=1}^{\infty} \frac{1}{k^2} \frac{6}{\pi^2} = 1,$$

so $\Pr[\Omega]$ has the proper normalization to 1.

Kolmogorov introduced these remarkably simple axioms of probability theory. There is a remarkable amount of information packed into the axioms **P1** and **P2**, as will become evident in the next few propositions.

Proposition 15.4. *Let Ω be a discrete sample space.*
(a) The impossible event $\emptyset$ satisfies $\Pr[\emptyset] = 0$.
(b) Let n be a positive integer and $E_1, E_2, \ldots, E_n$ be disjoint events in $P(\Omega)$. Then

$$\Pr[E_1 \cup E_2 \cup \cdots \cup E_n] = \sum_{k=1}^{n} \Pr[E_k].$$

(c) An event E satisfies $\Pr[E^{\complement}] = 1 - \Pr[E]$.
(d) The probability measure is monotonic, meaning that if A and B are events such that $A \subseteq B$, then $\Pr[A] \leqslant \Pr[B]$.

Proof.
(a) Let $E_1 = \Omega$ and $E_k = \varnothing$ for all $k \geq 2$. Then these events are pairwise disjoint. By countable additivity, we have
$$1 = \Pr[\Omega] = \sum_{k=1}^{\infty} \Pr[E_k].$$
As $\Pr[E_1] = 1$, we must have $\Pr[E_k] = 0$ for all $k \geq 2$. So $\Pr[\varnothing] = 0$.
(b) In axiom **P2**, if we choose $E_k = \varnothing$ for all $k \geq n+1$, then
$$\Pr[E_1 \cup E_2 \cup \cdots \cup E_n] = \Pr\left[\bigcup_{k=1}^{\infty} E_k\right] = \sum_{k=1}^{\infty} \Pr[E_k] = \sum_{k=1}^{n} \Pr[E_k],$$
where we used part (a) in the last equality.
(c) We have $\Pr[E] + \Pr[E^\complement] = \Pr[E \cup E^\complement] = \Pr[\Omega] = 1$. This implies the claim.
(d) The set B is the disjoint union of A and $A^\complement \cap B$. Therefore,
$$\Pr[B] = \Pr[A \cup (A^\complement \cap B)] = \Pr[A] + \Pr[A^\complement \cap B].$$
It follows that $\Pr[A] \leq \Pr[B]$.
$\square$

The next proposition gives a very simple yet often useful bound on the probability of union of events. We call it the **union bound**, but it is also known as **Boole's inequality** or simply as **subadditivity** of the probability measure.

Proposition 15.5 (Union Bound). *Let $E_1, E_2, \ldots$ be a finite or countably infinite number of events. Then*
$$\Pr\left[\bigcup_k E_k\right] \leq \sum_k \Pr[E_k].$$

Proof. The proof is simple but instructive, see Exercise 15.1. $\square$

Another consequence is a simple form of the inclusion-exclusion principle.

Proposition 15.6. *Let E and F be events. Then*
$$\Pr[E \cup F] = \Pr[E] + \Pr[F] - \Pr[E \cap F],$$
This fact can be convenient when calculating probabilities.

Proof. We can decompose $E \cup F$ into the disjoint union of the three events $E \backslash (E \cap F)$, $E \cap F$, and $F \backslash (E \cap F)$. Therefore, we have
$$\Pr[E \cup F] = \Pr[E \backslash (E \cap F)] + \Pr[E \cap F] + \Pr[F \backslash (E \cap F)]$$
$$= \Pr[E] + \Pr[F \backslash (E \cap F)] + (\Pr[E \cap F] - \Pr[E \cap F])$$
$$= \Pr[E] + \Pr[F] - \Pr[E \cap F],$$
where the second equality holds, since $0 = \Pr[E \cap F] - \Pr[E \cap F]$. $\square$

15.2 Combinatorial Probability

Exercises

15.1. Let $E_1, E_2, \ldots$ be a finite or countably infinite number of events that are not necessarily disjoint. Show that the probability measure is subadditive, meaning that

$$\Pr\left[\bigcup_k E_k\right] \leq \sum_k \Pr[E_k].$$

15.2. Let Ω be a sample space, and let ω be a fixed element of Ω. Show that $\Pr: P(\Omega) \to [0,1]$ given by

$$\Pr[A] = \begin{cases} 1 & \text{if } \omega \in A, \\ 0 & \text{otherwise.} \end{cases}$$

is a probability measure.

15.3. Let $E_1, E_2, \ldots, E_n$ be events. Show that

$$\Pr\left[\bigcup_{k=1}^n E_k\right] + \Pr\left[\bigcap_{k=1}^n E_k^C\right] = 1.$$

15.4. Let n be an integer satisfying $n \geq 2$. Let $E_1, E_2, \ldots, E_n$ be events. Show that the general inclusion-exclusion principle

$$\Pr\left[\bigcup_{k=1}^n E_k\right] = \sum_{k=1}^n \Pr[E_k] - \sum_{i<j} \Pr[E_i \cap E_j] + \sum_{i<j<k} \Pr[E_i \cap E_j \cap E_k]$$
$$+ \cdots + (-1)^{n+1} \Pr[E_1 \cap E_2 \cap \cdots \cap E_n]$$

holds. Use a proof by induction.

15.5. Let Ω be a countably infinite sample space. Show that one cannot define a uniform probability measure on Ω such that $\Pr[\{x\}] = \Pr[\{y\}]$ for all x, y in Ω.

15.2 Combinatorial Probability

In this section, we confine ourselves to a particularly simple situation. We consider experiments that have a finite number of different outcomes that are all equally likely. The prototypical example is the flipping of a fair coin that has the same odds of showing heads as it has showing tails.

In general, suppose that the experiment has outcomes that belong to a finite set Ω that we will call the sample space. The probability $\Pr[E]$ that the outcome of the experiment will belong to a subset E of Ω is given by

$$\Pr[E] = \frac{|E|}{|\Omega|}.$$

We call Pr the uniform probability measure.

Since the uniform probability measure is nothing but a normalized counting measure, it is not surprising that determining $\Pr[E]$ often boils down to a counting argument.

Marbles. Suppose that you have a bag full of marbles. There are 23 red marbles and 27 blue marbles in the bag, so a total of 50 marbles overall. If you grab without replacement 8 marbles overall, what is the probability that you get 5 red marbles?

We can answer this question by carefully considering the number of ways in which we can end up with precisely five red marbles when we grab eight marbles overall. We contrast this with the number of ways we can grab eight marbles from the bag. The quotient of these two counting numbers then gives us the probability that we are looking for.

There are $\binom{23}{5}$ ways to select 5 red marbles from the 23 available red marbles. Let us not forget that the $8 - 5 = 3$ other marbles must be blue. There are $\binom{27}{3}$ ways to select the remaining 3 blue marbles from the 27 available blue marbles. In other words, there are $\binom{23}{5}\binom{27}{3}$ ways to end up with precisely five red marbles when we grab eight marbles from the bag.

On the other hand, there are $\binom{50}{8}$ ways to choose 8 marbles from the 50 marbles in the bag. So this time we counted every possibility without restricting the number of red or blue marbles in any way.

Therefore, the probability to get precisely five red marbles when grabbing eight marbles from the bag is given by

$$\Pr[\text{get 5 red marbles from the bag}] = \frac{\binom{23}{5}\binom{27}{3}}{\binom{50}{8}}.$$

The numerical evaluation of this probability is

$$\Pr[\text{get 5 red marbles from the bag}] = \frac{33649 \cdot 80730}{536878650} \approx 0.183.$$

Pairing Up Socks. Mr. Bocks is quite enamored with his socks. He has n pairs in a drawer in his walk-in closet. None of the pairs are the same, so he got quite the variety.

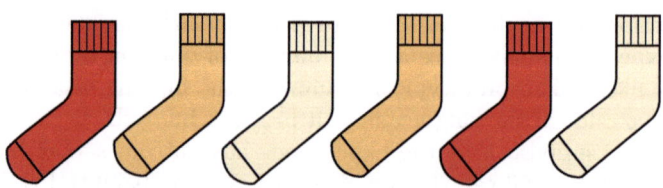

Recently, the light bulb in his closet burned out, so it is completely dark in there. Regrettably, Mr. Bocks has the unfortunate habit that he does not pair up his socks, but simply tosses them into the drawer after washing. How likely

15.2 Combinatorial Probability

is it that Mr. Bocks will get at least one matching pair of socks if he randomly grabs r socks?

Since the drawer contains $2n$ socks, he will get at least one pair with certainty if he grabs $r \geq n+1$ socks. However, if the number n of socks is large, then this seems to be a bit unreasonable.

So let's assume that Mr. Bocks is a bit more adventurous and chooses at most $r \leq n$ socks. Of course, in this case it might happen that he comes out of the closet with r different socks, and none of them match up to a pair. What are the odds of that event?

Let us first count the number of ways to pick r socks from the drawer. There are $2n$ ways to pick the first sock, $2n-1$ to pick the second, and so on. So there is a total of $2n(2n-1)(2n-2)\cdots(2n-r+1)$ ways to pick r socks.

Next we count the number of ways to pick r socks from the drawer so that we do not get any matching pair. There are $2n$ ways to pick the first sock. The second sock should be different from the first sock that we chose, so there are $2n-2$ ways to choose such a sock. For the next sock, we need to carefully avoid the two socks that will give a pair, but can choose any of the remaining $2n-4$ socks. The number of ways to pick r socks that cannot be matched in pairs is $2n(2n-2)(2n-4)\cdots(2n-2r+2)$.

Therefore, the probability that none of the r socks match is given by

$$\Pr[\text{none of the } r \text{ socks match}] = \frac{2n(2n-2)(2n-4)\cdots(2n-2r+2)}{2n(2n-1)(2n-2)\cdots(2n-r+1)}.$$

Our original goal was to find the probability that we *do* get a matching pair of socks. This is the probability of the complementary event. Since you either do or do not get a pair of matching socks, the two probabilities need to add up to one. Thus, the probability that he does get at least one pair of matching socks is given by

$$\Pr[\text{get at least one matching pair}] = 1 - \frac{2n(2n-2)(2n-4)\cdots(2n-2r+2)}{2n(2n-1)(2n-2)\cdots(2n-r+1)}.$$

For example, if Mr. Bocks has $n=10$ pairs of socks in his drawer, and he randomly grabs $r=6$ socks, then the probability that he gets a matching pair is

$$\Pr[\text{get at least one matching pair}] = 1 - \frac{20 \cdot 18 \cdot 16 \cdot 14 \cdot 12 \cdot 10}{20 \cdot 19 \cdot 18 \cdot 17 \cdot 16 \cdot 15} \geq 0.65.$$

So in this case it is more likely that he will end up with a matching pair of socks than without one.

Ducks in a Row. Every year, the small town of Canard organizes a duck race. There are n teams entering the race. Each team must flip a fair coin to decide whether to enter a *Baseball Duck* or a *Duck in Black*.

Now everyone knows that it is difficult to keep your ducks in a row. It is nearly impossible when two Ducks in Black are next to each other, since they quip so much in fowl language that this delays the start of the race. So the organizers want to know: How likely is it that the starting lineup does not contain two adjacent Ducks in Black?

There are two choices for each of the n positions, so there are a total of 2^n different starting lineups.

How many of these starting lineups do not contain adjacent Ducks in Black? If there is just one team participating, then both possible starting lineups do not contain adjacent Ducks in Black. If there are $n = 2$ teams, then there are exactly three starting lineups that do not contain two adjacent Ducks in Black, namely

If there are $n = 3$ teams, then there are five starting lineups that do not contain two adjacent Ducks in Black, since any of the $8 = 2^3$ possible starting lineups work with the exception of the following three choices:

.

We could keep going, but it starts to become too tedious to list all configurations. Let us denote by S_n the number of start lineups that do not contain two adjacent teams. We already determined that

$$S_1 = 2, \quad S_2 = 3, \quad S_3 = 5.$$

For larger values of n, we can set up a recurrence relation. If the n-th team puts in a Baseball Duck, then there are S_{n-1} lineups on the first $n-1$ positions that avoid adjacent Ducks in Black. If the n-th team puts in a Duck in Black, then there must be a Baseball Duck on position $n-1$, and any of the S_{n-2} starting lineups on the first $n-2$ positions that avoid adjacent Ducks in Black. In summary, we have

$$S_n = S_{n-1} + S_{n-2}$$

for all integers n such that $n \geqslant 3$. So $S_4 = 8$, $S_5 = 13$, $S_6 = 21$, and so on. In other words, the number S_n of starting line ups that do not contain two adjacent Ducks in Black is given by the Fibonacci number $S_n = f_{n+2}$.

Therefore, the probability that the duck selection of the n teams will create a starting lineup without two adjacent Duck in Black is given by

$$\Pr[\text{no adjacent Ducks in Black}] = \frac{S_n}{2^n} = \frac{f_{n+2}}{2^n}.$$

So asymptotically, we have

$$\Pr[\text{no adjacent Ducks in Black}] \sim \frac{1}{\sqrt{5}} \frac{\phi^{n+2}}{2^n} = \frac{\phi^2}{\sqrt{5}} \left(\frac{\phi}{2}\right)^n,$$

15.3 Conditional Probabilities

where $\phi = (1+\sqrt{5})/2$ is the golden ratio. Since the fraction $\phi/2$ is a positive real number less than 1, we have $(\phi/2)^n \to 0$ as $n \to \infty$.

Therefore, the probability that we get a starting lineup with no two adjacent Ducks in Black is very small when a large number of teams enter. Knowing this fact, the organizers allow for plenty of time until the race gets started.

Exercises

15.6. What is the sample space when drawing socks from a drawer containing n pairs of socks?

15.7. Suppose that you flip a fair coin three times.
(a) Determine the sample space of this random experiment (use H to denote heads and T to denote tails).
(b) What is the probability to get all heads?
(c) What is the probability to get two heads?
(d) What is the probability to get one head?
(e) What is the probability to get at least two heads?
(f) What is the probability to get at most one head?

15.8. Suppose that we flip a fair coin five times. We record heads by H and tails by T. So (H, T, T, H, H) means that we got heads in the first try, followed by tails in the next two tries, and the last two flips showed both heads.
(a) Formally describe the sample space of this experiment.
(b) Describe the event that there are more heads than tails in the five coin flips.
(c) What is the probability of the event that there are more heads than tails?

15.9. Suppose that you roll a pair of dice.
(a) What is the sample space of this random experiment?
(b) What is the event that the sum of the two face values is 5?
(c) What is the probability of the event that the sum of the face values is 5?
(d) What is the probability of the event that the sum of the face values is at least 5?

15.10. Suppose that you have 50 marbles in a bag, where 23 marbles are red and 27 marbles are blue. If you grab eight marbles from the bag without replacement, what is the probability to get at least six red marbles. Explain your result.

15.3 Conditional Probabilities

Let E and F be events over a sample space Ω such that $\Pr[F] > 0$. The **conditional probability** $\Pr[E\,|\,F]$ of the event E given F is defined by

$$\Pr[E\,|\,F] = \frac{\Pr[E \cap F]}{\Pr[F]}.$$

The value $\Pr[E\,|\,F]$ is interpreted as the probability that the event E occurs, assuming that the event F has occurred. By definition, $\Pr[E \cap F] = \Pr[E\,|\,F]\Pr[F]$, and this simple multiplication formula often turns out to be useful.

Example 15.7. Suppose that we have a spinner, and the hand of the spinner can land in one of 36 fields. The spinner is subdivided into three different areas that contain 12 fields each. The probabilities to land on a field in a given area are given by

$$\Pr[\text{area 1}] = \frac{1}{3}, \quad \Pr[\text{area 2}] = \frac{1}{3}, \quad \Pr[\text{area 3}] = \frac{1}{3}.$$

Suppose now that someone tells us that the hand of the spinner landed in an area above the dashed line.

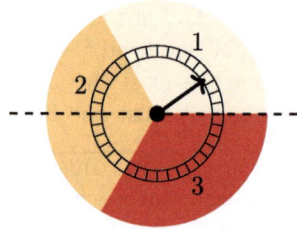

It is immediately clear that the conditional probability to land on area 3 is given by

$$\Pr[\text{area 3} \mid \text{above line}] = 0,$$

as area 3 is entirely below the dashed line. If the spinner landed above the dashed line, then

$$\Pr[\text{area 1} \mid \text{above line}] = \frac{2}{3}, \quad \Pr[\text{area 2} \mid \text{above line}] = \frac{1}{3}.$$

In other words, the spinner is twice as likely to be in area 1 than in area 2 given that it landed above the dashed line.

We can obtain these results by straightforward calculation. For example, the joint event of landing above the dashed line and in area 2 has probability $\Pr[\text{area 2} \cap \text{above line}] = 1/6$ and the probability to land above the line is $\Pr[\text{above line}] = 1/2$, whence

$$\Pr[\text{area 2} \mid \text{above line}] = \frac{\Pr[\text{area 2} \cap \text{above line}]}{\Pr[\text{above line}]} = \frac{1}{3},$$

confirming our aforementioned claim.

We have to be careful, though, as our intuition may fail us when considering conditional probabilities.

15.3 Conditional Probabilities

Example 15.8. Suppose that someone rolls a die twice. Before showing us the result, she reveals that in one of the two rolls the die showed a face value of six. What is the probability that the result will be a pair of sixes?

The outcome of the experiment is a pair of face values. Let P denote the event to get a pair of sixes and S the event that the pair of dices shows at least one six. Then our goal is to determine the conditional probability $\Pr[P \mid S]$, which is the conditional probability to roll a pair of sixes given that at least one die shows the value 6.

The event P of rolling a pair of sixes is given by the set $P = \{(⚅,⚅)\}$. The event $S =$ "at least one die shows a six" consists of the following 11 pairs:

$$S = \{(⚅,⚀), (⚅,⚁), (⚅,⚂), (⚅,⚃), (⚅,⚄), (⚅,⚅),$$
$$(⚄,⚅), (⚃,⚅), (⚂,⚅), (⚁,⚅), (⚀,⚅)\}.$$

The probability $\Pr[P]$ to get a pair of sixes is given by $\Pr[P] = 1/36$ and the probability to roll at least one six is given by $\Pr[S] = 11/36$. Since the event P is a subset of the event S, we have $S \cap P = P$. Therefore, we have

$$\Pr[P \mid S] = \frac{\Pr[P \cap S]}{\Pr[S]} = \frac{\Pr[P]}{\Pr[S]} = \frac{1}{11}.$$

It is essential to carefully calculate conditional probabilities, as it can be difficult to develop the right intuition for their value.

The next result shows how to express $\Pr[A \mid B]$ in terms of $\Pr[B \mid A]$.

Proposition 15.9 (Bayes' Theorem). *Let A and B be events with $\Pr[B] > 0$. Then*

$$\Pr[A \mid B] = \frac{\Pr[B \mid A]\Pr[A]}{\Pr[B]}.$$

Proof. By definition, we have

$$\Pr[A \mid B]\Pr[B] = \Pr[A \cap B] = \Pr[B \cap A] = \Pr[B \mid A]\Pr[A].$$

Dividing by $\Pr[B]$ yields the claim. □

Bayes' Theorem is at the heart of many algorithms in machine learning, e-mail spam detection, and countless other applications. There are a number of variations of this theorem that use the next proposition or some of its generalizations.

Proposition 15.10 (Law of Total Probability, Simple Version). *Let Ω be a sample space, and let A and B be events. Then*

$$\Pr[B] = \Pr[B \cap A] + \Pr[B \cap A^c]$$
$$= \Pr[B \mid A]\Pr[A] + \Pr[B \mid A^c]\Pr[A^c]$$

Proof. The events A and $A^\complement$ are disjoint and satisfy $\Omega = A \cup A^\complement$. Therefore, we have
$$\Pr[B] = \Pr[B \cap A] + \Pr[B \cap A^\complement].$$
The second equality follows directly from the definition of conditional probability. □

We can use the law of total probability to express the denominator $\Pr[B]$ in Bayes' theorem in a different form. Rather than memorizing the next result, you should memorize how to derive it from the previous two propositions.

Corollary 15.11. *Let A and B be events with $\Pr[B] > 0$. Then*
$$\Pr[A \mid B] = \frac{\Pr[B \mid A]\Pr[A]}{\Pr[B \mid A]\Pr[A] + \Pr[B \mid A^\complement]\Pr[A^\complement]}.$$

Example 15.12. We illustrate the virtue of conditional probabilities with the help of the notorious car and goats problem, which got famous through the Monty Hall game show. At the end of this show, a contestant was shown three closed doors. She was told that behind one door is a new car and behind the other two are goats. If the contestant chooses the door hiding the car, then she can keep the car. Once she has made her choice, the game show host—knowing which door conceals the car—opens one of the other two doors to reveal a goat. Monty then asks her whether she would like to switch doors. The question is: Should she switch?

Without loss of generality, let us assume that the contestant has chosen door 1, Monty has opened door 2, and now the contestant has to choose between doors 1 and 3. Let C_1 denote the event that the car is behind door 1, C_3 the event that the car is behind door 3, and M_2 the event that Monty opened door 2, hence contains a goat.

It is apparent that $\Pr[C_1] = 1/3$ and $\Pr[C_3] = 1/3$. Assuming that Monty will choose a door at random if both doors conceal goats, we get $\Pr[M_2 \mid C_1] = 1/2$. We certainly have $\Pr[M_2 \mid C_3] = 1$, because Monty has no choice in this case. Recall that our goal is to compare the conditional probabilities $\Pr[C_1 \mid M_2]$ and $\Pr[C_3 \mid M_2]$. We can use Bayes' rule to determine these probabilities. Indeed,
$$\Pr[C_1 \mid M_2] = \frac{\Pr[M_2 \mid C_1]\Pr[C_1]}{\Pr[M_2 \mid C_1]\Pr[C_1] + \Pr[M_2 \mid C_3]\Pr[C_3]} = \frac{1/6}{1/6 + 1/3} = 1/3.$$

Similarly, $\Pr[C_3 \mid M_2] = 2/3$. In conclusion, if she sticks with her decision, then the probability to get the car is $1/3$. If she switches, then the probability is $2/3$. This means that it is advisable that she switches doors.

There are many websites dedicated to this problem, and one finds heated discussions about the Monty Hall problem on the internet. You will notice that there exist different solutions, depending on the exact assumptions about Monty's knowledge and his strategy.

15.3 Conditional Probabilities

Exercises

15.11. Let Ω be a discrete sample space and A an event in $P(\Omega)$ such that $\Pr[A] > 0$. Show that the conditional probability $\Pr[\,\cdot\,|A]$ is a probability measure on A.

15.12. A mother has two children. At birth, each child has an equal chance to be a boy or a girl. Given that one child is a girl, what are the chances that the other child will be a girl as well?

15.13. Suppose that five fair coins are tossed. If at least one coin shows heads, what is the probability that precisely three of the five coins show heads?

15.14. A doctor suspects that his patient suffers from snark disease. The disease is rare, just 1 in 10,000 people carry the disease. The doctor administers a test that has a low probability for false negatives, namely there is a 1% probability of a carrier testing negative. The probability for a false positive is a bit higher but still very low, as there is a 2% probability of a non-carrier testing positive. The patient tests positive. What are the chances that the patient suffers from snark disease?

15.15. Suppose that $F_1, \ldots, F_n$ are events that partition the sample space Ω such that $\Pr[F_k] > 0$ holds for all k in the range $1 \leqslant k \leqslant n$. Show that

$$\Pr[E] = \sum_{k=1}^{n} \Pr[E\,|F_k]\Pr[F_k]$$

holds for any event E. This fact is attributed to Reverend Thomas Bayes.

15.16. Suppose that $A_1, \ldots, A_n$ are events that partition the sample space Ω. Let B be an event with $\Pr[B] > 0$. Show that

$$\Pr[A_1\,|B] = \frac{\Pr[B|A_1]\Pr[A_1]}{\sum_{k=1}^{n} \Pr[B|A_k]\Pr[A_k]}.$$

This fact is due to Reverend Thomas Bayes.

15.17. Suppose that there are three bags of marbles. The first one contains 30 red and 20 yellow marbles. The second bag contains 35 red and 15 green marbles. The third bag contains 40 red and 10 blue marbles. If you choose one of the bags uniformly at random and then randomly select one marble from the bag, what is the probability that the marble will be red?

15.18. We can often use conditional probabilities to calculate the probabilities of joint events. Let $A_1, A_2, \ldots, A_n$ be events satisfying $\Pr[A_1 \cap A_2 \cap \cdots \cap A_{n-1}] > 0$. Then

$$\Pr[A_1 \cap A_2 \cap \cdots \cap A_n] = \Pr[A_1]\Pr[A_2\mid A_1]\Pr[A_3\mid A_1 \cap A_2]\cdots$$
$$\cdots\Pr[A_n\mid A_1 \cap A_2 \cap \cdots \cap A_{n-1}].$$

15.19. Show that if A and B are events such that $\Pr[B] > 0$, then
$$\Pr[A \mid B] \leq \frac{\Pr[A]}{\Pr[B]}.$$

15.20. The Sure Thing Principle states that if A, B, and C are events such that $\Pr[A \mid C] \geq \Pr[B \mid C]$ and $\Pr[A \mid C^c] \geq \Pr[B \mid C^c]$, then $\Pr[A] \geq \Pr[B]$. Prove the Sure Thing Principle.

15.4 Independence

Two events A and B are called **independent** if and only if
$$\Pr[A \cap B] = \Pr[A]\Pr[B]$$
holds. We call the event A and B **dependent** if and only if they are not independent.

If the probabilities $\Pr[A]$ and $\Pr[B]$ are nonzero, this means that
$$\Pr[A \mid B] = \frac{\Pr[A \cap B]}{\Pr[B]} = \Pr[A] \quad \text{and} \quad \Pr[B \mid A] = \frac{\Pr[B \cap A]}{\Pr[A]} = \Pr[B].$$

In other words, two events are independent if the occurrence of one event does not change the probability of the other.

Example 15.13. Suppose that we roll a die twice. Let A be the event that in first roll the die shows a 5 or 6. Let B be the event that in the second roll the dies shows 4, 5, or 6. The probabilities of these events are
$$\Pr[A] = \frac{1}{3} \quad \text{and} \quad \Pr[B] = \frac{1}{2}.$$

The joint event $A \cap B$ is given by

$A \cap B = \{ (\boxdot,\boxdot), (\boxdot,\boxdot), (\boxdot,\boxdot), (\boxdot,\boxdot), (\boxdot,\boxdot), (\boxdot,\boxdot) \}.$

We have
$$\Pr[A \cap B] = \frac{6}{36} = \frac{1}{3} \cdot \frac{1}{2} = \Pr[A]\Pr[B],$$
so the two events A and B are independent.

We can extend the notion of independence to entire families of events. Let I be an arbitrary index set and A_k, $k \in I$, a family of events. This family of events is called **independent** if and only if for every finite subset F of the index set I, we have
$$\Pr\left[\bigcap_{k \in F} A_k\right] = \prod_{k \in F} \Pr[A_k].$$

It should be stressed that even if the index set I is infinite, the product formula must only hold for all finite subsets of the index set.

15.4 Independence

Example 15.14. Four students fill out a questionnaire about the likes and dislikes of the subjects: mathematics, history, and language arts. A student checks L if they like the subject and D if they dislike the subject. Alice likes only mathematics, so she writes (L, D, D). Beatrix likes only history, so she writes (D, L, D). Claire is into language arts but dislikes the others, so she writes (D, D, L). Dave likes all three subjects, so he writes (L, L, L). They put their answers into a hat and draw uniformly at random questionnaires. Then

$$\Pr[\text{likes math}] = \frac{1}{2}, \Pr[\text{likes history}] = \frac{1}{2}, \Pr[\text{likes language arts}] = \frac{1}{2}.$$

The probability to like two different subjects is $1/4$, as only Dave likes different subjects. We have

$$\Pr[\text{likes math} \cap \text{likes history}] = \frac{1}{4} = \Pr[\text{likes math}] \Pr[\text{likes history}].$$

So the events "likes math" and "likes history" are independent. In fact, any pair of subjects are liked independently. However, these three events are dependent, as the probability to like all three is given by

$$\Pr[\text{likes math} \cap \text{likes history} \cap \text{likes language arts}] = \frac{1}{4},$$

but this is not equal to the product

$$\Pr[\text{likes math}] \Pr[\text{likes history}] \Pr[\text{likes language arts}] = \frac{1}{8}.$$

This shows that it does not suffice to check pairwise independence, but one needs to verify the product law for all finite subsets of these three events.

We conclude this section by exploring the behavior of a sequence of events such as an infinite sequence of coin tosses. We might be interested in the event $E_n = \{H\}$ that the coin turns up heads at the n-th coin toss. However, let us assume that the coin will get increasingly more biased, say, $\Pr[E_n] = 1/n^a$ for some fixed exponent $a > 0$. So for large n, the probability that the coin in the n-th coin toss comes up tails approaches 1. How likely is it that an infinite number of heads turn up in this infinite sequence of coin tosses? The next result shows that the probability of this event is either 0 or 1.

Let $E_1, E_2, \ldots$ be events on a discrete sample space Ω. Then we can form the event

$$E^* = \bigcap_{n=1}^{\infty} \bigcup_{k=n}^{\infty} E_k.$$

An outcome ω is contained in E^* if and only if ω is contained in $\bigcup_{k=n}^{\infty} E_k$ for all n. In other words, ω is contained in E^* if and only if ω is contained in E_k for an infinite number of indices k.

Even though it might seem that deriving the probability of the limiting event E^* is difficult, the next result gives remarkably simple conditions that can establish $\Pr[E^*]$.

Proposition 15.15 (Borel-Cantelli). *Let $E_1, E_2, \ldots$ be events and denote by $E^* = \limsup_{n\to\infty} E_n$ the event that infinitely many of the events E_n occur.*

(a) If $\sum_{k=1}^{\infty} \Pr[E_k] < \infty$, then $\Pr[E^*] = 0$. In other words, with probability 1, only finitely many events E_n will occur.

(b) If $\sum_{k=1}^{\infty} \Pr[E_k] = \infty$ and $E_1, E_2, \ldots$ are independent events, then $\Pr[E^*] = 1$.

Proof. (a) We notice that

$$\Pr[E^*] = \Pr\left[\bigcap_{n=1}^{\infty}\bigcup_{k=n}^{\infty} E_k\right] \leq \Pr\left[\bigcup_{k=n}^{\infty} E_k\right] \leq \sum_{k=n}^{\infty} \Pr[E_k] =: R_n.$$

Since the sum $\sum_{k=1}^{\infty} \Pr[E_k]$ converges, the right-hand side terms $R_n \to 0$, as $n \to \infty$. We can conclude that $\Pr[E^*] = 0$.

(b) We denote by F_n^C the event

$$F_n^C = \bigcap_{k=n}^{\infty} E_k^C,$$

which expresses that none of the events E_k with $k \geq n$ occur. The complement of the event E^* is given by

$$(E^*)^C = \bigcup_{n=1}^{\infty}\bigcap_{k=n}^{\infty} E_k^C = \bigcup_{n=1}^{\infty} F_n^C.$$

Therefore, it suffices to show that $\Pr[F_n^C] = 0$.

Let us look at the probability of finite approximations to F_n^C. The independence of the events E_k implies the independence of the events E_k^C. Therefore, we get

$$\Pr\left[\bigcap_{k=n}^{m} E_k^C\right] = \prod_{k=n}^{m} \Pr[E_k^C] = \prod_{k=n}^{m}(1 - \Pr[E_k]) \leq \exp\left(-\sum_{k=n}^{m} \Pr[E_k]\right),$$

where in the last inequality we used that $1 - x \leq e^{-x}$ holds for all real numbers x. We can conclude that

$$\Pr[F_n^C] = \lim_{m \to \infty} \Pr\left[\bigcap_{k=n}^{m} E_k^C\right] \leq \exp\left(-\sum_{k=n}^{\infty} \Pr[E_k]\right).$$

As the sum diverges, it follows that

$$\Pr[F_n^C] = 0.$$

By the union bound, we can conclude that

$$\Pr[(E^*)^C] = \Pr\left[\bigcup_{n=1}^{\infty} F_n^C\right] \leq \sum_{k=1}^{\infty} \Pr[F_n^C] = 0,$$

15.4 Independence

so $\Pr[E^*] = 1 - \Pr[(E^*)^C] = 1$, as claimed.

$\square$

The independence of events in part (b) is necessary, as Exercise 15.30 shows. A fun consequence is the occurrence of any text in an infinite string of letters generated uniformly at random, see Exercise 15.31.

Exercises

15.21. Consider a standard deck of 52 cards. Let A be the event to draw an ace, and B the event to draw a card of diamonds. Are the events A and B independent? Prove your result.

15.22. Two events A and B are disjoint if and only if $A \cap B = \emptyset$. When are A and B independent?

15.23. Let $\Omega = \{H, T\} \times \{H, T\} \times \{H, T\}$ be the sample space of three successive independent coin flips. Suppose that the probability to get heads in a single toss is p, and the probability to get tails is $1 - p$. Consider the events

$$A = \{(H, H, H), (H, H, T), (H, T, H), (T, H, H)\},$$

and $B = \{(H, H, H), (T, T, T)\}$. For what choice of the parameter p in the range $0 \leq p \leq 1$ are the two events A and B independent? Prove your result.

15.24. Let A and B be events such that $\Pr[A] \neq 0$ and $\Pr[B] \neq 0$. Show that if $\Pr[A \mid B] > \Pr[A]$, then $\Pr[B \mid A] > \Pr[B]$.

15.25. Let Ω be a finite sample space. Let A be an event in $P(\Omega)$. Show that if A is independent of all events $B \in P(\Omega)$, then $\Pr[A] = 0$ or $\Pr[A] = 1$.

15.26. Let A and B be independent events. Show that
(a) A^C and B are independent events.
(b) A^C and B^C are independent events.

15.27. Let n be an integer such that $n \geq 2$. Let $E_1, E_2, \ldots, E_n$ be independent events. Show that
(a) the events $E_1^C, E_2, \ldots, E_n$ are independent,
(b) the events $E_1^C, E_2^C, \ldots, E_n^C$ are independent.

15.28. Let n be an integer such that $n \geq 2$. Let $E_1, E_2, \ldots, E_n$ be independent events. Show that the probabilistic inclusion-exclusion formula for independent events can be written in the form

$$\Pr[E_1 \cup E_2 \cup \cdots \cup E_n] = 1 - \prod_{k=1}^{n}(1 - \Pr[E_k]).$$

This is much simpler to prove than the general formula.

15.29. Let n be an integer such that $n \geq 2$. Let $E_1, E_2, \ldots, E_n$ be independent events. Show that

$$\Pr\left[\bigcup_{k=1}^{n} E_k\right] \geq 1 - \exp\left(\sum_{k=1}^{n} \Pr[E_k]\right).$$

15.30. Show that if we omit the independence of the events in part (b) of the Borel-Cantelli proposition, then the limiting event E^* can occur with probability p such that $p \neq 0$ and $p \neq 1$. Give a simple example.

15.31. Lev Nicolayevich Tolstoy's epic "War and Peace" has 3,230,047 characters. Suppose that a Monkey types an infinite string of letters uniformly at random on a standard typewriter with 88 characters. A biologist observes that the Monkey's written text contains the complete "War and Peace" epic exactly—character by character—not just once, but several times. Since the probability to write Tolstoy's epic is exceedingly small (after all, the probability $1/88^{3,230,047} < 10^{-1,000,000}$), the biologist concludes that the Monkey must be a genius. Show that the result is not surprising at all, and the result can be reproduced by a simple machine that depresses the keys uniformly at random. How often will "War and Peace" occur in the infinite stream of letters?

15.5 Random Variables

In this section, we are going to introduce random variables, which are a convenient way to specify events. A random variable is a function that associates a numerical value to each outcome of an experiment. For instance, if we roll a pair of dice, then the sum of the two face values is a random variable.

Let Ω be a discrete sample space. A **random variable** X is a function from the sample space Ω to the set of real numbers. In essence, the random variable translates cumbersome specifications of events in $P(\Omega)$ to the familiar language of subsets of real numbers.

Example 15.16. Consider rolling a die twice. The sample space Ω contains 36 pairs of face values. Let X denote the random variable that gives the sum of face values. Then the event that the face values sum to 8 or 9 is specified by $X^{-1}([8,9])$. This is more concise than writing the event in the form

$$\{(\boxdot,\boxdot),(\boxdot,\boxdot),(\boxdot,\boxdot),(\boxdot,\boxdot),(\boxdot,\boxdot),$$
$$(\boxdot,\boxdot),(\boxdot,\boxdot),(\boxdot,\boxdot),(\boxdot,\boxdot)\}.$$

Associating numerical values with events can have other benefits as well, as we will see shortly.

Probability theorists often use the notation $X \in B$ to express the event

$$X^{-1}(B) = \{\omega \in \Omega \mid X(\omega) \in B\}.$$

15.5 Random Variables

The reason for this notation is simple: for instance, writing $X \in \{8, 9\}$ looks less cluttered than $X^{-1}(\{8, 9\})$. If $B = \{x\}$ is a singleton set, then we often write $X = x$ as a shorthand for $X^{-1}(\{x\})$.

Example 15.17. Let $\Omega = \{H, T\} \times \{H, T\} \times \{H, T\}$ be the sample space of tossing a coin three times. Let $X \colon \Omega \to \{0, 1, 2, 3\}$ be the random variable counting the number of heads. Then $X = 2$ denotes the event

$$\{(T, H, H), (H, T, H), (H, H, T)\},$$

and $X = 3$ the event $\{(H, H, H)\}$.

Among the simplest examples of random variables are indicator random variables, which have merely two different values.

Example 15.18. Let Ω be a discrete sample space. For an event $A \in P(\Omega)$, we define the function

$$I_A(x) = \begin{cases} 1 & \text{when } x \in A, \\ 0 & \text{when } x \notin A. \end{cases}$$

Then I_A is called the **indicator random variable** of the event A. Suppose that $p = \Pr[A]$ is the probability of the event A. Then

$$\Pr[I_A = 1] = p \quad \text{and} \quad \Pr[I_A = 0] = 1 - p.$$

More elaborate random variables are often composed of weighted sums or products of indicator random variables.

Example 15.19. Consider the sample space $\Omega = \{H, T\} \times \{H, T\} \times \{H, T\}$ of a sequence of three coin tosses. Let $I_{H_1}, I_{H_2}, I_{H_3}$ be the indicator random variables of the events that the first, second, or third coin toss shows heads, respectively. Then the sum $X = I_{H_1} + I_{H_2} + I_{H_3}$ of these three indicator random variables is the random variable counting the total number of heads that we have seen in Example 15.17.

A random variable $X \colon \Omega \to \mathbf{R}$ on a discrete sample space Ω has a finite or countably infinite set of values. Therefore, the values attained by X can be indexed by positive integers

$$\{X(\omega) \mid \omega \in \Omega\} = \{\alpha_1, \alpha_2, \alpha_3, \ldots\}.$$

We can assume without loss of generality that the values α_k are pairwise distinct. Let us denote the probability of the event $X = \alpha_k$ by

$$p_k = \Pr[X = \alpha_k].$$

The probabilities p_k allow us to express the probability of the event $X \in B$ for any subset B of the set of real numbers. Indeed, since the events $X = \alpha_k$ and $X = \alpha_\ell$ are disjoint when $k \neq \ell$, we obtain

$$\Pr[X \in B] = \Pr[X \in B \cap \{\alpha_1, \alpha_2, \ldots\}] = \sum_{\alpha_k \in B} p_k,$$

where the sum extends over all indices $k \in \mathbf{N}_1$ such that $\alpha_k \in B$.

We define the **distribution function** $F_X(x)$ of the random variable X by

$$F_X(x) = \Pr[X \in (-\infty, x]] = \Pr[X \leq x],$$

where $X \leq x$ is the commonly used shorthand for $X \in (-\infty, x]$. Sometimes this is also called the **cumulative distribution function** of the random variable X.

Example 15.20. Let us consider once more the indicator random variable I_A of an event A in a discrete sample space. Suppose that $p = \Pr[A]$ is the probability of the event A, whence

$$\Pr[I_A = 1] = p \quad \text{and} \quad \Pr[I_A = 0] = 1 - p.$$

The distribution function $F_{I_A}(x)$ of the random variable I_A is given by

$$F_{I_A}(x) = \begin{cases} 0 & \text{when } x < 0, \\ 1-p & \text{when } 0 \leq x < 1, \\ 1 & \text{when } x \geq 1. \end{cases}$$

One can reconstruct the probability mass function $p(x) = \Pr[I_A = x]$ of the indicator random variable from the distribution function.

We call two random variables X and Y **independent** if and only if the events $X = x$ and $Y = y$ are independent for all values x and y, meaning

$$\Pr[X = x, Y = y] = \Pr[X = x]\Pr[Y = y].$$

Exercises

15.32. The nomenclature "random variable" was introduced before the modern definition of probability theory. Use the definition of the random variable to explain why the slogan "A random variable is neither random nor a variable" is correct.

15.33. Consider the measurable space $(\Omega, P(\Omega))$, where the sample space describes the outcomes of three successive coin tosses, so

$$\Omega = \{H, T\} \times \{H, T\} \times \{H, T\}.$$

Let X be the random variable counting the number of heads. Explicitly describe the events given by
(a) $X \geq 2$,
(b) $X = 1$,
(c) $X < 1$.

15.6 Expectation

15.34. Let X denote the number of heads on three independent coin flips of a biased coin that turns up heads with probability p. Give the distribution function F_X of X.

15.35. Suppose that two dice are rolled. Let X be the random variable that gives the larger of the two face values. Determine $\Pr[3 \leq X \leq 4]$.

15.36. Suppose that we toss a fair coin until the coin shows heads. Let X denote the number of coin tosses. So we have $X = 1$ when the coin shows head after the first toss, $X = 2$ when the coin showed tails and then heads in the second try.
(a) Determine $\Pr[X = n]$.
(b) Determine the distribution function F_X.

15.6 Expectation

Random variables specify events using the values of the random variable. The expected value of a random variable is essentially an average of its values. Despite being an average, the expected value often gives us some insights about the behavior of the probability distribution of a random variable.

Let X be a random variable on a discrete sample space Ω. The **expectation value** of X is defined to be

$$\mathrm{E}[X] = \sum_{\alpha \in X(\Omega)} \alpha \Pr[X = \alpha],$$

when this sum is unconditionally convergent in $\overline{\mathbf{R}}$, the extended real numbers. The expectation value is also called the **mean** of X. If X is a random variable with nonnegative integer values, then the expectation can be calculated by

$$\mathrm{E}[X] = \sum_{x=1}^{\infty} \Pr[X \geq x],$$

which is often convenient. If X and Y are two arbitrary discrete random variables, then

$$\mathrm{E}[aX + bY] = a\,\mathrm{E}[X] + b\,\mathrm{E}[Y],$$

that is, the expectation operator is linear. This is an extremely useful result.

If X and Y are independent discrete random variables, then

$$\mathrm{E}[XY] = \mathrm{E}[X]\,\mathrm{E}[Y].$$

Caveat: If X and Y are not independent, then this is in general false.

We suggest that you formally prove the last three claims, see Exercises 15.37–15.39.

The next example illustrates how linearity of expectation can ease computations, especially when a random variable can be expressed as sum of indicator random variables.

Example 15.21. Suppose that n persons give their hats to the hat check girl. She is distraught and hands the hats back uniformly at random. We want to answer the following question: On average, how many persons get their own hat back?

We take $\Omega = \{1, \ldots, n\}$ as sample space and allow all subsets of Ω as events, $\mathcal{F} = 2^\Omega$. Let $X_k \colon \Omega \to \mathbf{R}$ be the random variable defined by $X_k(k) = 1$ and $X_k(x) = 0$ for all $x \neq k$. Hence, $X_k = 1$ denotes the event that person k receives her own hat. The probability that the kth person receives her own hat back is $\Pr[X_k = 1] = 1/n$, since she will receive one of n possible hats.

Let $X = X_1 + \cdots + X_n$ denote the number of persons receiving their own hats. We have

$$\mathrm{E}[X] = \sum_{k=1}^{n} \mathrm{E}[X_k] = \sum_{k=1}^{n} 1 \cdot \Pr[X_k = 1] = n(1/n) = 1,$$

by linearity of expectation, and by definition of the expectation value. This means that on average one person gets her own hat back.

The expectation can be used to bound probabilities, as the following simple but fundamental result shows:

Theorem 15.22 (Markov's Inequality). *If X is a nonnegative random variable and t a positive real number, then*

$$\Pr[X \geq t] \leq \frac{\mathrm{E}[X]}{t}.$$

Proof. Let Y denote the random variable

$$Y(\omega) = \begin{cases} 0 & \text{if } X(\omega) < t, \\ 1 & \text{if } X(\omega) \geq t, \end{cases}$$

hence $Y = 1$ denotes the event $X \geq t$. The expectation value of X satisfies

$$\mathrm{E}[X] \geq \mathrm{E}[tY] = t\,\mathrm{E}[Y] = t\,\Pr[X \geq t],$$

which proves the claim. $\square$

Variance. The **variance** $\mathrm{var}[X]$ of a discrete random variable X is defined by

$$\mathrm{var}[X] = \mathrm{E}[(X - \mathrm{E}[X])^2] = \mathrm{E}[X^2] - \mathrm{E}[X]^2,$$

whenever this expression is well-defined. The variance measures the squared deviation from the expected value $\mathrm{E}[X]$.

It is easy to see that variance is *not* a linear operator, since

$$\mathrm{var}[X + X] = 4\,\mathrm{var}[X]$$

15.6 Expectation

holds, to mention just one example. Moreover, $\mathrm{var}[aX + b] = a^2 \mathrm{var}[X]$ for all $a, b \in \mathbf{R}$. If X and Y are independent random variables, then the variance satisfies

$$\mathrm{var}[X + Y] = \mathrm{var}[X] + \mathrm{var}[Y]. \tag{15.1}$$

The random variable X will rarely deviate from the expectation value if the variance is small. This is a consequence of the Chebyshev's useful inequality:

Theorem 15.23 (Chebyshev's Inequality). *Let Y be an integrable random variable. Then*

$$\Pr[|Y - \mathrm{E}[Y]| \geq \beta] \leq \frac{\mathrm{var}[Y]}{\beta^2}. \tag{15.2}$$

Proof. Define a nonnegative random variable X by setting $X = (Y - \mathrm{E}[Y])^2$. Then

$$\Pr[|Y - \mathrm{E}[Y]| \geq \beta] = \Pr[X \geq \beta^2] \leq \frac{\mathrm{E}[X]}{\beta^2} = \frac{\mathrm{var}[Y]}{\beta^2},$$

where we used Markov's inequality. □

Bernoulli Distribution. Tossing a biased coin can be described by a random variable X that takes the value 1 if the outcome of the experiment is **head**, and the value 0 if the outcome is **tail**. Assume that $\Pr[X = 1] = p$ and $\Pr[X = 0] = 1 - p$ for some real number $p \in (0, 1)$. The random variable X is said to have the Bernoulli distribution with parameter p. We can compute the expectation value and the variance as follows:

$$\mathrm{E}[X] = p, \quad \mathrm{var}[X] = \mathrm{E}[X^2] - \mathrm{E}[X]^2 = p - p^2 = p(1 - p).$$

Binomial Distribution. Let $X_1, \ldots, X_n$ denote independent identically distributed random variables, all having a Bernoulli distribution with parameter p. Then the random variable $X = X_1 + \cdots + X_n$ describes the number of heads in a sequence of n coin flips. The expectation of X can be immediately computed by linearity of expectation, and, since the random variables X_k are independent, we can compute the variance using (15.1):

$$\mathrm{E}[X] = np, \quad \mathrm{var}[X] = np(1 - p).$$

The probability of the event $X = x$, for integers in the range $0 \leq x \leq n$, is

$$\Pr[X = x] = \binom{n}{x} p^x (1 - p)^{n - x}.$$

Indeed, choose x positions in a sequence of length n. The probability that the sequence will show heads at exactly these positions is $p^x(1-p)^{n-x}$. The result follows, since there are $\binom{n}{x}$ ways to choose x positions in a sequence of length n.

Uniform Distribution. Let X be a random variable that takes integer values in $\{1, \ldots, n\}$. Such a random variable is said to be uniformly distributed if $\Pr[X = k] = 1/n$ for all integers k in the range $1 \leqslant k \leqslant n$. The expectation value and the variance of X are respectively given by

$$\mathrm{E}[X] = \frac{n+1}{2}, \qquad \mathrm{var}[X] = \frac{n^2 - 1}{12}.$$

The expectation value follows from the definition. We can verify the variance by noting that

$$\mathrm{E}[X^2] = \sum_{k=1}^{n} \frac{1}{n} k^2 = \frac{1}{n} \frac{n(1+n)(1+2n)}{6} = \frac{(1+n)(1+2n)}{6},$$

hence $\mathrm{var}[X] = \mathrm{E}[X^2] - \mathrm{E}[X]^2 = \dfrac{(1+n)(1+2n)}{6} - \dfrac{(n+1)^2}{4} = (n^2 - 1)/12$.

Geometric Distribution. Suppose we keep tossing a biased coin, which has the Bernoulli distribution with parameter p, until the event **head** occurs. Let the random variable X denote the number of coin flips needed in this experiment. We say that X is geometrically distributed with parameter p. The density function of X is given by

$$p_X(x) = \Pr[X = x] = p(1-p)^{x-1}$$

for $x = 1, 2, \ldots$, and $p_X(x) = 0$ otherwise. The expectation value and the variance of X are given by

$$\mathrm{E}[X] = \frac{1}{p}, \qquad \mathrm{var}[X] = \frac{1-p}{p^2}.$$

It is possible to derive these facts directly from the definitions. For the expectation value, this can be done without too much effort, but for the variance, this is cumbersome. In the next section, we will introduce a tool that can significantly simplify such calculations.

Negative Binomial Distribution. Let $X_1, \ldots, X_n$ be independent random variables, all having geometric distribution with parameter p. The random variable $X = X_1 + \cdots + X_n$ describes the number of coin flips that are necessary until n heads occur, when heads has probability p. The random variable X is said to have negative binomial distribution with parameters n and p. Linearity of expectation and additivity of variance for independent random variables show that

$$\mathrm{E}[X] = \sum_{k=1}^{n} \mathrm{E}[X_k] = \frac{n}{p}, \qquad \mathrm{var}[X] = \sum_{k=1}^{n} \mathrm{var}[X_k] = \frac{n(1-p)}{p^2}.$$

15.6 Expectation

The probability of the event $X = k$ is

$$\Pr[X = k] = \binom{k-1}{n-1} p^n (1-p)^{k-n}, \qquad k \geq n.$$

Indeed, a sequence of k coin flips that contains exactly n heads at specified positions has probability $p^n(1-p)^{k-n}$. By specification, the last position is a head, hence there are $\binom{k-1}{n-1}$ other positions that can be chosen for the heads.

Poisson Distribution. A random variable X with non-negative integer values is said to be Poisson distributed with parameter $\lambda > 0$ if

$$\Pr[X = k] = \frac{\lambda^k}{k!} e^{-\lambda}, \qquad k = 0, 1, 2, \ldots.$$

The Poisson distribution can be used to approximate the Binomial distribution if n is large and p is small. Indeed, suppose that $\lim_{n \to \infty} n p_n = \lambda$, then

$$\lim_{n \to \infty} \binom{n}{k} p_n^k (1-p_n)^{n-k} = e^{-\lambda} \frac{\lambda^k}{k!}.$$

This formula is frequently used when the evaluation of the binomial distribution is not feasible.

Exercises

15.37. Let X be a discrete random variable that has nonnegative integer values. Show that

$$\mathrm{E}[X] = \sum_{n=1}^{\infty} \Pr[X \geq n].$$

15.38. Let X and Y be random variables on discrete sample spaces. Show that expectation is a linear operator, that is, for all real numbers a and b, we have

$$\mathrm{E}[aX + bY] = a\,\mathrm{E}[X] + b\,\mathrm{E}[Y].$$

15.39. Let X and Y be two independent random variables on discrete sample spaces. Show that $\mathrm{E}[XY] = \mathrm{E}[X]\,\mathrm{E}[Y]$. [Note that some dependent random variables are also multiplicative, but the relation does not hold in general.]

15.40. Suppose that you have a biased coin that produces heads with probability p, where $0 < p < 1$, but unfortunately this probability is not known to you. Von Neumann showed that it is possible to use such a biased coin to construct a source for fair coin flips. Derive a scheme such that the expected number of biased coin flips does not exceed $1/(p(1-p))$. [Hint: Consider consecutive pairs of biased coin flips.]

15.41. Alice is a compulsive coupon collector. Currently, she is collecting charming Harry Potter characters that are contained in overpriced cereal boxes. There are n different characters, and each box contains one character. She wants to get the complete set. Show that she must expect to buy nH_n cereal boxes, where H_n denotes the Harmonic number $H_n = 1 + 1/2 + 1/3 + \cdots + 1/n$.

15.42. Let us continue the previous exercise. Let X denote the random variable counting the number of boxes required to collect at least one character of each type. Calculate the variance $\text{var}[X]$.

15.43. A geometric random variable describes how many times we have to flip a biased coin until we obtain heads. Suppose that we have n such coins and we toss all of them once in one round. Let X denote the random variable counting the number of rounds until heads have occurred at least once for all n coins. Show that

$$\text{E}[X] = \sum_{m=0}^{\infty} \left(1 - \left(1 - (1-q)^m\right)^n\right) = \sum_{m=0}^{\infty} \left(1 - \left(1 - p^m\right)^n\right). \qquad (15.3)$$

Here q denotes the probability that the coin shows heads and $p = (1-q)$.

15.44. We continue the previous exercise. Derive from equation (15.3) the alternate expression

$$\text{E}[X] = \sum_{k=1}^{n} \binom{n}{k} (-1)^{k+1} \frac{1}{1 - (1-q)^k}.$$

15.7 The Probabilistic Method

Suppose that we want to prove the existence of a combinatorial object that has certain properties. In the probabilistic method, we approach this problem by defining a sample space of combinatorial objects and showing that a randomly chosen element of this space has the desired properties with positive probability. Paul Erdős showed that this simple idea can lead to spectacular results. An interesting aspect is that one can apply the probabilistic method to solve problems that do not seem to call for probabilistic solutions. We discuss some well-known results that nicely illustrate this method.

Ramsey Numbers. What is the *smallest* number $n = R(a,b)$ such that in *any* set of n people there must be
(a) a mutually acquainted people or
(b) b mutual strangers?
The numbers $R(a,b)$ are called **Ramsey numbers**.

In general, computing $R(a,b)$ can be remarkably difficult. Among n people, there are $2^{\binom{n}{2}}$ possible acquaintance/stranger relations. To show that $n = R(a,b)$, we must verify that all $2^{\binom{n}{2}}$ choices satisfy the conditions (a) and (b)

15.7 The Probabilistic Method

above. Furthermore, we must show that there is one particular set of $n-1$ people that satisfies neither (a) nor (b).

We can translate this problem into graph-theoretic terms. We can model the set of n people by a complete graph. We can use a coloring of the graph with two colors red and blue. However, this time we will color the edges rather than the vertices. We color an edge (i,j) *red* (and depict it by a solid arc) if i and j are acquainted, and color it *blue* (and depict it by a dashed arc) if i and j are strangers. In this language, the Ramsey number $R(a,b)$ is the smallest integer n such that in *any* edge-coloring of K_n with the two colors *red* and *blue*, there exists

(a) an induced *red* (solid) K_a subgraph or

(b) an induced *blue* (dashed) K_b subgraph.

We can gain some insight into this problem by having a look at a few simple instances. Let us begin with one of the simplest cases.

Proposition 15.24. *Let n be an integer satisfying $n \geqslant 2$. Then the Ramsey number $R(2,n)$ is given by*
$$R(2,n) = n.$$

Proof. Any coloring of K_n either has (a) one or more red (=solid) edges, so it contains a red K_2, or (b) it does not contain any red edges, but then it contains a blue (=dashed) K_n.

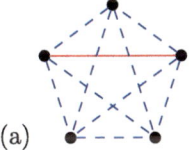 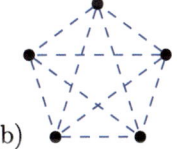

(a) (b)

We can also formulate it as follows. At a party with n people, there are either two people knowing each other or they are all mutual strangers. □

Exchanging acquaintances and strangers, we get $R(n,2) = n$. The next larger case is already more difficult to argue, so we start with just a lower bound.

Proposition 15.25. *The Ramsey number $R(3,3) > 5$.*

Proof. Consider the graph K_5 with the edge-coloring:

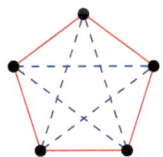

In other words, this graph does not have an induced red K_3 subgraph, nor an induced blue K_3 subgraph. □

Loosely speaking, the previous proposition asserts that in a particular party of five people, it can happen that there are no three people that are mutually acquainted and no three people that are mutually strangers. The next proposition shows that this can never happen in a party of six people.

Proposition 15.26. *The Ramsey number $R(3,3) = 6$.*

Proof. It suffices to show that $R(3,3) \leqslant 6$. Let $G = (V, E)$ be the red induced subgraph of the complete graph K_6, and recall that red models acquaintance. Let $u \in V$ be an arbitrary vertex. Then there are two cases:
(a) Suppose that the set $N(u) = \{v \in V | (u, v) \in E\}$ has at least 3 elements. Then either $N(u)$ is an independent set of strangers and the claim holds, or we have two adjacent vertices $v_1, v_2 \in N(u)$, in which case $\{u, v_1, v_2\}$ is a clique of mutually acquainted people and the claim also holds.
(b) Suppose that the set $N(u) = \{v \in V | (u, v) \in E\}$ has at most two elements. Then there are three vertices x, y, z in $V \setminus N[u]$. If those three are mutually acquainted, then we are done. Otherwise, if two of the vertices x, y, z are not acquainted, let us say x and y are strangers, then $\{u, x, y\}$ are mutually strangers, and the claim also holds.

In any case, we have established that $R(3,3) \leqslant 6$, as claimed. □

One can show that $R(4,4) = 18$, but determining the exact value of $R(5,5)$ is already an open problem. The reason is that the corresponding graphs get large and enumerating all colorings becomes infeasible.

Erdős proved the following remarkable lower bound on the Ramsey numbers.

Proposition 15.27 (Erdős). *If $\binom{n}{k} 2^{1-\binom{k}{2}} < 1$, then $R(k, k) > n$.*

Proof. Consider K_n and a random 2-coloring on its edges, namely we color an edge *red* with probability $1/2$, and *blue* with probability $1/2$. For any k-subset S of the set of vertices, let M_S be the event that the induced subgraph on S is monochromatic. Then,

$$\Pr[M_S] = \Pr[S\ \text{red}] + \Pr[S\ \text{blue}] = \frac{1}{2^{\binom{k}{2}}} + \frac{1}{2^{\binom{k}{2}}} = 2^{1-\binom{k}{2}}.$$

By the union bound, the probability that some k-subset forms a monochromatic subgraph is at most $\binom{n}{k} 2^{1-\binom{k}{2}}$.

Since $\binom{n}{k} 2^{1-\binom{k}{2}} < 1$, there exists some 2-coloring for which there is no monochromatic K_k. Therefore, we can conclude that the Ramsey number $R(k, k) > n$. □

Corollary 15.28. *For all integers k such that $k \geqslant 3$, the Ramsey number*

$$R(k, k) > 2^{k/2}.$$

15.7 The Probabilistic Method

Proof. Given an integer k, we define $n = \lfloor 2^{k/2} \rfloor$. The simple estimate $\binom{n}{k} \leq n^k/k!$ yields

$$\binom{n}{k} 2^{1-\binom{k}{2}} \leq \frac{n^k}{k!} 2^{1-\frac{k(k-1)}{2}} \leq \frac{\left(2^{k/2}\right)^k}{k!} 2^{1-\frac{k^2}{2}+\frac{k}{2}}.$$

Simplifying the right-hand side gives the bound

$$\binom{n}{k} 2^{1-\binom{k}{2}} \leq \frac{2^{1+\frac{k}{2}}}{k!}.$$

For $k \geq 3$, the right-hand side satisfies $2^{1+\frac{k}{2}}/k! < 1$, see Exercise 15.46. It follows from the previous proposition that $R(k,k) > \lfloor 2^{k/2} \rfloor$. Since $R(k,k)$ is an integer, it even follows that $R(k,k) > 2^{k/2}$. □

Large Cuts. Given an undirected graph $G = (V, E)$, a partition of the vertex set V into disjoint subsets A and B is called a **cut**. In other words, two subsets A and B form a cut of G if and only if

$$A \cap B = \emptyset \quad \text{and} \quad A \cup B = V.$$

The **size** of the cut (A, B) in the graph G is given by the number of edges crossing the cut (A, B). In other words,

$$\text{size}(A, B) = |\{(u, v) \in E \mid u \in A, v \in B\}|.$$

Finding a cut of maximum size is a computationally hard problem. However, we can use the probabilistic method to prove the existence of large cuts.

The key in the argument is the following probabilistic version of the pigeonhole principle.

Proposition 15.29. *The values of a discrete random variable cannot always be less than its expected value.*

Proof. Seeking a contradiction, suppose that X is a discrete random variable that has values always less than $\mu = \mathrm{E}[X]$. Then

$$\mathrm{E}[X] = \sum_{\alpha \in X(\Omega)} \alpha \Pr[X = \alpha] < \sum_{\alpha \in X(\Omega)} \mu \Pr[X = \alpha] = \mathrm{E}[X],$$

contradiction. □

Similarly, a random variable cannot always be larger than its expected value.

Proposition 15.30. *Given an undirected graph $G = (V, E)$ with m edges, there exists a partition of V into two disjoint sets A and B such that at least $m/2$ edges cross the cut (A, B).*

Proof. For each vertex, flip a fair coin and put the vertex in A if the coin shows heads, and put the vertex in B if the coin shows tails. Let $e_1, e_2, \ldots, e_m$ be an enumeration of the edges in E. Define the indicator random variable X_k by

$$X_k = \begin{cases} 1 & \text{if edge } k \text{ crosses the cut } (A, B), \\ 0 & \text{otherwise} \end{cases}$$

The probability that the edge crosses the cut (A, B) is $1/2$; hence,

$$\mathrm{E}[X_k] = \frac{1}{2}.$$

Let $\mathrm{size}(A, B)$ denote the size of the cut (A, B). Then

$$\mathrm{E}[\mathrm{size}(A,B)] = \mathrm{E}\left[\sum_{k=1}^m X_k\right] = \sum_{k=1}^m \mathrm{E}[X_k] = \frac{m}{2}.$$

By Proposition 15.29, there exists a cut (A, B) of size $m/2$ or larger. □

Exercises

15.45. Determine the Ramsey number $R(2, 2)$.

15.46. For any integer k such that $k \geqslant 3$, prove that

$$2^{1+\frac{k}{2}} \leqslant k!$$

by induction on k.

15.47. Show that the Ramsey numbers $R(a, b)$ satisfy the bound

$$R(a, b) \leqslant R(a-1, b) + R(a, b-1)$$

for all integers a and b such that $a \geqslant 3$ and $b \geqslant 3$.

15.48. Use the previous exercise to show that the Ramsey number $R(a, b)$ satisfies the upper bound

$$R(a, b) \leqslant \binom{a+b-2}{a-1}$$

for all integers a and b such that $a \geqslant 2$ and $b \geqslant 2$.

15.49. Use the previous exercise to show that the diagonal Ramsey number $R(k, k)$ is bounded from above by

$$R(k, k) \leqslant 4^{k-1}.$$

15.50. Suppose that n players play a tennis tournament, where each player plays against all other players. Players either win or loose. The tournament graph contains a vertex for each player and an edge (a, b) if player a won against player b. A tournament has property P_k if and only if for every set of k players, there exists one player who beats them all. Show that if $\binom{n}{k}(1 - 2^{-k})^{n-k} < 1$, then there exists a tournament of n players that has property P_k.

15.51. Use the probabilistic method to show that every graph $G = (V, E_G)$ has a bipartite subgraph $H = (V, E_H)$ such that $|E_H| \geq \frac{1}{2}|E_G|$.

15.8 Notes

There exist an abundance of good introductory text on probability theory. The book by Chung and AitSahlia [13] is a very well-written introduction to elementary probability theory that also briefly discusses some more advanced topics. Feller [27] is a wonderful classic. The book by Grimmett and Stirzaker [31] gives a well-rounded introduction and contains numerous excellent exercises. The book by Ross [65] is an easily readable introduction, which contains a well-chosen number of interesting applications.

If you want to make further progress, then you need to learn more about measure-theoretic aspects of probability theory. The books by Gut [32], Jacod and Protter [40], Klenke [46], Rosenthal [64], Williams [80] all give solid but accessible measure-theoretic introductions to probability theory. These books will give you a good start, and all contain at least a brief discussion of the Lebesque integral and related topics. All of the previous books are accessible to undergraduate students.

Probability generating functions, and much more, are discussed in Graham et al. [30]. The book by Alon and Spencer [5] contains numerous applications illustrating the probabilistic method.

Bibliography

[1] A.V. Aho, J.E. Hopcroft, and J. Ullman. *The Design and Analysis of Computer Algorithms*. Addison-Wesley Longman Publishing Co., Inc., Boston, MA, USA, 1st edition, 1974.

[2] M. Aigner. *Combinatorial theory*. Classics in Mathematics. Springer-Verlag, Berlin, 1997. Reprint of the 1979 original.

[3] M. Aigner. *A course in enumeration*, volume 238 of *Graduate Texts in Mathematics*. Springer, Berlin, 2007.

[4] M. Aigner. *Discrete mathematics*. American Mathematical Society, Providence, R.I., 2007.

[5] N. Alon and J. Spencer. *The Probabilistic Method*. John Wiley & Sons, New York, 2nd edition, 2000.

[6] T. Andreescu and V. Crişan. *Mathematical Induction – A powerful and elegant method of proof*. XYZ Press, 2017.

[7] P. Bachmann. *Die Analytische Zahlentheorie*. Teubner, 1894.

[8] R.A. Beeler. *How to count*. Springer, Cham, 2015. An introduction to combinatorics and its applications.

[9] A.T. Benjamin and J.J. Quinn. *Proofs that really count*, volume 27 of *The Dolciani Mathematical Expositions*. Mathematical Association of America, Washington, DC, 2003. The art of combinatorial proof.

[10] G. Boole. *A treatise on the calculus of finite differences*. Dover Publications, New York, 2nd edition, 1960.

[11] G. Cantor. *Gesammelte Abhandlungen mathematischen und philosophischen Inhalts*. Springer-Verlag, Berlin-New York, 1980. Reprint of the 1932 original.

[12] N. Caspard, B. Leclerc, and B. Monjardet. *Finite ordered sets*, volume 144 of *Encyclopedia of Mathematics and its Applications*. Cambridge University Press, Cambridge, 2012. Concepts, results and uses.

[13] K.L. Chung and F. AitSahlia. *Elementary Probability Theory: With Stochastic Processes and an Introduction to Mathematical Finance.* Undergraduate Texts in Mathematics. Springer New York, 4th edition, 2010.

[14] P. M. Cohn. *Universal algebra*, volume 6 of *Mathematics and its Applications*. D. Reidel Publishing Co., Dordrecht-Boston, Mass., second edition, 1981.

[15] P.M. Cohn. *Basic Algebra.* Springer London, 2003.

[16] T.H. Cormen, C.E. Leiserson, R.L. Rivest, and C. Stein. *Introduction to Algorithms.* The MIT Press, 3rd edition, 2009.

[17] B. A. Davey and H. A. Priestley. *Introduction to lattices and order.* Cambridge University Press, New York, second edition, 2002.

[18] J.-M. De Koninck and A. Mercier. *1001 problems in classical number theory.* American Mathematical Society, Providence, RI, 2007. Translated from the 2004 French original by De Koninck.

[19] O. Deiser. *Einführung in die Mengenlehre.* Springer-Lehrbuch. [Springer Textbook]. Springer-Verlag, Berlin, second edition, 2004. Die Mengenlehre Georg Cantors und ihre Axiomatisierung durch Ernst Zermelo. [The set theory of Georg Cantor and its axiomization by Ernst Zermelo].

[20] U. Dudley. *Elementary Number Theory.* Dover Books on Mathematics. Dover Publications, 2nd edition, 2012.

[21] H.-D. Ebbinghaus. *Einführung in die Mengenlehre.* Spektrum Academischer Verlag, Heidelberg, fourth edition, 2003.

[22] H.D. Ebbinghaus, J. Flum, and W. Thomas. *Mathematical Logic.* Undergraduate Texts in Mathematics. Springer New York, 1996.

[23] P.J. Eccles. *An introduction to mathematical reasoning.* Cambridge University Press, Cambridge, 1997. Numbers, sets and functions.

[24] H.B. Enderton. *Elements of set theory.* Academic Press [Harcourt Brace Jovanovich, Publishers], New York-London, 1977.

[25] H.B. Enderton. *A mathematical introduction to logic.* Harcourt/Academic Press, Burlington, MA, second edition, 2001.

[26] L. Euler. Solution d'une question curieuse que ne paroit soumise à aucune analyse. *Mémoire de l'Academie des Sciences de Berlin*, 15:310–337, 1759 (published 1766).

[27] William Feller. *An Introduction to Probability Theory and Its Applications*, volume 1. Wiley, 1968.

[28] J.H. Gallier. *Logic for Computer Science – Foundations for Automatic Theorem Proving*. John Wiley and Sons, 1987.

[29] D. Goldrei. *Propositional and Predicate Calculus – A Model of Argument*. Springer, 2005.

[30] R.L. Graham, D.E. Knuth, and O. Patashnik. *Concrete mathematics: a foundation for computer science*. Addison-Wesley, Reading, Mass., 2nd edition, 1994.

[31] G. Grimmett and D. Stirzaker. *Probability and Random Processes*. Oxford University Press, Oxford, 3rd edition, 2001.

[32] A. Gut. *Probability: A Graduate Course*. Springer, 2005.

[33] P.R. Halmos. *Naive set theory*. Springer-Verlag, New York-Heidelberg, 1974. Reprint of the 1960 edition, Undergraduate Texts in Mathematics.

[34] G.H. Hardy. *Orders of infinity, the 'infinitärcalül' of Paul Du Bois-Reymond*. Cambridge University press, Cambridge, 1910.

[35] J. Herman, R. Kučera, and J. Šimša. *Equations and Inequalities – Elementary Problems and Theoremsin Algebra and Number Theory*. Springer-Verlag New York, 2000.

[36] K. Hrbacek and T. Jech. *Introduction to set theory*, volume 220 of *Monographs and Textbooks in Pure and Applied Mathematics*. Marcel Dekker, Inc., New York, third edition, 1999.

[37] M. Huth and M. Ryan. *Logic in Computer Science: Modelling and Reasoning About Systems*. Cambridge University Press, New York, NY, USA, 2nd edition, 2004.

[38] K. Ireland and M. Rosen. *A Classical Introduction to Modern Number Theory*. Graduate Texts in Mathematics. Springer, 1990.

[39] N. Jacobson. *Basic Algebra I. Basic Algebra*. Dover Publications, 2nd edition, 2009.

[40] J. Jacod and P. Protter. *Probability Essentials*. Springer-Verlag, Berlin, 2000.

[41] T. Jech. *Set theory*. Springer Monographs in Mathematics. Springer-Verlag, Berlin, 2003. The third millennium edition, revised and expanded.

[42] K. Jordán. *Calculus of finite differences*. Chelsea Pub. Co., New York, 3rd edition, 1965.

[43] S. Jukna. *Extremal combinatorics*. Texts in Theoretical Computer Science. An EATCS Series. Springer-Verlag, Berlin, 2001. With applications in computer science.

[44] V.G. Kac and P. Cheung. *Quantum calculus.* Universitext. Springer, New York, 2002.

[45] I. Kaplansky. *Set theory and metric spaces.* Chelsea Publishing Co., New York, second edition, 1977.

[46] A. Klenke. *Probability Theory: A Comprehensive Course.* Universitext. Springer, 2nd edition, 2013.

[47] D.E. Knuth. Big omicron and big omega and big theta. *SIGACT News*, 8(2):18–24, 1976.

[48] D.E. Knuth. *Fundamental Algorithms*, volume 1 of *The Art of Computer Programming*. Addison-Wesley, Reading, Massachusetts, third edition, 1997.

[49] D.E. Knuth. *Seminumerical Algorithms*, volume 2 of *The Art of Computer Programming*. Addison-Wesley, Reading, Massachusetts, third edition, 1997.

[50] D.E. Knuth. *Sorting and Searching*, volume 3 of *The Art of Computer Programming*. Addison-Wesley, Reading, Massachusetts, second edition, 1998.

[51] D.E. Knuth. *Combinatorial Algorithms: Part 1*, volume 4A of *The Art of Computer Programming*. Addison-Wesley, Reading, Massachusetts, first edition, 2011.

[52] M. Koecher. *Klassische Elementare Analysis.* Springer Basel AG, 1987.

[53] Thomas Koshy. *Catalan numbers with applications.* Oxford University Press, Oxford, 2009.

[54] K. Kunen. *Set theory*, volume 34 of *Studies in Logic (London)*. College Publications, London, 2011.

[55] E. Landau. *Handbuch der Lehre von der Verteilung der Primzahlen.* Teubner, Leipzig, 1909. 2 volumes. Reprinted by Chelsea, New York, 1953.

[56] E. Landau. *Foundations of Analysis.* AMS Chelsea Publishing Series. American Mathematical Society, 2001.

[57] C.H.C. Little, K.L. Teo, and B. Van Brunt. *The Number Systems of Analysis.* World Scientific, 2003.

[58] J. Loeckx and K. Sieber. *The Foundations of Program Verification.* Wiley and Teubner, 1984.

[59] Z. Manna. *Mathematical Theory of Computation.* McGraw-Hill, 1974.

[60] E. Mendelson. *Introduction to Mathematical Logic*. Chapman & Hall/CRC, 6th edition, 2015.

[61] L.M. Milne-Thomson. *The calculus of finite differences*. AMS Chelsea Pub., Providence, R.I., 2000.

[62] V.H. Moll. *Numbers and Functions: From a Classical-experimental Mathematician's Point of View*. Student mathematical library. American Mathematical Society, 2012.

[63] S. Roman. *Lattices and ordered sets*. Springer, New York, 2008.

[64] J.S. Rosenthal. *Rigorous Probability Theory: A First Look*. World Scientific, 2nd edition, 2006.

[65] S.R. Ross. *Introduction to Probability Models*. Academic Press, San Diego, 7th edition, 2000.

[66] R. Schindler. *Logische Grundlagen der Mathematik*. Springer, Berlin, 2009.

[67] R.M. Smullyan. *Forever Undecided: A Puzzle Guide to Gödel*. Oxford University Press, 1987.

[68] R.M. Smullyan. *First-order logic*. Dover Publications, Inc., New York, 1995. Corrected reprint of the 1968 original.

[69] R.M. Smullyan. *A beginners guide to mathematical logic*. Dover, 2014.

[70] R.M. Smullyan and M. Fitting. *Set theory and the continuum problem*, volume 34 of *Oxford Logic Guides*. The Clarendon Press, Oxford University Press, New York, 1996. Oxford Science Publications.

[71] J. Spencer. *Asymptopia*. Student Mathematical Library. American Mathematical Society, 2014.

[72] R.P. Stanley. *Enumerative combinatorics. Vol. 1*, volume 49 of *Cambridge Studies in Advanced Mathematics*. Cambridge University Press, Cambridge, 1997. With a foreword by Gian-Carlo Rota, Corrected reprint of the 1986 original.

[73] R.P. Stanley. *Enumerative combinatorics. Vol. 2*, volume 62 of *Cambridge Studies in Advanced Mathematics*. Cambridge University Press, Cambridge, 1999. With a foreword by Gian-Carlo Rota and appendix 1 by Sergey Fomin.

[74] R.P. Stanley. *Catalan numbers*. Cambridge University Press, New York, 2015.

[75] J. Stopple. *A primer of analytic number theory: from Pythagoras to Riemann*. Cambridge University Press, 2003.

[76] W.T. Trotter. *Combinatorics and partially ordered sets*. Johns Hopkins Series in the Mathematical Sciences. Johns Hopkins University Press, Baltimore, MD, 1992. Dimension theory.

[77] D.J. Velleman. *How to prove it*. Cambridge University Press, Cambridge, second edition, 2006. A structured approach.

[78] W. Walter. *Analysis I*. Springer, Berlin, 1985.

[79] M.H. Weissman. *An Illustrated Theory of Numbers*. American Mathematical Society, 2017.

[80] D. Williams. *Probability with Martingales*. Cambridge University Press, 1991.

Index

Symbols
k-combination, 274
k-permutation, 273
n-ary predicate, 35
p-adic valuation, 177

A
accumulation point
    lower, 248
    upper, 247
adjacent, 351
antichain, 146
antidifference, 225
antisymmetric, 70
arity, 123
assertion, 11
asymmetric, 70
asymptotically equal, 237
asymptotically tight bound, 250
asymptotic lower bound, 259
asymptotic upper bound, 254
axioms, 12

B
base set, 123
Big Theta, 250
bi-implication, 15
bijective, 75
Binet's formula, 327
binomial coefficient, 99, 219
binomial series, 312
binomial theorem, 99

bipartite, 355
block, 134
Boolean formulas, 15
Boolean function
    monotone, 165
Boolean variable, 15
Boole's inequality, 384
bound, 36
bounded above, 157, 247
bounded below, 158, 249

C
canonical disjunctive normal
    form, 34
cardinality
    equal, 82
    less than or equal to, 82
Cartesian product, 68
Catalan number, 343
ceiling, 171
centered hexagonal number, 321
certain event, 382
chain, 146
characteristic polynomial, 336
choice function, 77
chord, 375
chromatic number, 368
clique, 370
clique number, 370
closed neighborhood, 351
codomain, 73

color class, 371
comparable, 146
complementary
     counting, 271
complete bipartite graph, 356
complete graph, 355
complete lattice, 166
component number, 360
composite number, 44
composition, 72, 74
conclusion, 14, 26
conditional probability, 389
congruence modulo n, 140
conjunction, 13
constructor, 123
contrapositive, 21
converse, 22
cosine function, 313
countable set, 84
cover graph, 151
covers, 149
critical graph, 374
cumulative distribution
     function, 400
cut, 409
cut-vertex, 379
cycle, 6
cycle graph, 355

D
degree, 351
dependent, 394
derangement, 294
difference operator, 211
Diophantine equation, 191
discrete partial order, 146
disjoint sets, 60
disjunction, 13
distance, 361
distribution function, 400
    cumulative, 400
divides, 43
division of integers, 185
divisor, 185
domain, 73

E
edges, 5
element, 4, 52
elementary events, 381
embedding, 164
empty graph, 354
empty set, 52
equipotent, 82
equivalence class, 134
equivalence relation, 133
equivalent, 15
Euler's totient function, 294
even integer, 42
event, 381
existential quantifier, 36
expectation value, 401
exponential function, 312

F
faces, 365
factorial, 219
factorial function, 99
falling factorial power, 217
Fermat's Little Theorem, 101
Fibonacci numbers, 95
floor, 171
forest, 363
formal derivative, 307
formal integral, 309
formal system, 28
forward difference operator, 211
free, 127
free variables, 36
function, 73
    bijective, 75
    eventually nonzero, 252
    injective, 74
    one-to-one, 74
    onto, 75
    preimage, 74
    restriction, 74
    surjective, 75
functionally complete set, 34
functional property, 76
Fundamental Theorem of Difference
     Calculus, 221

INDEX

G
generating function, 301
geometric series, 97
graph, 5
    bridge, 361
    Cartesian product, 357
    circuit, 358
    connected, 359
    cycle, 358
    diameter, 361
    order, 349
    path, 358
    size, 349
    trail, 358
    walk, 358
graph complement, 356
greatest common divisor, 187
greatest element, 157
greatest lower bound, 159
Grinberg equation, 377

H
Hamiltonian-connected, 379
Hamiltonian cycle, 6, 374
Hamiltonian graph, 374
Hamiltonian path, 6, 374
happened-before, 148
harmonic number, 228
Hasse diagram, 151
homogeneous linear recurrence relation, 320
hypercube graph, 357
hypotheses, 26
hypothesis, 14

I
ideal, 195
identity map, 74
implication, 14
impossible event, 382
incident, 351
incomparable, 146
indefinite sum, 225
independence number, 370
independent, 394
independent set, 370
indicator random variable, 399
induction hypothesis, 91
inductively defined, 123
infimum, 159
inflationary, 164
inhomogeneous linear recurrence relation, 320
initial segment, 119
injective, 74
input size, 260
internal node, 362
internal vertex, 362
intersection, 60
inverse function, 75
inverse relation, 71
irreflexive, 70
isomorphism, 164

K
knight graph, 5
knight moves, 3
knight's tour, 4
    closed, 4

L
lattice, 166
leaf, 362
least element, 81, 158
least upper bound, 159
left-shifted sequence, 307
limit, 41
limit inferior, 244
limit superior, 244
linear congruence equation, 196
linear order, 146
linear recurrence relation, 320
logical consequence, 25
logically equivalent, 21
lower bound, 158
lower envelope, 243

M
maximal, 159
maximal degree, 352
maximal planar graph, 366
maximum independent set, 370

mean, 401
Mersenne prime, 49
minimal, 106, 159
minimal degree, 351
modus ponens, 26
monotonic, 164
multinomial coefficient, 286
multiple, 185
multiplicative inverse, 304

N
natural map, 135
negation, 14
neighborhood, 351
nodes, 5

O
odd integer, 42
operator
  difference, 211
  shift, 214
ordered k-selection with
    repetition, 284
ordered pair, 67
order extension, 161

P
Padovan sequence, 320
palindromic polynomial, 332
partially ordered set, 145
  height, 146
  width, 153
partial order, 145
partite classes, 355
partition, 134
path, 6
path graph, 354
perfect square, 43
planar graph, 364
plane graph, 365
polynomial, 217
  constant term, 217
  degree, 217
  leading coefficient, 217
  leading term, 217
power, 309

power set, 56
predicate, 35
predicate of degree n, 35
premises, 26
prime, 104
prime gap, 44
prime number, 44
principal ideal, 195
probability measure, 382
product of power
    series, 304
proper subset, 54
proposition, 11
propositional calculus, 28
propositional
    function, 35

Q
quotient, 103
quotient set, 135

R
Ramsey numbers, 406
random variable, 398
  independent, 400
  indicator, 399
range, 73
rational numbers, 143
reciprocal polynomial, 331
recurrence relation, 320
  degree, 320
recursively defined, 123
reflected polynomial, 331
reflexive, 70
reflexive and transitive
    closure, 150
regular, 353
relation, 69
  binary, 69
  wellfounded, 106
remainder, 103
right-shifted sequence, 307

S
sample space, 381
satisfiable, 21

INDEX 423

set, 4, 52
    countable, 84
    finite, 83
    inductive, 79
    transitive, 81
    uncountable, 84
set complement, 64
set difference, 63
set equality, 52
set intersection, 62
set union, 60, 62
shift operator, 214
sine function, 313
singleton set, 53
size of a cut, 409
spanning subgraph, 363
spanning tree, 363
square-free, 202
star graph, 356
statement, 11
Stirling numbers
    of the second kind, 222
strict asymptotic lower
    bound, 259
strict asymptotic upper
    bound, 256
strictly ordered set, 147
strict order, 147
subadditivity, 384
subgraph, 363
subset, 5, 54
substitution technique, 23
successor, 79
sum of power
    series, 304
supremum, 159
surjective, 75
symmetric, 71
symmetric difference, 65

T
tautology, 21
telescoping sum, 96
total order, 146
transitive, 71
trapdoor function, 203
tree, 362
triangle, 354
Tribonacci numbers, 321

U
unimodal sequence, 283
union bound, 384
unit, 44
universal quantifier, 36
universe, 35
universe of discourse, 35
unordered k-selection with
    repetitions, 285
upper bound, 157
upper envelope, 243

V
vacuously true, 14
valid, 26
valuation, 20
variance, 402
vertex coloring, 368
vertices, 5

W
walk, 5
    length, 358
well-ordered set, 161
worst-case running time, 261

GPSR Compliance

The European Union's (EU) General Product Safety Regulation (GPSR) is a set of rules that requires consumer products to be safe and our obligations to ensure this.

If you have any concerns about our products, you can contact us on ProductSafety@springernature.com

In case Publisher is established outside the EU, the EU authorized representative is:

Springer Nature Customer Service Center GmbH
Europaplatz 3
69115 Heidelberg, Germany

Batch number: 08115475

Printed by Printforce, the Netherlands